E. Lüscher / P. Korpiun
H. J. Coufal / R. Tilgner (Eds.)

Photoacoustic Effect

Edgar Lüscher/Peter Korpiun
Hans-Jürgen Coufal/Rainer Tilgner (Eds.)

Photoacoustic Effect
Principles and Applications

Proceedings of the First International Conference
on the Photoacoustic Effect in Germany

Held on February 23–26, 1981 in Bad Honnef (FRG)
Sponsored by the Stiftung Volkswagenwerk and the
Deutsche Physikalische Gesellschaft

With 227 Figures

Springer Fachmedien Wiesbaden GmbH

CIP-Kurztitelaufnahme der Deutschen Bibliothek

Photoacoustic effect, principles and applications:
proceedings of the 1. Internat. Conference on the
Photoacoust. Effect in Germany, held on February
23–26, 1981 in Bad Honnef (FRG)/Edgar Lüscher ...
(eds.). Sponsored by the Stiftung Volkswagenwerk
and the Dt. Physikal. Ges. — Braunschweig; Wiesbaden:
Vieweg, 1984.
    ISBN 978-3-528-08573-5   ISBN 978-3-663-06820-4 (eBook)
    DOI 10.1007/978-3-663-06820-4
NE: Lüscher, Edgar [Hrsg.]: International Conference
on the Photoacoustic Effect ⟨01, 1981, Honnef⟩

1984

Produced by IVD, Industrie- u. Verlagsdruck, Walluf

Preface

The interest in the photoacoustic effect as a base of new experimental techniques increased appreciately in the last ten years. Originally, the effect was used in the optic spectroscopy to investigate very weak optical absorption of gases. Now more and more publications report on applications to an increasing number of fields.

This volume presents the contributions to the first international conference on the photoacoustic effect held in Germany.

Corresponding to the present situation in photoacoustics, contributions on basic principles of the effect still represent the relative majority of the papers. Spectroscopy and detection of minute concentrations as well as monitoring the different partners in chemical reactions have remained up to now the main field of analytical application. There is, however, a growing number of new problems mostly unknown in photoacoustics still at the end of the seventies. Calorimetric applications to phase transitions, energy conversion processes as well as monitoring of photochemical reactions have gained remarkable attention within the photoacoustic community.

It should be expected photoacoustics to become still more interdisciplinary in character. Applications of the photoacoustic effect to fields, at first sight so far distant, as electron paramagnetic resonance as well as nondestructive testing of materials are hints to this direction. Contributions on instrumentation rounded up the exchange of many field's knowledge.

One value of this conference was to make participants more aware of this evolutionary interaction. It offered the important opportunity to promote the spreading of that powerful technique into a still growing field of applications.

The idea to hold such a meeting in Germany for the first time
was due to Prof. H. Pelzl. Its realization was greatly supported
by financial help of the Stiftung Volkswagenwerk and the Deutsche
Physikalische Gesellschaft, which is gratefully acknowledged.

The Editors

Contents

2. Spectroscopic Applications

3. Photoacoustic Detection and Monitoring

4. Thermal Applications

5. Photochemistry

6. Nondestructive Testing

7. Magnetic Resonance

8. Instrumentation

1. Basic Principles

PHOTOACOUSTIC EFFECT IN CONDENSED MATTER -
HISTORICAL DEVELOPMENT

Edgar Lüscher
Physik-Department, Technische Universität München, D-8046 Garching

1. TO THE HISTORY

The concept on which photoacoustic effects and spectroscopy are based was first reported in 1880 by Alexander Graham Bell[1]. One of the transmitters of Bell's so-called "photophone", which he built together with Sumner Tainter, consisted of a mirror, which was activated by the sound waves of a voice, Fig. 1. A receiver section was made up of a hearing tube. It was mounted in that way that a hard rubber diaphragm was in the focus of a lens, Fig. 2. To explain the experimental observations Bell suggested that the illuminated front side of the diaphragm is heated and cooled periodically more than the rare side. The varying thermal expansion along the diaphragm leads to a periodically varying bending. This explanation was supported by calculations of Lord Rayleigh[2]. The intensity of a beam of sunlight also was modulated

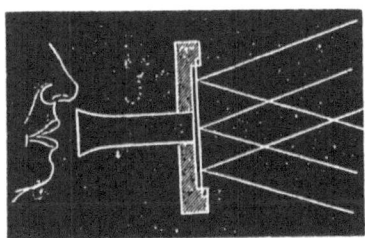

FIGURE 1   Bell's photophonic emmitter. The intensity distribution of a light beam reflected on a mirror forming the diaphragm of a microphone is modulated by voice (mirror: silvered mica or microscope glass). Figure taken from Bell's work[1].

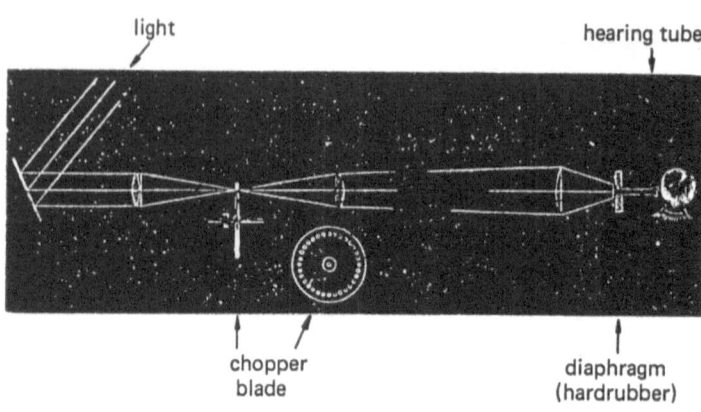

FIGURE 2   Bell's sketch of a photophonic arrangement using a hearing
tube as receiver. Light intensity modulated by a mechanical
chopper is focused to the hard rubber diaphragm of the hearing tube. Taken from ref. 1.

by the mirror. Bell wrote to his father on the behave of this discover:
"I have heard articulate speech produced by sunlight: I have heard
a ray of the sun laugh and cough and sing! I have been able to
hear a shadow, and I have even perceived by ear the passage of a
cloud across the sun's disk...
Can imagination picture what the future of this invention is to
be...."

Bell also prophesied:

    "I recognize the fact that the spectrosphere must ever remain a
    mere adjunct to the spectroscope: but I anticipate that it has
    a wide and independent field of usefulness in the investigation
    of absorption-spectra in the ultra-red."

Not only today but even at that time the newspapers were not always
friendly towards science. In their issue of 30 August 1880, the "New
York Times" wrote:

    "The ordinary man ... may find a little difficulty in comprehending
    how sunbeams are to be used. Does Professor Bell intend to connect
    Boston and Cambridge ... with a line of sunbeams hung on telegraph
    posts, and, if so, what diameter are the sunbeams to be, and how is
    he to obtain them of the required size? What will become of the

sunbeams after the sun goes down? Will they retain their power to communicate sound, or will it be necessary to insulate them, and protect them against weather by a thick coating of gutta-percha? The public has a great deal of confidence in Scientific Persons, but until it actually sees a man going through the streets with a coil of No. 12 sunbeams on his shoulder and suspending it from pole to pole, there will be a general feeling that there is something about Professor Bell's photophone which places a tremendous strain on human credulity.".

The ambivalence of scientific achievements is not restricted to modern science but even Bell looked for military applications of his "photophone", as illustrated in Fig. 3.

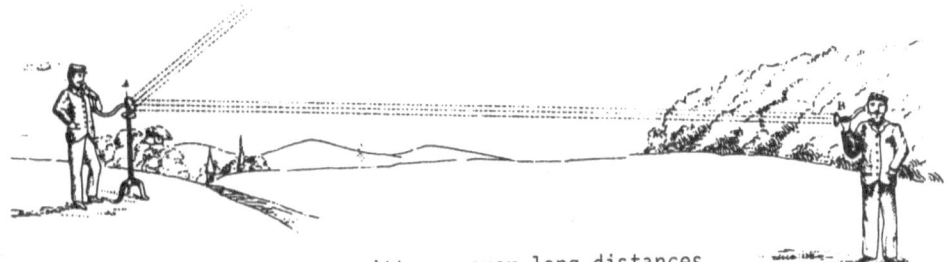

FIGURE 3   Photophonic transmittance over long distances. Taken from Bell[1].

A number of publications on studies using spectrophones by Mercadier[3], Tyndall[4], Preece[5], Roentgen[6], and Rayleigh[2] followed 1881 Bell's first work. Mercadier[3] investigated solid and liquid samples, gases and vapours. He was convinced that the effect is caused by absorption of light in the gas or vapour but not by a solid or liquid sample. A lot of experimental evidence led Tyndall[4] to a similar conclusion. Preece[5] recognised that the gas plays an important part in the generation of sound. From his experimental observations he concluded that the bending of the sample (diaphragm) does not contribute essentially to the sound. At about the same time Roentgen[6] reported acoustical observations of pressure oscillations due to the absorption of light in various gases. For city gas he found that the intensity of sound (amplitude) decreased with increasing modulation frequency of the light intensity; in his own words: "The phenomenon is, on the whole, rather complicated, because, besides the capability of the gases to absorb light, the specific heat also, as well as the ability of the gases to equalize more or less rapidly any differences of temperature that may be present, play a part ...". Therefore, it is surprising that Lord Rayleigh[2] despite of the whole background knowledge of experimental evidence explained the phenomena of sound generation merely by thermal bending of the sample.

During a visit in Europe 1880 to 1881 Bell learned all about the works of the European scientists mentioned above. He started some experiments at Paris. Bell summarized the results of his and S. Tainter's investigations in an extensive publication[7] where he also gave descriptions of the various apparatus. Graham Bell's "spectrophone" is shown in Fig. 4. The eye-piece of a commercial spectroscope is removed and a strong light absorbing substance is placed in the focal plane behind an opaque diaphragm containing a slit. The sound generated by the absorbing substance can be detected by a hearing tube. It has been used to investigate the "audibility" of different substances in the spectrum[7]. The driver of the light-chopper was taken from a sewing machine, Fig. 5. A double beam spectrophone is shown in Fig. 6. The focal lengths of both lenses A and B are identical and are illuminated with equal intensities. They are modulated by the chopping disk C. If the optical absorption of the two samples in the vessels D and E differs the sound intensity can be equalized by changing the position of the vessel D on the support whereas E is fixed. Using a pendulum L that screens alternately the light beams from the lenses A and B the intensity of sound from the samples in the vessels D and E can be heard separatly with one hearing tube H to be compared.

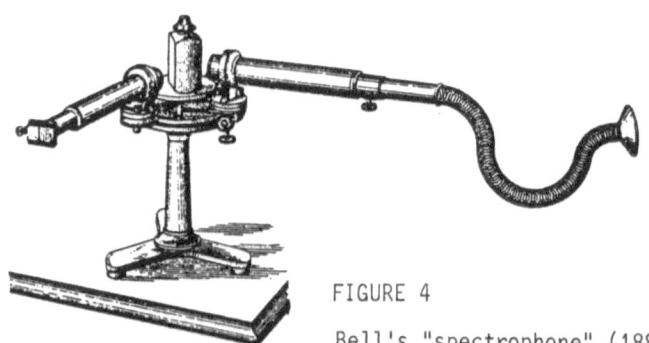

FIGURE 4

Bell's "spectrophone" (1881) with hearing tube[7].

FIGURE 5

Mechanical chopper constructed by Bell[7].

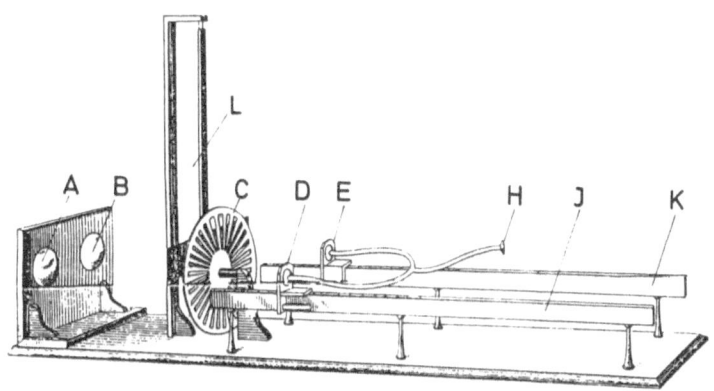

FIGURE 6   Bell's double beam spectrophone (1881). A,B: lenses; C: inter-
rupting-disk; D,E: vessels containing samples to be compared,
connected by flexible tubes of equal length to a hearing-
tube H. I, K: graduated supports; L: pendulum with screen
interrupting alternately the two light beams. (In the original work of Bell the letters have been unfortunately ommitted,
see footnote, in Ref.7, page 523.)

    In the following years the interest in the photoacoustic effect
ended soon. In none of the works published in the eighties of the last
century a quantitative satisfactive description of the phenomenon has
been given. The photoacoustic method was then replaced by optical spectroscopic methods and was forgotten until Veingerov[8] started in the
thirties of this century an extensive series of experiments on analysis
of gas mixtures. Gas analysis became probably the most widespread and

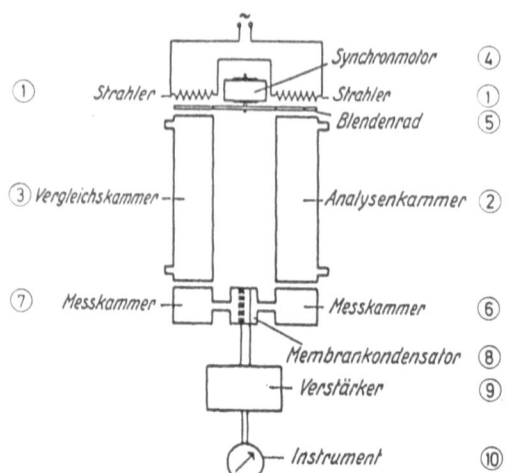

FIGURE 7   IR recording gas analyser, Luft 1943[9]
            1: IR source, filament; 2: vessel for gas mixture to be ana-
            lysed with inlet and outlet for continuous use; 3: vessel
            with reference gas (air); 4: chopper driver with 5: chopper
            blade; 6,7: differential photoacoustic gas cell; 8: membrane
            capacitor; 9: amplifier; 10: detecting and recording unit.

important application of photoacoustics[9-11]. Luft[9] published 1943 a
detailed paper on an apparatus for continuous gas analysis in the infrared that had been developed in the years before the world war. The
scheme of the apparatus is shown in Fig. 7. The source of ultrared radiation is an electrically heated filament (1). One light beam passes
a cell (2) containing the gas or gas mixture to be analysed. A second
light beam traverses a reverence gas (air) in cell (3). A mechanical
chopper (4,5) modulates the light intensity of both beams with equal
phase angle. A modulation frequency of 50/8 = 6.25 Hz was used. Both
beams enter different parts of a differential photoacoustic gas cell
(6) and (7) respectively seperated by a membrance capacitor (8). Both
parts of this cell are filled with the pure gas that should be analysed
in the gas mixture. Therefore the difference of the amplitude of the
pressure variation in (6) and (7) measured by the capacitance is determined by the different light attenuation by the gases in the cells
(2) and (3). The advantage of separating of the chamber (2) with the
gas to be analysed from the photoacoustic cell (6) is that measurements
can be performed continously on flowing gas without disturbing the
membrane by the pressure fluctuations. The sensivity was estimated to
be in the order of 1 ppm.

A very interesting technique was developed by Kreuzer and Patel[12]; they were able to detect very low concentrations of contaminants in a gas by tuning an infrared laser to a vibrational frequency of the contaminant and recording the microphonic signal resulting from the absorption.

The first theoretical analysis (to our knowledge) of the processes in the spectrophone cell on the basis of a two-state gas model was presented by R. Kaiser[13].

Turning now to photoacoustic effects in condensed matter: the first application of this was made by E. Hey and K. Gollnick at the Max-Planck-Institut für Kohleforschung in Mühlheim[14], but only the abstract of this contribution was published, due to the early death of E. Hey. (See also the contribution of Gollnik in this volume.) For the optoacoustic effects in very weakly absorbing materials we would like to refer to the contribution of Patel in this volume. The first publication of optoacoustic spectra of several solid samples in 1973 was by Harshbarger and Robin[15] at the Bell Laboratories; the strong promoter of photoacoustic spectra of many inorganic and organic solids was Al Rosencwaig[16] who was stimulated by Harshbarger and Robin. At that time Rosencwaig was also working at the Bell laboratory. It was thanks to the activities of Rosencwaig that there was a world-wide rebirth of interest in optoacoustic methods and, since then, there has been a real boom in publications in this field. The qualitative curve of the number of publications in this field is shown in Fig. 8, but the experience in any field shows that the curve representing enthusiasm in the physics aspects of it is better shown by the dashed line.

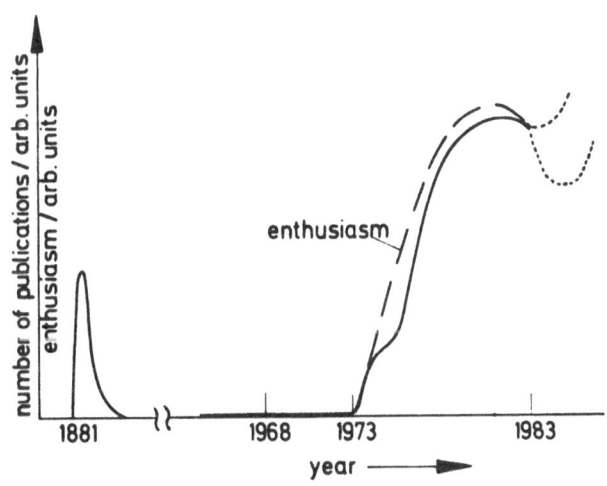

FIGURE 8

Interest in opto- or photoacoustics as a function of time. Number of publications (solid line), enthusiasm (dashed line).

2. FUNDAMENTALS OF THE OPTOACOUSTIC EFFECT IN CONDENSED MATTER

Light absorbed by condensed matter is converted, partially or totally, into heat by non-radiative transitions. The acoustic signal in
a gas-microphone cell is due to a periodic heat flow from the solid
sample to the surrounding gas, represented in the four steps of Fig. 9.
A typical experimental set up is shown in Fig. 10.

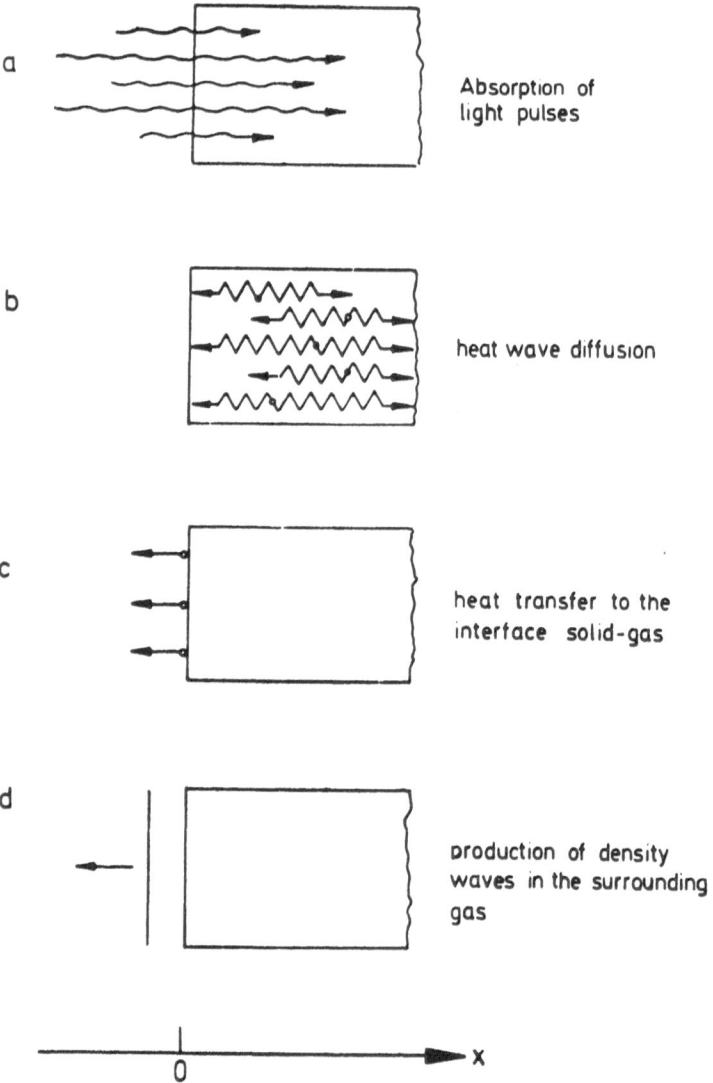

FIGURE 9   Four step process from light absorption to pressure variation
            in the gas.

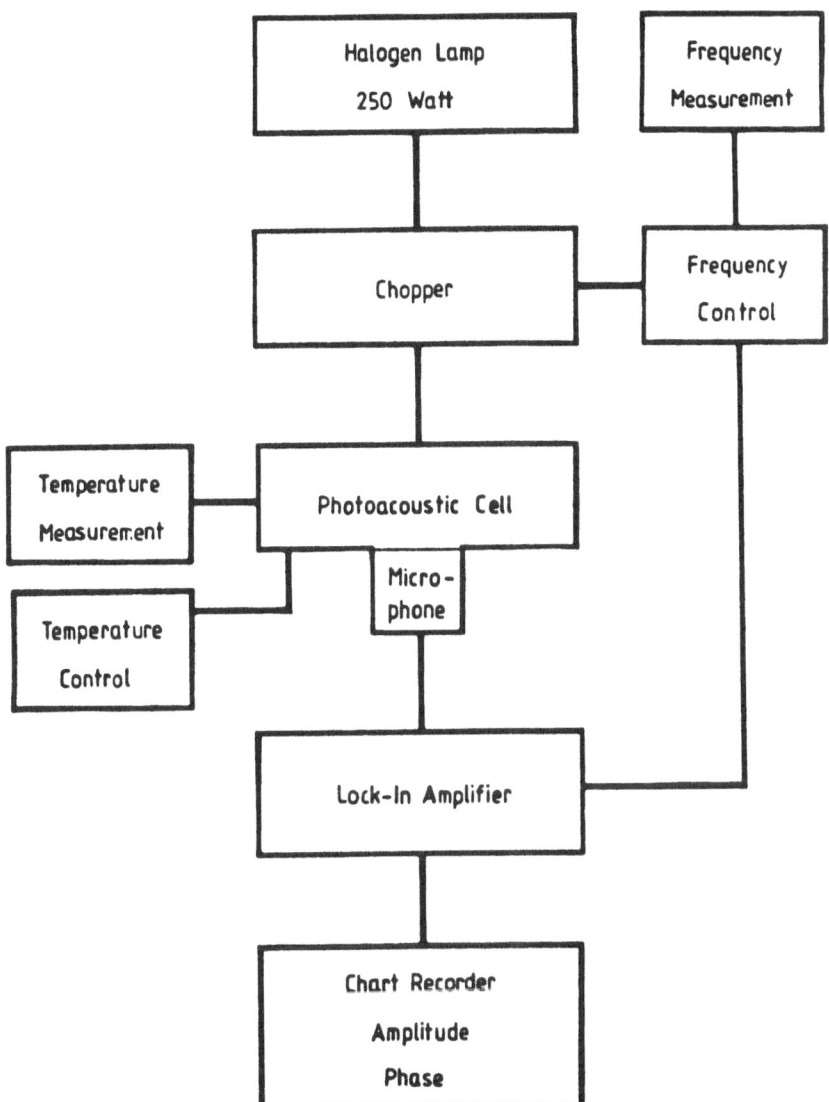

FIGURE 10 Typical experimental set-up for PAS measurements.

For the generation of heat waves, many excitation mechanisms can be used, providing that a change in energy-states is produced and that the relaxation to a lower state has a radiationless component. Examples for such generative mechanisms are:
periodically varying or pulsed electrical heating, exothermal or endo-thermal chemical reactions, external induced change of conformation, periodical external mechanical strain, etc.

The most widely used model for describing the optoacoustic or photoacoustic effect on condensed samples in a gas-microphone cell was developed by Rosencwaig and Gersho[17]; valid for acoustic wavelengths much greater than the dimensions of the sample and the gas column. The absorption of the periodically varying electromagnetic radiation, following Beer's Law, is described in a linear model in terms of change of intensity in an element dx:

$$dI = -\beta \ I \ dx \tag{1}$$

where I = intensity of radiation and b = optical absorption coefficient.

Integration of eq. (1) leads to the well known relation:

$$I = I_0 \ e^{-\beta x} \tag{2}$$

where I_0 is the intensity of the incident radiation.

If ω is the modulation frequency of the incident radiation, the incident a.c. intensity I_0^{ac} is:

$$I_0^{ac} = \frac{1}{2} \ I_0 \ [1 - \cos \omega t] \tag{3}$$

The heat-diffusion equation has to be solved for the sample, the gas and the backing[17]. For the sample it has the form

$$\frac{\partial^2 T}{\partial x^2} = \frac{1}{\alpha_s} \frac{\partial T}{\partial t} - \frac{\beta \ \sigma_{rt} \ I_0}{2 \varkappa_s} \ exp[-\beta x] \cdot (1 - \cos \omega t) \tag{4}$$

with T = temperature, $\varkappa_s$ = thermal conductivity, σ_{rt} = probability for radiationless transitions of the sample.
α_s is the thermal diffusivity defined as:

$$\alpha_s = \frac{\varkappa_s}{\rho_s c_{ps}} \tag{5}$$

with ρ_s = density and c_{ps} = specific heat capacity of the sample.

The thermal diffusion length μ_s of the sample is defined as:

$$\mu_s = \frac{1}{a_s} = [\frac{2\alpha_s}{\omega}]^{1/2} \tag{6}$$

where a_s represents the thermal diffusion coefficient:

$$a_s = [\frac{\omega}{2\alpha_s}]^{1/2} \tag{7}$$

The periodic heat-diffusion generation produces a periodic temperature variation T^{ac} in the surrounding gas (Fig. 11), according to Rosencwaig and Gersho:

$$T^{ac}(x,t) = \exp[-a_g x][\theta_1 \cos(\omega t + a_g x) - \theta_2 \sin(\omega t + a_g x)] \tag{8}$$

with the complex amplitude of the periodic temperature variation:

$$\theta = \theta_1 + i\theta_2 \tag{9}$$

The temperature variation in the gas dies out at length $2\pi\mu g$, Fig. 11.

Depending on the geometrical position of the optical absorption length $\mu_\beta = \beta^{-1}$ and the thermal diffusion length μ_s, both in the sample and in the probe-support (backing), Rosencwaig and Gersho found the results for the photoacoustic signal in each of the following two cases:

1. optical transparent samples and
2. optical opaque samples;

as represented in Fig. 12.

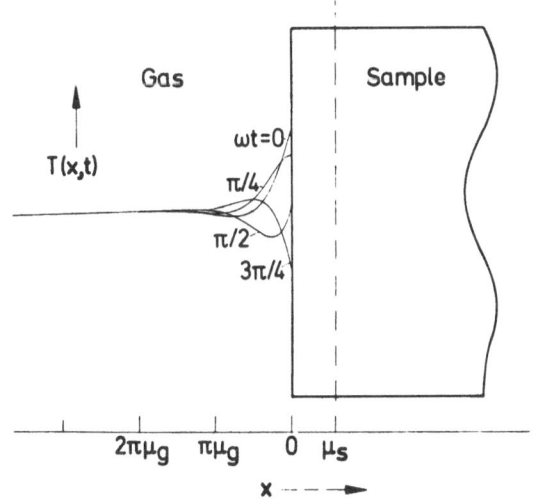

FIGURE 11   Instantaneous picture of the spatial temperature variation $T(x,t)$ in the gas near sample gas boundary. Temperature oscillation dies out with the thermal diffusion length in the gas, μ_g. Thermal diffusion length of sample = μ_s.

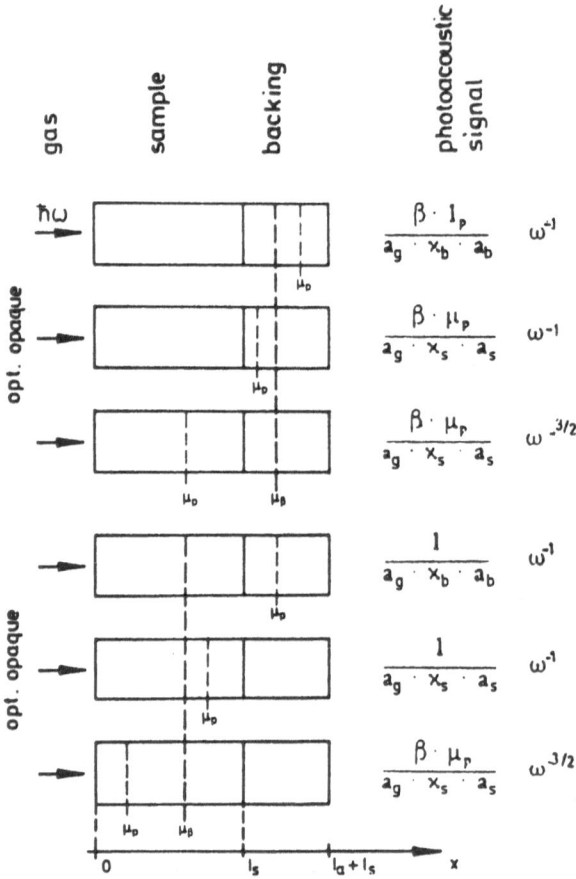

FIGURE 12   Amplitude of the photoacoustic signal for various ratios of
the thermal diffusion length $\mu_s = a_{\bar{s}}^1$, the optical absorption
length $\mu_\beta = \beta^{-1}$, and the sample length l_s.

The theoretical prediction for the dependence of the photoacoustic
signal from the frequency ω (ω^{-1} for $\mu_s > \mu_\beta$ and $\omega^{-3/2}$ for $\mu_s < \mu_\beta$) was
experimentally verified by Rosencwaig[18]; for example, in GaP - repre-
sented in Fig. 13. For the thick probe, the $\omega^{-3/2}$ dependence of the
photoacoustic signal is fulfilled for $\omega/2\pi \gtrsim 300$ Hz. A more advanced
theory must take into account the exact boundary conditions (e.g.
placement of detectors; possible Helmholtz resonances; three-dimensional heatflow; etc.). McDonald and Wetsel[19] solved the hydro-dyna-
mical equations for the sample and the surrounding gas volume.

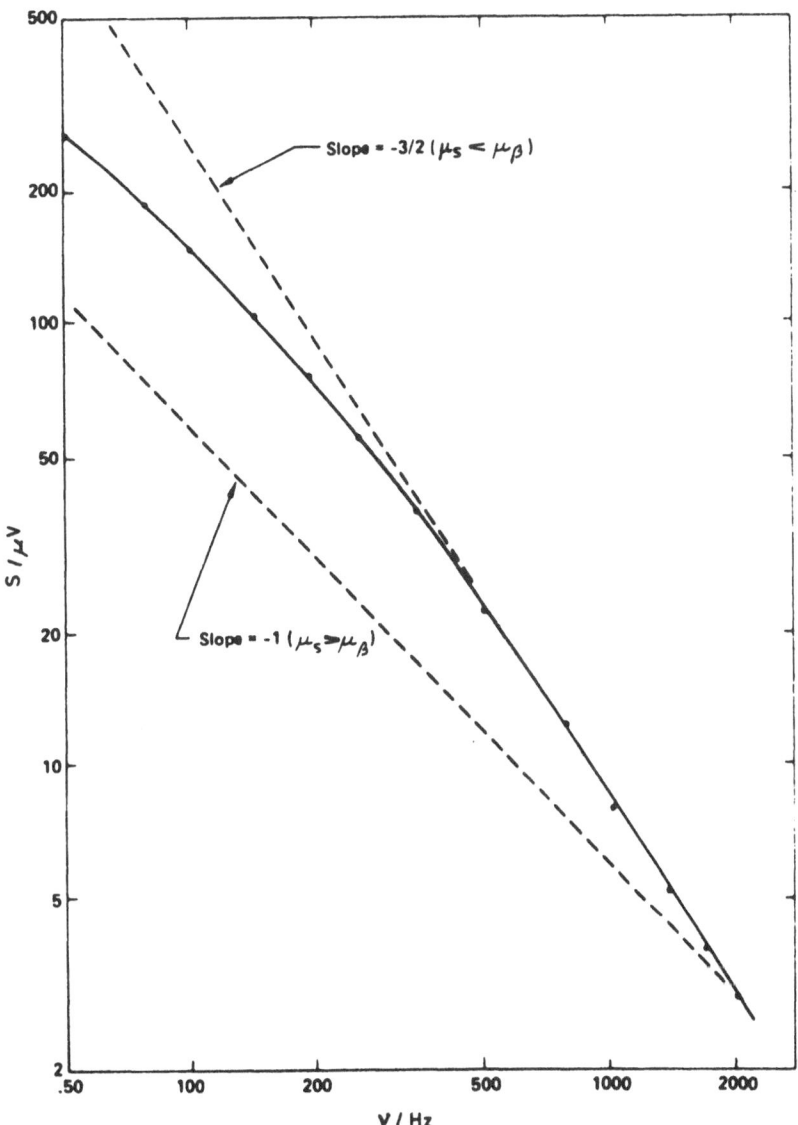

FIGURE 13   Amplitude S of the photoacoustic signal for GaP with 1 mm thickness at 524 nm versus modulation frequency ν from Rosencwaig[18]. Circles (solid line) : experimental data. Dashed lines : calculated.

3. EXPERIMENTAL METHODS

The whole frequency range, from microwaves up to X-rays, can be used in principle as electromagnetic radiation sources, but the most frequently used are Xenon lamps and lasers. For every kind of detection, an optimalisation of the signal/noise ratio is necessary.

Criteria for the sample chambers are:
1. excellent acoustical insulation from outside noise
   (vibrations, etc.)
2. minimisation of optoacoustic signals due to the interaction
   of incident radiation through walls, windows or detectors, etc.

A wide range of detectors are possible, depending on the frequency range chosen, e.g. microphone, piezocrystal, golay-cell, hot wire Schlieren technique. A typical "classical" experimental chamber realized by Rosencwaig[18] is represented in Fig. 14. Various cells are described also in other contributions to this book.

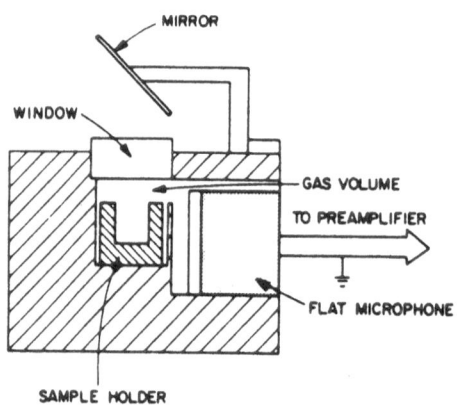

FIGURE 14   Photoacoustic gas-microphone cell (Rosencwaig[18]).

4. EXAMPLES OF APPLICATIONS

Gaining information through optoacoustic measurements is especially suited to the study of the physics of radiationless transitions, caloric properties, and phase transitions.

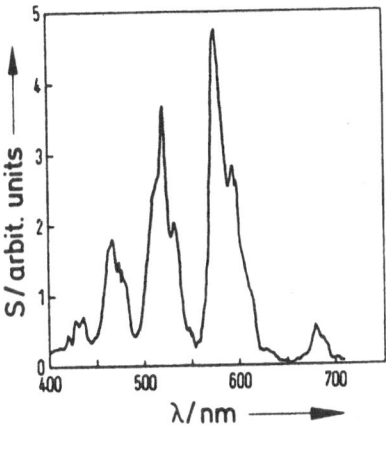
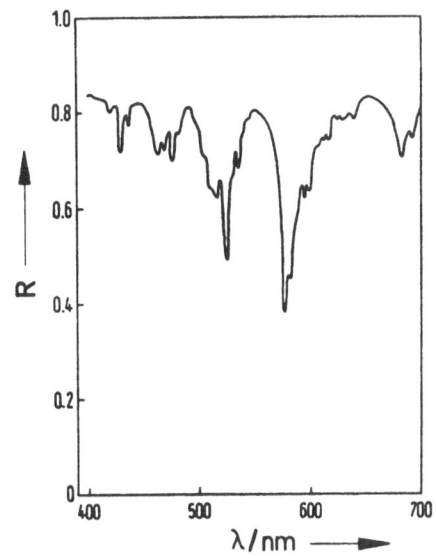

FIGURE 15   Comparison of PAS signal amplitude S with DRS relative re-
            emission R as function of wavelength λ of the light for
            Nd_2O_3. Modulation frequency 190 Hz (Tilgner and Lüscher[20]).

   In a detailed study, we compared optoacoustic spectroscopy to
conventional optical spectroscopy, more especially absorption and reflexion measurements[20]. As an example of our results the spectra of
Nd_2O_3 measured with PAS and diffuse reflectance spectroscopy (DRS) are
given in Fig. 15.

   For many applications there is· no particular advantage in using
the photoacoustic methods, provided that it is an experienced reflexion-
spectroscopist working with them. (See also article of Tilgner in this
volume.)

   A very sensitive parameter in photoacoustics is the phase-relation
between the incident pulsed light and the detected acoustic signal in
the surrounding gas, Fig. 16. Using the phase sensitive lock-in technique to measure the photoacoustic signal, it can be determined experimentally relatively easy. This becomes obvious from the first measurement of the photoacoustic effect at a first-order phase transition in
Gallium, H_2O and K_2SnCl_6 performed by Pelzl and his group in Bochum[21].
In Fig. 17 are shown the results of the measurement at the melting
point of Gallium. An example of a structural phase transition of VO_2 at

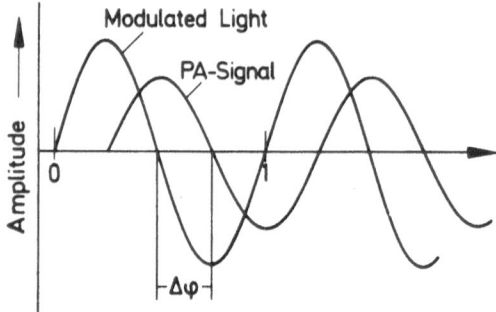

FIGURE 16   Characterisation of the photoacoustic signal with respect to a
             sinusoidally modulated light intensity by the shift of the
             phase angle, $\Delta\varphi$.

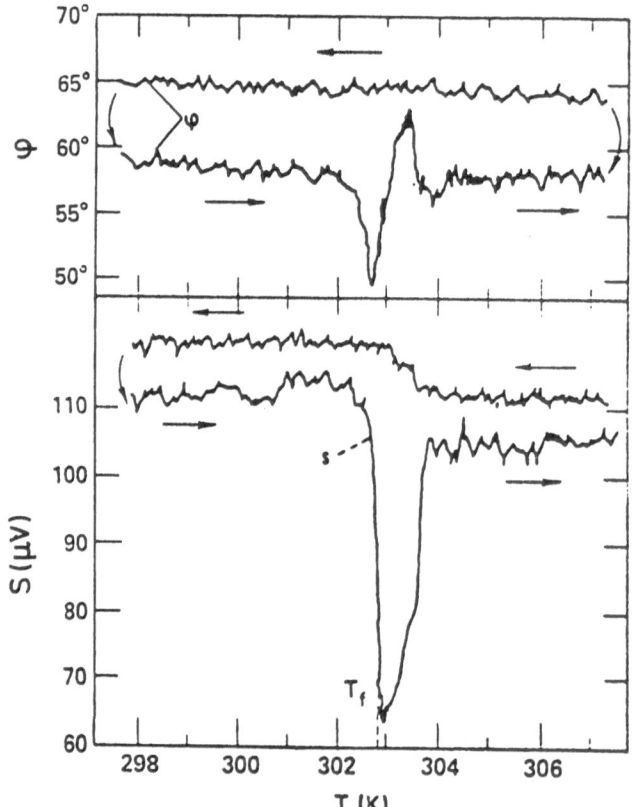

FIGURE 17   Amplitude S and phase angle φ of the PA signal at the melting
             point of gallium, T_f = 302.9 K, measured at increasing ($\rightarrow$)
             and decreasing ($\leftarrow$) temperature by Florian et al.[21].

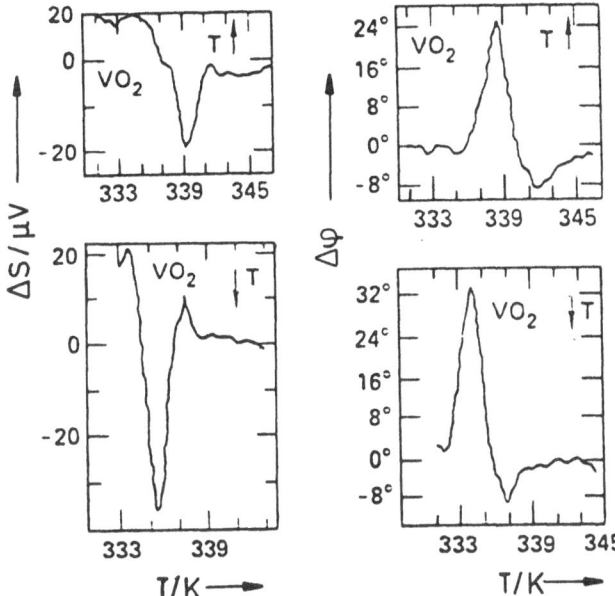

FIGURE 18   Variation of amplitude,ΔS, and phase angle,$\Delta \varphi$, of PAS signal at a structural phase transition in VO_2 near 340 K measured at increasing ($\uparrow$) and decreasing ($\downarrow$) temperature by Korpiun et al.[22].

T_c = 338.6 K is given in Fig. 18 (Korpiun et al.[22]). We found in contradiction to Pelzl and coworker that the temperature dependence of the amplitude and phase relation is reversible with respect to the temperature.

The PA detection of a first order phase transition in a biomolecular system by Knoll et al.[23] should be mentioned also. Chlorophyll a incorporated into dipalmitoyllecithin vesicles can aggregat into domains under certain conditions at a certain temperature. The aggregation modifies a first order phase transition near 43 °C of DPPC which depends on the concentration of chlorophyll a. From the results shown in Fig. 19 it becomes obvious that the phase angle seems to be more sensitive to the phase transition than the amplitude.

Korpiun and Tilgner[24] gave a model for the PAE in the transition region that explain very satisfactorily the experimental results. (See also the contribution of Korpiun in this volume.)

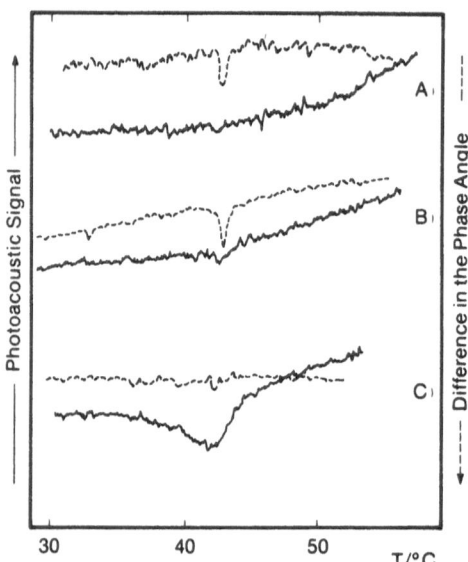

FIGURE 19   Amplitude (full line) and phase angle (dashed line) of DPPC
            dispersion (15 % wt/wt) without (A), with 0.33 mole % (B),
            and with 10 mole % (C) chlorophyll a. The phase transition
            occurs near 43 $^{\circ}$C (Knoll et al.[23]).

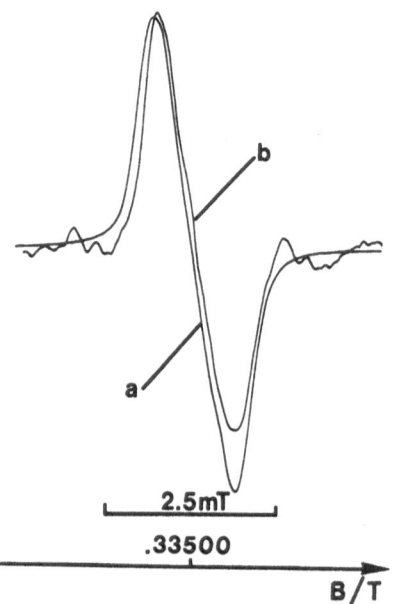

FIGURE 20   First derivative ESR-absorption spectra of DPPH recorded at
            a microwave frequency of 9394283 x 10^3 s^{-1} with a modulation
            frequency of 20 s^{-1} and a modulation amplitude of 1 mT:
            a) conventional detection, b) acoustical detection. Measured
            by Coufal[26].

Another application of the photoacoustic effect is the detection
of magnetic resonances. In the first publication on this field Numes
et al.[25] described the observation of ferromagnetic resonance in Fe
films. The results of ESR measurements on DPPA by Coufal[26] using a
microphone detector is shown in Fig. 20.

5. RESUMÉ

The range of applications for optoacoustic methods in solids is
very wide, but only for very special sample forms (gel, biological materials, etc.) one may obtain additional informations to conventional
optical methods.

REFERENCES

1 A.G. Bell, Am.J.Sci. 20, 305 (1880)

2 Rayleigh (Lord), Nature 23, 274 (1881)

3 M.E. Mercadier, Phil.Mag. 11, 78 (1881)

4 J. Tyndall, Proc.Roy.Soc. 31, 307 (1881)

5 W.H. Preece, Proc.Roy.Soc. 31, 506 (1881)

6 W.C. Roentgen, Ann.Phys.u.Chem. 12, 155 (1881)

7 A.G. Bell, Phil.Mag. 11, 510 (1881)

8 M.L. Veingerov, Dokl.Akad.Nank.USSR 19, 687 (1938)

9 K.F. Luft, Z.techn.Phys. 24, 97 (1943)

10 S.M. Luchin, Dokl.Akad.Nauk.USSR 49, 418 (1945)

11 W.G. Fastie, A.U. Pfund J.O.S.A. 37, 762 (1947)

12 L.B. Kreuzer, C.K.N. Patel, Science 173, 45 (1971)

13 R. Kaiser, Can.J.Phys. 37, 1499 (1959)

14 E. Hey, K. Gollnick, Ber. Bunsengesellschaft 72, 263 (1968)

15 W.R. Harshbarger, M.B. Robin, Acc.of Chem.Res. 6, 329 (1973),
    Chem.Phys.Lett. 21, 462 (1973)

16 A. Rosencwaig, Opt.Com. 7, 305 (1973)

17 A. Rosencwaig, A. Gersho, J.Appl.Phys. <u>47</u>, 64 (1976)

18 A. Rosencwaig, Rev.Sci.Instrum. <u>48</u>, 1133 (1977)

19 F.A. McDonald, G.C. Wetsel, Jr., J.Appl.Phys. <u>49</u>, 2313 (1978)

20 R. Tilgner, E. Lüscher, Z.f.Physikal.Chemie <u>111</u>, 19 (1978)

21 R. Florian, J. Pelzl, M. Rosenberg, H. Vargas, R. Wernhardt, phys. stat.sol. (a) 48, K35 (1978)

22 P. Korpiun, J. Baumann, E. Lüscher, E. Papamokos, R. Tilgner, phys.stat.sol. (a) 58, K13 (1980)

23 W. Knoll, J. Baumann, P. Korpiun, U. Theilen, Biochem.Biophys.Res. Comm. <u>96</u>, 968 (1980)

24 P. Korpiun, R. Tilgner, J.Appl.Phys. <u>51</u>, 6115 (1980)

25 O.A.C. Nunes, A.M.M. Monteiro, K.S. Neto, Appl.Phys.Lett.<u>35</u>, 656 (1979)

26 H. Coufal, Appl.Phys.Lett. <u>39</u>, 215 (1981)

OPTO-ACOUSTIC SPECTROSCOPY - A TOOL FOR THE STUDY OF OPTICAL SPECTRA OF VERY TRANSPARENT MATERIALS

C. K. N. Patel

Bell Laboratories
Murray Hill, New Jersey 07974

1. INTRODUCTION AND HISTORICAL PERSPECTIVE

The principle of opto-acoustic spectroscopy, namely, the measurement of optical absorption spectra through the detection of acoustical signal generated when the absorbed optical energy decays through nonradiative channels, has a long history. The first experiments can be traced to Bell,[1] who in 1880, showed that solar radiation, dispersed with a prism and chopped at an audio frequency is absorbed by different materials, e.g., cloth, carbon-black, etc., to varying amounts depending upon the wavelength and produced varying audio signals. The opto-acoustic spectroscopy is a variation even of an earlier scheme - the calorimetric spectroscopy where the temperature rise of the sample is used to measure the absorbed (and subsequently nonradiatively relaxed) radiation. The application of opto-acoustic detection to modern spectroscopy occurred with the availability of a variety of fixed frequency and tunable lasers having reasonable amounts of power outputs. Atwood and Kerr[2] used a capacitive manometer for measuring the pressure changes induced in a gas when the gas is irradiated by optical radiation which is absorbed by the gas (called a spectrophone instrument). Kreuzer[3] was one of the first to recognize the usefulness of sensitive acoustic microphones for detection of small optical absorption in gases. He used a fixed frequency He-Ne laser at 3.39 μm to measure absorption in methane. Spectroscopy studies using the sensitive opto-acoustic technique and tunable lasers were first reported by Kreuzer and Patel[4] who studied the weak absorptions in a dilute mixture of NO and H_2O in air at ~5.3 μm using tunable spin-flip Raman lasers. The continued use of the term "opto-acoustic" first used by Kreuzer for fixed frequency laser - electret microphone detection and by Kreuzer and Patel for tunable laser-electret microphone studies of gases is appropriate since the opto-acoustic spectroscopy of gases has changed very little over the last ten years in any fundamental fashion. The very sensitive opto-acoustic technique for measurements of very low absorptions in gaseous samples has found numerous scientific and practical applications.[5] Extension of the gas-phase opto-acoustic spectroscopy to condensed phase materials was first shown by Harshbarger and Robin[6] and by Rosencwaig[7] who coined the word "photo-acoustic" for the opto-acoustic studies of condensed phase samples where the optical absorption takes place in the condensed phase sample while the acoustic signal pick-up is by a gas phase microphone placed in a closed chamber (filled with an optically transparent gas) which contains the sample to be studied. Photo-acoustic spectroscopy has many positive aspects for spectroscopic studies of materials but the ability to measure small absorption coefficients is not one of them. The photo-acoustic spectroscopy, as it is now generally known, is generally useful only for the study of relatively opaque materials. Acoustic impedance mismatch between the condensed phase sample and the gas medium which transmits the acoustic signal to the microphone results in very little transmission of the acoustic signal generated in the condensed phase sample into the gaseous medium. The "acoustic" signal detected by the microphone arises from the periodic heating of the gas in contact with the condensed sample surface and thus the term "photo-acoustic" has been a misnomer. Such a spectroscopic technique is a "photo-thermal" one. (This heating of the gas in contact with illuminated surface has very important implications in applications involving depth profiling[8] but the chopping frequency dependence of the signal output[9] from the microphone makes quantitative studies of optical absorption at best very difficult if not impossible.) To get around the problems of impedance mismatch, schemes that do not use a gaseous medium for acoustic signal pickup rely on a piezoelectric transducer immersed in a liquid or attached to a solid. Since the piezoelectric transducers have very good high frequency response, the latter scheme generally is seen to work best with pulsed tunable lasers and the technique is called opto-acoustic spectroscopy[10] of condensed phase samples and the nomenclature is appropriate for distinguishing it from the earlier "photo-acoustic" technique. Earlier attempts to use chopped CW laser radiation for opto-acoustic spectroscopy of liquids[11] and solids [12] have met with only moderate success. Figure 1 puts the three currently used acoustic detection schemes - gas phase, solids, and liquid and solids - into perspective.

In the present paper, I will review the pulsed laser, immersed (for liquids) or attached (for solids) transducer gated detection opto-acoustic spectroscopy. We have shown that this form of spectroscopy is capable of measuring absorption coefficient - sample length ($\alpha\ell$) products as small as $\sim 10^{-7}$ over a wavelength that covers the region from ~ 450 nm to $\sim 1.6\ \mu$m. While the purpose of the paper is not to give complete details of the many studies that have been carried out to date, I will give a brief description of the experimental techniques and of the various studies to acquaint the readers with the status of what has turned out to be an extremely powerful optical technique for measuring optical spectra of very transparent materials.

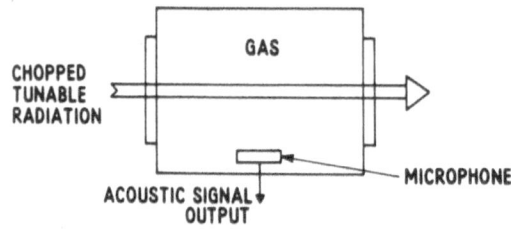

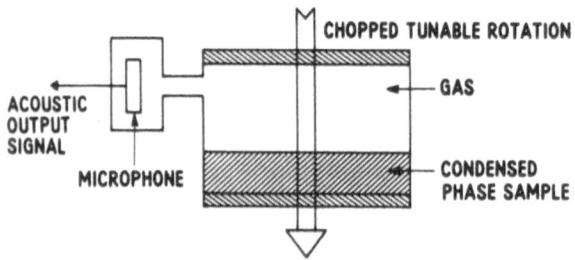

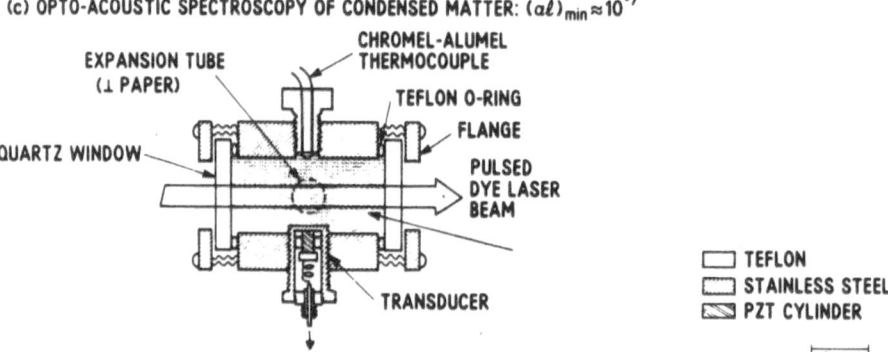

Fig. 1.   Schematic representation of opto-acoustic spectroscopy of gases, photo-acoustic spectroscopy and opto-acoustic spectroscopy of condensed matter.

2. EXPERIMENTAL DETAILS

1. Transducer

   All of the pulsed laser opto-acoustic spectroscopy studies to date have been carried out using piezoelectric ceramic transducers.[13] For opto-acoustic studies of transparent materials (as well as for all other studies), the piezo-electric material is enclosed in a stainless steel enclosure[14] (shown in Figure 2) which serves a number of very useful and essential purposes as opposed to using bare transducers. These are summarized below.

   a. In the case of liquid opto-acoustic studies, where the piezoelectric transducer is immersed in the liquid, it is necessary to ascertain that

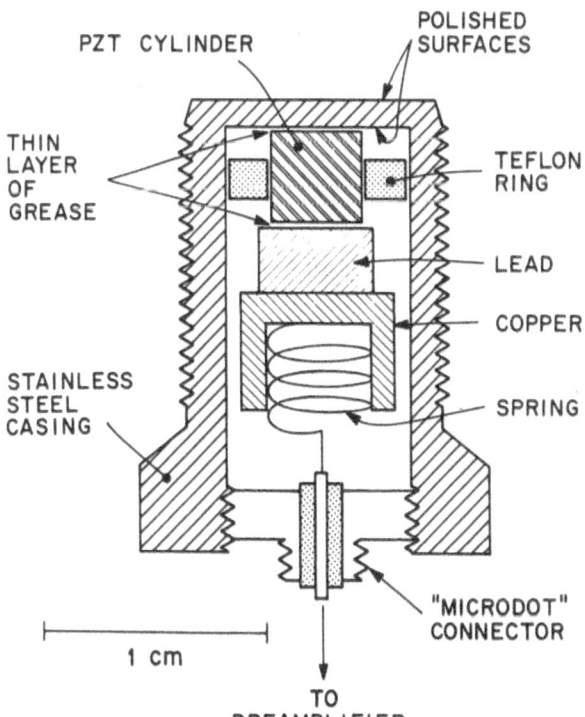

Fig. 2.

Cross-section of piezo-electric transducer assembly.

the liquid is not contaminated. The enclosed geometry assures us that only inert surface is in contact with liquid which is a prerequisite for minimum contamination. Bare piezoelectric ceramic transducers immersed in the liquid clearly pose significant danger of contamination and subsequent uncertainty regarding the origin of the observed opto-acoustic signal, i.e., whether the observed opto-acoustic signal arises from optical absorption in the liquid under investigation or from the contaminants introduced by the bare transducer.

b. Scattered light from the bulk can be intercepted by the transducer and absorbed, giving rise to false signals which would interfere with the opto-acoustic signals arising from bulk absorption in the sample. The bare transducers are nearly optically black resulting in a potentially serious problem from scattered light absorption by the transducer. The enclosed geometry, shown in Figure 2, helps minimize such interference signals since the stainless steel diaphragm separating the piezo-electric ceramic material and the material being investigated is highly polished on the side that faces the sample. Hence most if not all of the bulk scattered light intercepted by the transducer assembly is reflected. Thus, typically, the interference signals arising from the scattered light are reduced by a factor of about 20 to 50.

c. A practical experimental problem of electrical pickup when we use pulsed lasers for the studies is exacerbated with the use of bare transducers but is minimized or is nonexistent when the piezoelectric transducer is electrically shielded in the housing as shown in Figure 2.

2. Transducer preamplifier: Because of the low levels of signal output, we have used a special preamplifier to optimize signal-to-noise ratio. Details of the electronic circuit and its characteristic are given in Ref. 15. Further amplification was provided by commercial preamplifiers such as Ithaco 1120.

3. Lasers: Two types of tunable dye lasers have been used in the present studies. The first is a flash lamp pumped dye laser (Chromatix CMX-4) capable of producing laser pulses of energy ~ 1 mJ, in a pulse duration of $1-2$ μsec with a repetition rate of 10 Hz. Tunability of this laser extends from ~ 450 nm to ~ 700 nm (with a variety of dyes). The other laser, which has now become the workhorse of most of our studies, is a dye laser pumped with frequency doubled output from a Nd:YAG laser. This laser produced pulses of energy > 10 mJ, with a duration of ~ 7 nsec at a repetition rate of 10 Hz. Tunability of this primary dye laser extends from ~ 560 nm to ~ 800 nm. This laser has been further used to produce stimulated Raman scattering in high pressure hydrogen gas which extends the tunability range of the laser out to ~ 1.6 μm.

4. Electronics: The pulsed acoustic output from the piezoelectric transducer is in the form of a ringing waveform (due to mechanical resonances of the piezo-electric ceramic material). This is detected using boxcar detectors. Laser power, for normalization purposes, is measured using a broadband pyroelectric detector. Data collection and normalization is accomplished using a minicomputer as shown in Figure 3.

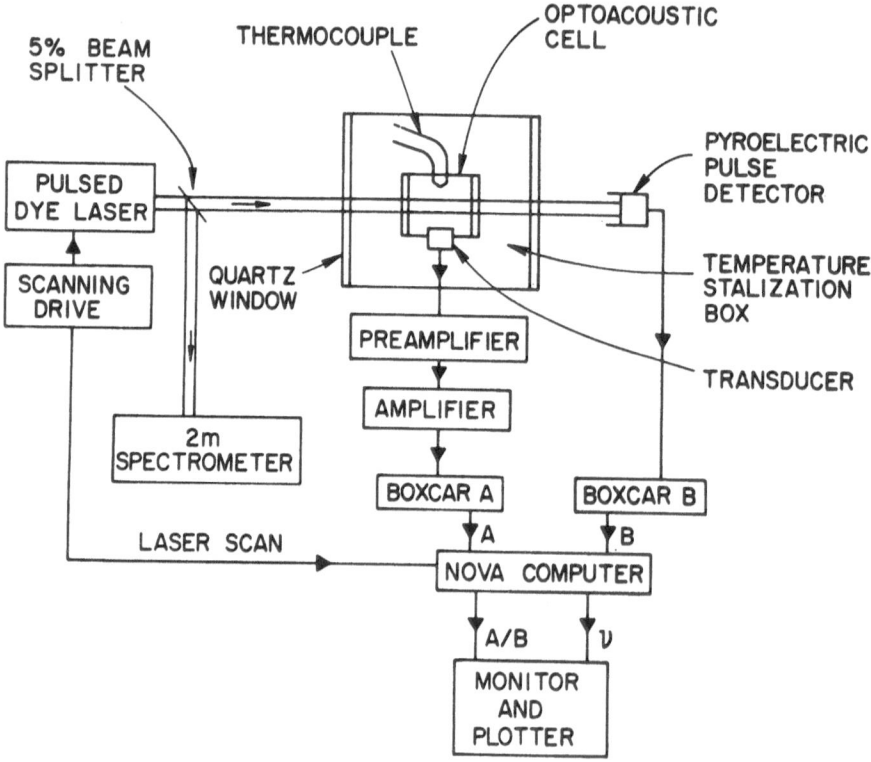

Fig. 3.   Experimental setup for opto-acoustic spectroscopy of condensed
           phase materials.

3. OPTO-ACOUSTIC RESPONSE

The voltage output from the transducer is directly related to the absorption

coefficient as shown in Refs. 10, 15 and 16. We will not rederive the

expressions but give them below. The crucial parameters that determine the

relationship of the acoustic voltage output on the various thermo-physical

parameters of the material being studied are the laser pulse length, τ_p, acoustic

transit time across the laser illuminated column of the material, τ_a, and the response time of the piezo-electric (or other) transducer, τ_r. The acoustic transit time is given by

$$\tau_a = \frac{r}{v_a} \tag{1}$$

where r is the radius of the laser beam in the medium and v_a is the acoustic velocity.

a. For $\tau_p > \tau_a, \tau_r$

$$\frac{V_{oa}}{E} = K_p \frac{\beta v_a}{C_p} \alpha \tag{2}$$

b. For $\tau_r \gg \tau_a$ and $\tau_a \gg \tau_p$

$$\frac{V_{oa}}{E} = K_a \frac{\beta v_a^2}{C_p} \left[\frac{Z_{abs}}{Z_{abs} + Z_{pzt}} \right] \alpha \tag{3}$$

c. For $\tau_a \gg \tau_r$ and $\tau_a \gg \tau_p$

$$\frac{V_{oa}}{E} = K_r \frac{\beta v_a^2}{C_p} \alpha \tag{4}$$

where V_{oa} is the voltage output from the transducer, E is the laser pulse energy, K_p, K_a, and K_r are constants which depend upon the specific geometry, transducer response, etc., β is the voltmetric expansion coefficient, C_p is the isobaric specific heat of the medium, and Z_{abs} and Z_{pzt} are acoustic impedances of the optical absorber and the piezoelectric transducer, respectively. α is the optical absorption coefficient. We have tacitly assumed here that $\alpha\ell \ll 1$ where ℓ is the length of the medium being studied.

Calibration of the experimentally observed V_{oa}/E to give a quantitative value of absorption coefficient requires an evaluation of K's. This is best done through comparison with a "doped" sample as discussed in detail in Ref. 17. However, a word of caution to the reader: when transferring the calibration from a known material to another, care should be exercised to use the appropriate relations from those given in Eqs. (2)-(4).

For the experimental situations encountered so far, $\tau_r \approx 1-2$ μsec for piezoelectric ceramic transducers used. For the flash lamp pumped dye laser, $\tau_p \approx 2$ μsec and $\tau_a \approx 0.9$ μsec (typically $r \approx 1.5$ mm and $v_a \sim 1.7 \times 10^5$ cm sec^{-1}). Thus Eq. (2) applies. For the frequency doubled Nd:YAG pumped dye laser (and its extended tuning range using stimulated Raman scattering in high pressure hydrogen), $\tau_p \approx 7$ nsec. Thus Eq. (3) applies. See Ref. 16 for details.

4. LIMITS TO OPTO-ACOUSTIC DETECTION SENSITIVITY

Presently achieved sensitivity for measuring small absorption coefficient is

$$(\alpha \ell)_{\min} \approx 10^{-7} \tag{5}$$

which is limited by the preamplifier noise. Notice that with the minimum $\alpha \ell$ described in Eq. (2) we can measure very small absorptions of the order of $\sim 10^{-7}$ cm^{-1} for reasonably long (~ 1 cm) samples or for strong absorbers, say $\alpha \sim 10^3$ cm^{-1}, we can carry out spectroscopy on very thin layers of $\leq$ monolayer. However, other sources of limitation will come into the picture as we improve the transducer responsivity and the characteristics of the preamplifier. These are 1) window absorption signal (in the case of liquid opto-acoustic spectroscopy), 2) bulk scattering signal, and 3) electrostriction signal. The first two can be minimized by time gating the acoustic voltage output from the transducer and by ingeneous schemes described in Refs. 15 and 18. The third, which has been the subject of considerable confusion,[19] does not become a problem for $\alpha\ell$ as small as $\sim 10^{-9}$ cm^{-1} as discussed in Ref. 15. Further, the electrostriction signal is largely wavelength independent and thus it corresponds to a constant background signal which can be subtracted electronically to bring out the wavelength dependent absorption peaks. Hence, considerable improvement in minimum $\alpha\ell$ is to be expected by merely increasing the laser pulse energy and repetition rate.

5. EXPERIMENTAL RESULTS

The demonstrated ability to measure $\alpha\ell$ as small as $\sim 10^{-7}$ has led to a number of exciting linear and nonlinear spectroscopic studies. In this paper I will not discuss the details of those studies that have already been published but merely list them with appropriate references.

1. Linear Spectroscopy

    a. Aromatic hydrocarbons: Measurements have been made of the weak 6th, 7th and 8th harmonics[20] of C-H stretch of C_6H_6 (see Figure 4). A highly corrosion resistant opto-acoustic cell[21] for rapid studies of weakly absorbing liquids has made routine measurements up to $\alpha\ell \lesssim 10^{-6}$ easy (see Figure 5). Measurements of the 6th harmonic of C-H stretch in substituted benzenes have been reported.[21]

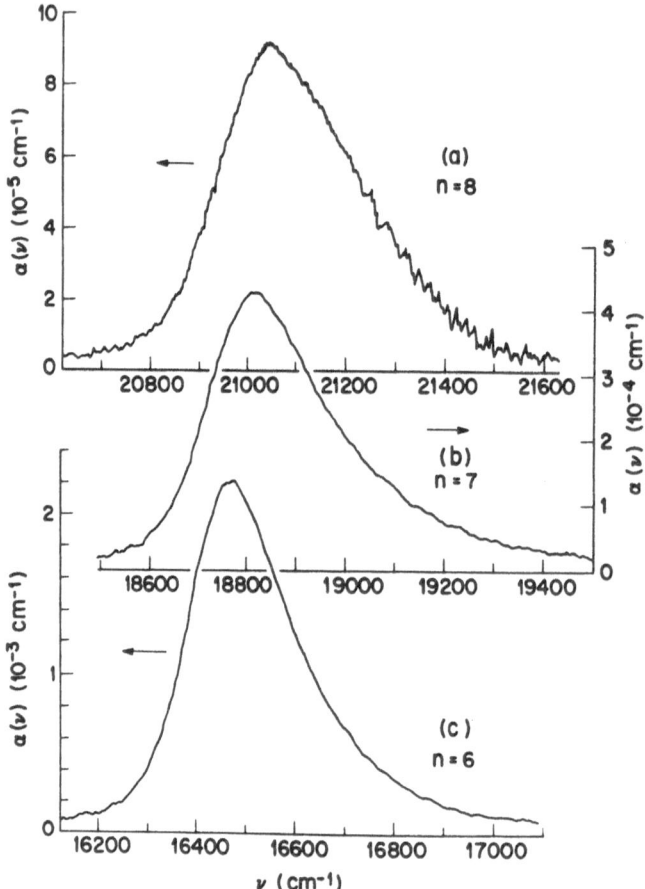

Fig. 4.   Absorption spectra of 6th, 7th, and 8th harmonics of C-H stretch in benzene measured by opto-acoustic spectroscopy.

b. H_2O and D_2O: Accurate and reproducible optical absorption measurements of H_2O and D_2O have now become possible.[14,22] A typical absorption spectrum of H_2O is seen in Figure 6.

c. Thin Liquid Film: Measurements of thin liquid film (1 μm–5 μm thick) of Ho_2O_3, etc., solution have been reported.[18] These results show that films as thin as monolayers can be measured for strongly absorbing liquids.

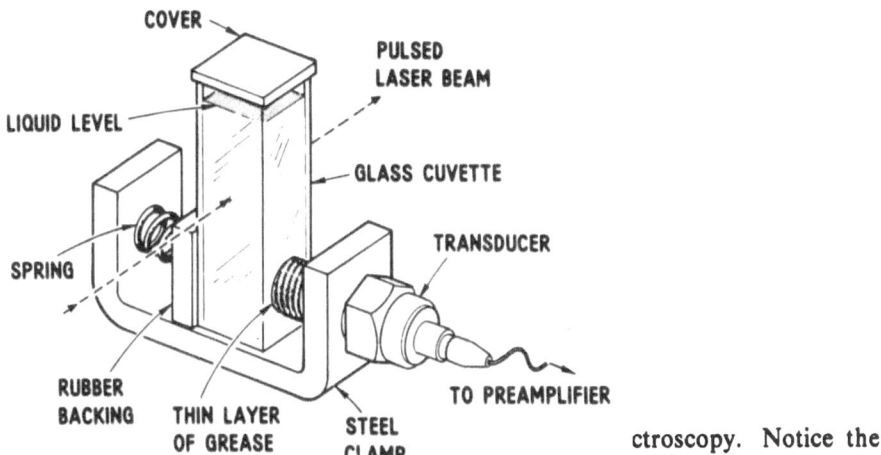

ctroscopy. Notice the easy interchangeability of sample couvetts facilitates rapid change of samples.

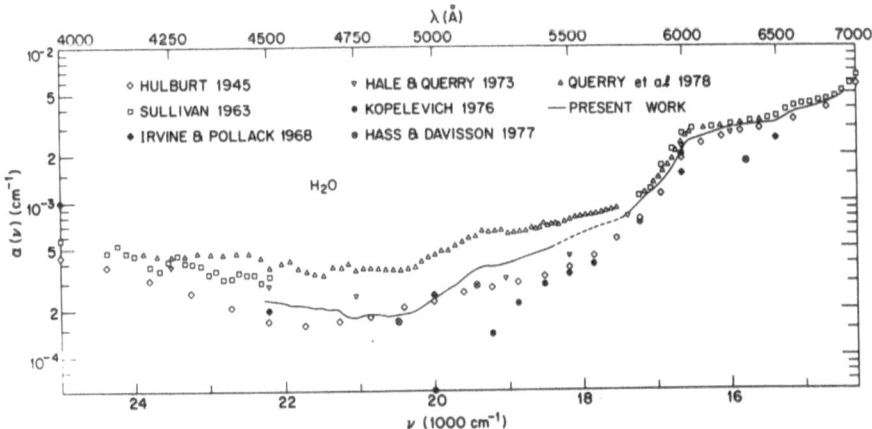

spectroscopy. (For details see Refs. 14 and 22.)

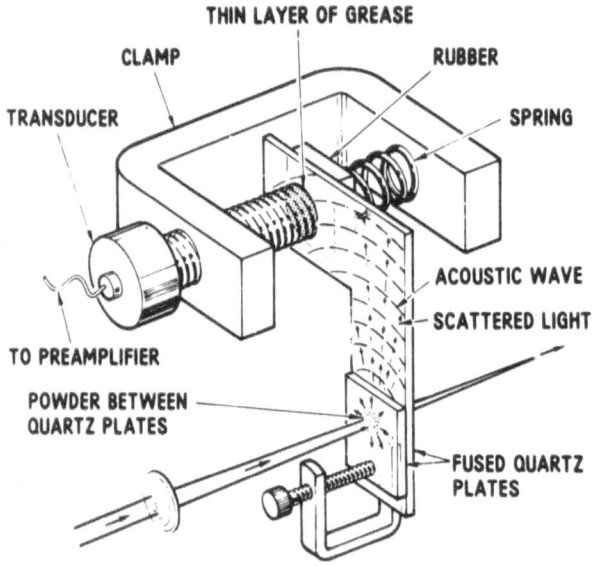

Fig. 7.

Opto-acoustic "cell" for

thin-films of liquids,

powders and solids.

d. Powders: An ingeneous experimental arrangement (see Figure 7) has allowed a study of powdered materials which are highly scattering. Spectra of Ho_2O_3, Er_2O_3, and Dy_2O_3 have been measured.[23] Path is now paved for measurements of small absorption coefficients in powders. This capability may turn out to be extremely crucial in exploratory studies[24] of low loss materials for long wavelength optical fibers.

e. Solids: Experiments are under way to study solid materials of technological interest such as preforms for low loss optical fibers and thin epitaxial films of semiconductors such as GaAs, InP, etc., using the experimental setup shown in Figure 8.

f. Low temperature studies: Since the piezo-electric transducers have no obvious low temperature limitations, the opto-acoustic spectroscopy scheme described above can be directly extended to

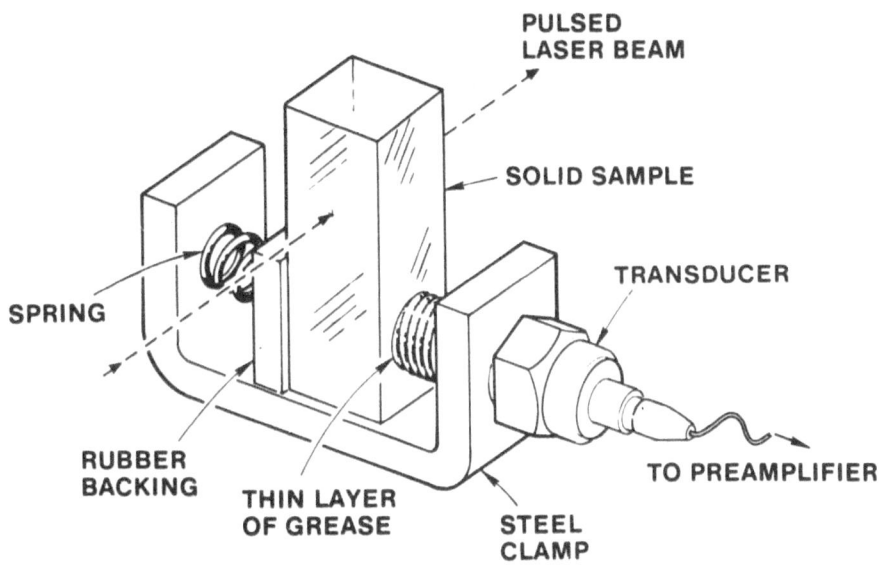

low temperatures. Cryogenic liquids such as liquid methane[25] and liquid ethylene[26,27] have already been measured and show a host of fascinating weak absorption features (see Figure 9 for a spectrum of methane). For details of why methane and ethylene studies are interesting, see Refs. 25 and 26.

g. Long wavelength opto-acoustic studies: For solids mentioned in (e) wavelength of interest is $\lambda > 0.8$ μm. The stimulated Raman scattering in high pressure hydrogen (described above) has allowed demonstration that opto-acoustic spectroscopy can be carried out[27] up to $\lambda \approx 1.6$ μm with the same sensitivity as that possible in the visible region (i.e., $(\alpha\ell)_{min} \approx 10^{-7}$).

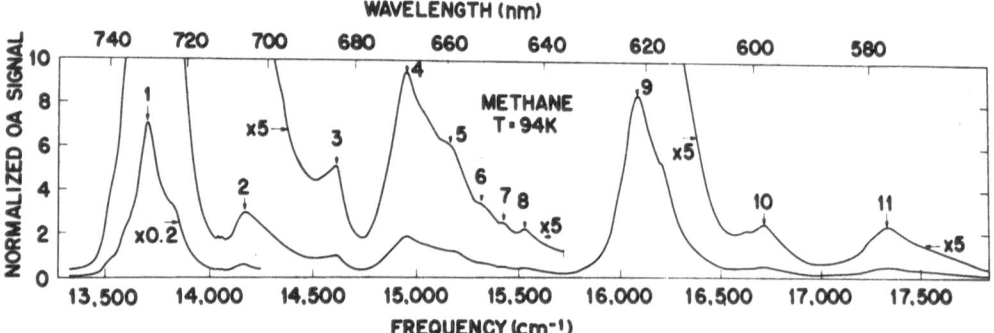

Fig. 9.   Absorption spectrum of liquid methane measured using opto-
           acoustic techniques.

2. Nonlinear Spectroscopy

    a. Two-photon absorption: Weak two photon absorption measurements
        using opto-acoustic techniques in C_6H_6 due to the $^1A_{1g}$–$^1B_{2u}$
        transition has been carried out.[28] First accurate, quantitative
        measurements of two photon absorption cross-section have been
        reported. Smallest two photon absorption cross-section that can be
        measured under present circumstances is estimated to be
        $\sigma_{2\omega} \approx 10^{-53}$ cm^4 sec photon^{-1} mol^{-1}.

    b. Raman gain: By using two laser radiations which can be tuned so
        that the difference frequency equals the Raman allowed frequency of
        the medium, opto-acoustic Raman gain spectroscopy (OARS) of
        liquids such as C_6H_6, 1,1,1 trichloroethane, acetone, toluene and n-
        hexane have been carried out.[29] Both polarized as well as
        depolarized parts of the Raman scattering tensor can be studied.
        Smallest Raman gain that can be measured so far (using the OARS
        technique) is $\sim 10^{-5}$ cm^{-1}.

6. FUTURE

The advantages of (a) high detection sensitivity described above, (b) use of pulsed lasers which make studies over broad wavelength regions easier (than using CW lasers), (c) use of time gating for discrimination of unwanted interference signals, (d) ability to obtain quantitative data on absorption coefficients, and (e) straightforward interpretation of the experimental data clearly make the pulsed laser, submerged (in the case of liquids) or attached (in the case of solids) transducer, gated detection opto-acoustic technique an exciting tool for optical spectroscopy of weakly absorbing condensed phase materials.

A large number of future studies are suggested from what has already been demonstrated. These include:

1. Studies of higher order Raman processes.

2. Materials testing (see also V.1.e and V.1.g).

3. Trace and impurity detection.

4. Studies of forbidden transitions.

5. Studies of excited states.

6. Monolayers.

7. Opto-acoustic microscopy.

8. Extensions to electron beam and X-ray excitation of the medium.

It is clear that the future of opto-acoustic spectroscopy of condensed phase materials will be limited only by the imaginative and creative applications of the technique.

REFERENCES

1.   A. G. Bell, *Proc. Am. Assoc, Adv. Sci.* **29,** 115 (1880).

2.   E. L. Kerr and J. G. Atwood, *Appl. Opt.* **7,** 915 (1968).

3.   L. B. Kreuzer, *J. Appl. Phys.* **42,** 2934 (1971).

4.   L. B. Kreuzer and C. K. N. Patel, *Science,* **173,** 45 (1971).

5.   C. K. N. Patel, *Science,* **202,** 157 (1978) and references cited therein; C. K. N. Patel, R. J. Kerl and E. G. Burkhardt, *Phys. Rev. Lett.* **38,** 1204 (1977); C. K. N. Patel, *Phys. Rev. Lett.* **40,** 535 (1978).

6.   W. R. Harshbarger and M. B. Robin, *Acc. Chem. Res.* **6,** 329 (1973).

7.   A. Rosencwaig, *Opt. Comm.* **7,** 305 (1973).

8.   M. J. Adams and G. F. Kirkbright, *Analyst,* **102,** 678 (1977).

9.   A. Rosencwaig and A. Gersho, *J. Appl. Phys.* **47,** 64 (1976).

10.   C. K. N. Patel and A. C. Tam, *Appl. Phys. Lett.* **34,** 467 (1979).

11.   Y. Kohanzadeh, J. R. Whinnery, and M. M. Carroll, *J. Acoust. Soc. Am.* **57,** 67 (1975).

12.   A. Hordvik, *Appl. Opt.* **16,** 2827 (1977).

13.   Lead zirconate titanate (PLZT) material LTZ-2 obtained from Transducer Products, Conn. U.S.A.

14.   C. K. N. Patel and A. C. Tam, *Nature,* **280,** 302 (1979).

15.   C. K. N. Patel and A. C. Tam, *Rev. Mod. Physics* (to appear in July 1981 publication).

16. E. T. Nelson and C. K. N. Patel, *Opt. Lett.* (to appear in July 1981 publication).

17. A. C. Tam, C. K. N. Patel and R. J. Kerl, *Opt. Lett,* **4,** 81, (1979).

18. C. K. N. Patel and A. C. Tam, *Appl. Phys. Lett.* **36,** 7 (1980).

19. S. R. J. Brueck, H. Kildal, and L. J. Belanger, *Opt. Comm.* **34,** 199 (1980).

20. C. K. N. Patel, A. C. Tam, and R. J. Kerl, *J. Chem. Phys.* **71,** 1470 (1979).

21. A. C. Tam and C. K. N. Patel, *Opt. Lett.* **5,** 27 (1980).

22. A. C. Tam and C. K. N. Patel, *Appl. Opt.* **18,** 3348 (1979).

23. A. C. Tam and C. K. N. Patel, *Appl. Phys. Lett.* **35,** 843 (1979).

24. C. K. N. Patel, *Proceedings of SPIE Symposium,* (February 12, 1981, Los Angeles, Calif.).

25. C. K. N. Patel, E. T. Nelson and R. J. Kerl, *Nature,* **286,** 368 (1980).

26. E. T. Nelson and C. K. N. Patel, *Proc. Nat. Acad. Sci.* **78,** 702 (1981).

27. E. T. Nelson and C. K. N. Patel, (to be published).

28. A. C. Tam and C. K. N. Patel, *Nature,* **280,** 304 (1979).

29. C. K. N. Patel and A. C. Tam, *Appl. Phys. Lett.* **34,** 760 (1979).

THERMODYNAMIC MODELS OF THE PHOTOACOUSTIC EFFECT

Peter Korpiun

Physik-Department, Technische Universität München, D-8046 Garching

I. INTRODUCTION

The source of the acoustic signal is the heat, in whatever manner it may be created from the energy of light absorbed by the sample. The whole process, beginning with the created heat δQ and ending when the pressure signal δP arrives at the acoustical detector, can be described using the formalism of thermodynamics.

In the present contribution the considerations are restricted to the thermal aspects of the photoacoustic effect (PAE) in gas - microphone cells. It should be possible to study with the PAE in principle those thermodynamic properties that govern the creation of heat, its diffusion through the sample to the gas and its influence on the pressure of the gas. Such thermodynamic properties are the thermal conductivity, the specific heat capacity, the latent heat at a first-order phase transition, the density, the thermal expansion coefficient, and the compressibility. The samples of condensed matter that may be investigated are solid, amorphous and liquid samples.

The first theoretical treatment of the PAE on condensed samples have been published by Parker[1]. It was followed by the model of Rosencwaig and Gersho[2] that nowadays is used mostly to interpret the experimental results. Later on, Aamodt, Murphy and Parker[3] have worked out the model of Parker more precisely. Based on the same ideas McDonald and Wetsel[4] have developed a fairly general description of the PAE. In the present contribution the main features of the models of Rosencwaig and Gersho and of McDonald and Wetsel will be presented.

In the model of Rosencwaig and Gersho[2] the pressure variation δP in the gas surrounding the sample is only due to the temperature variation treated by the transport of heat δQ from the sample to the gas. The temperature distribution in the sample and the gas is obtained as solutions of the differential equation of conduction of heat.

Following the ideas of Parker[1], McDonald and Wetsel[4] take into account the pressure as a variable originating from the thermal expansion. They start from a set of two equations. The first one is the acoustic wave equation, the second one is the differential equation of conduction of heat. Both are modified by terms concerning the thermal expansion.

2. MODEL OF ROSENCWAIG AND GERSHO (RG MODEL)

The scheme of a gas-microphone cell with a condensed sample is shown in Fig. 1. It consists of the sample with length l_s fixed to the bottom of the cell, the backing with length l_b and the gas, the coupling medium between sample and microphone, with length l_g. It is closed by an optical window with thickness l_w. The intensity of the light is modulated sinusoidally. It is at the sample surface $x = 0$, Fig. 1,

$$J = \frac{J_0}{2} (1+e^{j\omega t}) \quad . \tag{1}$$

If the absorption coefficient of the sample is β the energy per unit area absorbed along the distance from x to $x + dx$ is

$$\beta J e^{\beta x} dx \quad . \tag{2}$$

It is equal to the heat power created in the sample if the efficiency for the conversion to heat is equal to unity[2].

2.1 Temperature Distribution

The heat expressed by Eq. (2) warms up sample, gas and backing by an amount of $\phi (x,t)$ above the ambient temperature T_0,

$$\phi(x,t) = T(x,t) - T_0 \quad . \tag{3}$$

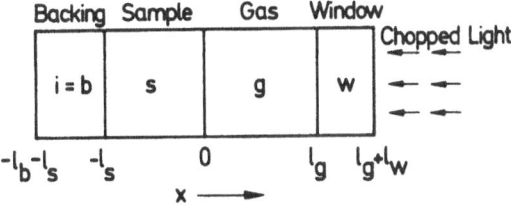

FIGURE 1   Scheme of a PA gas-microphon cell. s = sample, g = gas,
            b = backing, w = window. The window is not considered in
            the RG model.

The temperature field in the sample cell assembly is described by the differential equation of conduction of heat. For the sample $-l_s \leq x \leq 0$ it has the form

$$\frac{\partial^2 \phi_s}{\partial x^2} - \frac{1}{\alpha_s} \frac{\partial \phi_s}{\partial t} = - \frac{\beta J_0}{2\lambda_s} e^{\beta x} (1+e^{j\omega t}) \qquad (4)$$

with a term for heat sources according to Eq. (2). The thermal diffusivity is defined by

$$\alpha_s = \lambda_s / c_{ps} \rho_s \quad , \qquad (5)$$

where λ_s is the thermal conductivity, c_{ps} the specific heat capacity and ρ_s the density of the sample.

For the gas and the backing material without heat sources, there hold the equations

$$\frac{\partial^2 \phi_g}{\partial x^2} - \frac{1}{\alpha_g} \frac{\partial \phi_g}{\partial t} = 0 \quad , \qquad 0 \leq x \leq l_g \quad , \qquad (6)$$

and

$$\frac{\partial^2 \phi_s}{\partial x^2} - \frac{1}{\alpha_b} \frac{\partial \phi_s}{\partial t} = 0 \quad , \qquad -l_s - l_g \leq x - l_s \quad . \qquad (7)$$

For the solutions of the equations (4), (6) and (7) Rosencwaig and Gersho made the following ansatz:

$$\phi_s(x,t) = e_{1s} + e_{2s}x + de^{\beta x} + (U_s e^{\sigma_s x} + V_s e^{-\sigma_s x} - E_1 \beta t)e^{j\omega t} \quad , \qquad (8)$$

in the sample $-l_s \leq x \leq 0$,

$$\phi_g(x,t) = e_{1g} + e_{2g}x + V_g e^{-\sigma_g x + j\omega t} \qquad (9)$$

in the gas $0 \leq x \leq l_g$, and

$$\phi_b(x,t) = e_{1b} + e_{2b}x + V_b e^{\sigma_b x + j\omega t} \quad , \qquad (10)$$

in the backing material $-l_s - l_b \leq x \leq -l_s$.

Here is

$$\sigma_i = (1+j) a_i \quad , \qquad (11)$$

where

$$a_i = \sqrt{\frac{\omega}{2\alpha_i}} \qquad\qquad (12)$$

is the thermal diffusion coefficient of medium labelled by i. The term
with amplitude U_s represents a temperature wave running to the left,
the term with V_s a temperature wave running to the right. The waves are
damped to e^{-1} along the distance of a thermal diffusion length μ_i.
That is $1/2\pi$ of the wave length λ_{Ti} of the temperature wave or

$$\lambda_{Ti} = 2\pi\,\mu_i \quad . \qquad\qquad (13)$$

The gas and the backing material are assumed to be thermally very thick,
that means

$$a_g l_g \gg 1 \quad , \quad a_b l_b \gg 1 \quad . \qquad\qquad (14)$$

Therefore the waves are damped out at $x = l_g$ and $x = -l_b$ respectively
and no temperature wave are reflected.
The limitation of the RG model to thermally thick gas columns, $a_g l_g \gg 1$,
is an essential restriction of its applicability to experimental situations. As it is shown below, the signal increases with decreasing
length of the gas column.

The terms linear in x describe stationary temperature gradients.
They become important, if processes are investigated that depend on
temperature, e.g. phase transitions[5].

The real constants e_{hi} and the complex constants U_i and V_i are
determined by the continuity of temperature and heat flow at the
boundaries.

As long as one is not interested in temperature dependent problems,
it is sufficient to know U_i and V_i. Both constants depend on the thermal
properties of the sample cell assembly, mentioned above and the cell
geometry. Rosencwaig and Gersho have shown that important informations
are obtained from measurements of the PA signal, that means the temperature amplitude as a function of the modulation frequency. Plots of
the amplitude over frequency are characterized by: 1. the ratio of the
sample length to the optical absorption length, βl_s; 2. the ratio of l_s
to the thermal diffusion length a_s^{-1} of the sample, $a_s l_s$; 3. the ratio
of the thermal diffusion length to the optical absorption length, β/a_s.

Rosencwaig and Gersho have presented a full discussion of this problem. Therefore, the reader should be referred to Sect. V in the original work[2].

2.2 Pressure Temperature Relation

It is evident that the temperature of the gas is related to the pressure by an equation of state. In the region of audio frequencies, $\lesssim 10^4$ Hz, the wave length of the sound wave is much greater than the length l_g of the gas column. Therefore, the pressure of the gas does not depend on x, Fig. 1. The problem is to determine the pressure that corresponds to a spatial temperature distribution in the gas in the form of Eq. (9), if the pressure is assumed to be constant all over the volume.

Rosencwaig and Gersho have solved this problem by introducing a spatially averaged temperature. They define an average temperature variation $<\delta T>$ in a gas column with the length of a temperature wave length $2\pi\mu_g$, Eq. (13), in the form of

$$<\delta T> = \frac{1}{2\pi\mu_g} \int_0^{2\pi\mu_g} V_g \, e^{-\sigma_g x + j\omega t} \, dx = \frac{\sqrt{2}}{4\pi} V_g \, e^{j(\omega t - \frac{\pi}{4})} \quad , \tag{15}$$

where V_g is the complex amplitude of Eq. (9). This average temperature is one of the mean features of the RG model.

The relation between $<\delta T>$ and the pressure variation δP is obtained from the assumption that the thermodynamic process is performed in two steps.

In the first step the temperature variation $<\delta T>$ varies isobarically the volume of the gas column with the initial length $2\pi\mu_g$ by the amount (Fig. 2)

$$\delta V = V \, \beta_T \, <\delta T> = 2\pi\mu_g \, A \, \beta_T \, <\delta T> \quad , \tag{16}$$

using the well known relation

$$(\frac{\partial V}{\partial T})_P = V \, \beta_T \quad . \tag{17}$$

Here, β_T is the thermal cubic expansion coefficient of the gas and A the cross section of the gas column.

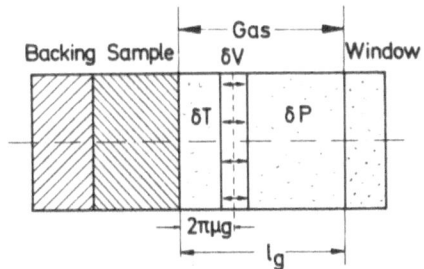

FIGURE 2   Piston model of Rosencwaig and Gersho for a cylindrical
            assembly with cross section A. Two step process
            $<\delta T> \rightarrow \delta V \rightarrow \delta P$

In the second step, the displacement δV, Eq. (16) is considered as a
"piston" acting on the remaining gas column and leading to an adiabatic
variation of pressure (Fig. 2),

$$\delta P = \frac{\delta V}{\varkappa_S \, A \, l_g} \quad . \tag{18}$$

It is obtained from the relation

$$\left(\frac{\partial V}{\partial P}\right)_S = \varkappa_S \, V \quad , \tag{19}$$

where $\varkappa_S$ is the adiabatic compressibility and $V \approx A \, l_g$ is the volume
of the gas column with the assumption $l_g \gg 2\pi\mu_g$.
With Eq. (16) the pressure temperature relation of RG becomes

$$\delta P = 2\pi\mu_g \, \beta_T \, <\varepsilon T>/\varkappa_S \, l_g \quad . \tag{29}$$

For an ideal gas $\beta_T = 1/T_o$ and $\quad_S = 1/\gamma P, \gamma = c_p/c_v$.
The pressure variation is related to the temperature at the sample to
gas boundary using Eq. (15) by

$$\delta P = \frac{\gamma\mu_g \, V_g \, P}{\sqrt{2} \, l_g \, T_o} \, e^{j(\omega t - \frac{\pi}{4})} \quad . \tag{21}$$

It is proportional to the temperature amplitude V_g.
The pressure variation decreases with increasing volume, pressure and
temperature of the gas. From Eq. (21) becomes obvious that in an experiment the length of the gas column should be made as small as possible.
On the other hand , the validity of solutions in the form given by
Rosencwaig and Gersho is restricted to gas lengths that are large compared to the thermal diffusion length, $a_g l_g \gg 1$. One condition is in
contradiction to the other.

2.3 Final Remarks

As just pointed out, one problem of the RG model is that the limiting lower frequency of its applicability cannot be defined. From hand waving arguments it can be estimated to be in the order of a thermal diffusion length. This has been shown experimentally by Aamodt et al.[3] and by Tam and Wong[6]. These authors found also experimentally that for length of the gas columns smaller than a diffusion length, the PA-amplitude decreases with decreasing l_g, Fig.3. Tam and Wong[6] have explained this signal behaviour at small gas lengths with a "residual volume". It is essentially the volume in the tube or channel connecting the gas region in front of the sample with the microphone. Another point of the RG model one should think about is the idea of the "piston". It is based on the assumption that a subvolume of the gas expands isobarically, while the pressure of the remaining volume changes. Therefore, the pressure should vary over the volume. But as long as the wave length of the pressure wave is large compared to l_g the pressure is constant. Starting from the basic ideas of Rosencwaig and Gersho it should be possible to take into account all phenomena mentioned above in a proper way and develop a modified RG

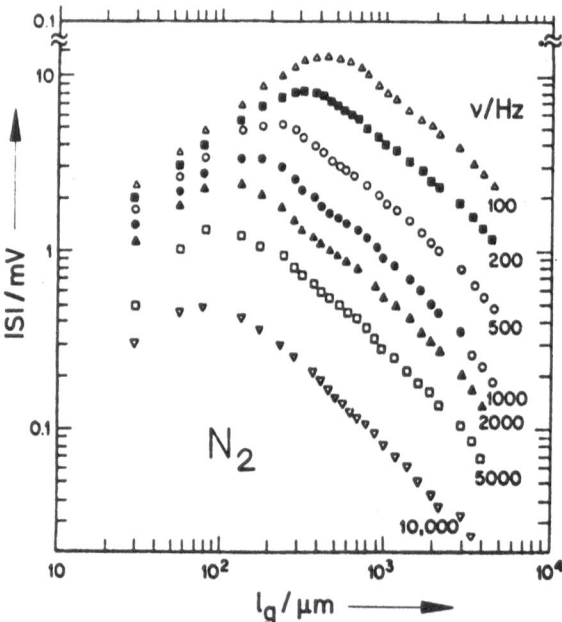

FIGURE 3   Amplitude of the PA signal $|S|$ as a funtion of the length of the N_2 gas column measured by Tam and Wong[6] at various frequencies. Sample: Silicon.

model even for $l_g \to 0$. For not too small optical absorption and for low modulation frequencies the more complicate model of McDonald and Wetsel
should not describe better experimental results than such a modified RG
model.

3. MODEL OF MCDONALD AND WETSEL (MW-MODEL)

Light absorbed periodically by a liquid sample varies not only the
temperature as assumed in the RG model, but also the pressure. That
means, beside a temperature there is also an elastic wave proceeding to
the gas. In the interpretation of the most experimental results this
acoustic wave can be neglected, but its contribution to the pressure
signal of the gas increases with decreasing optical absorption.

3.1 Pressure and Temperature Waves

McDonald and Wetsel[4] start from a system of linear differential
equations that couple thermal and acoustical processes:

$$\nabla^2 P_i - \rho_i \varkappa_{Ti} \frac{\partial^2 P_i}{\partial t^2} = -\rho_i \beta_{Ti} \frac{\partial^2 \phi_i}{\partial t^2} \quad , \tag{22}$$

$$\lambda_i \nabla^2 \phi_i - \frac{\lambda_i}{\alpha_i} \frac{\partial \phi_i}{\partial t} + Q_i = -T_0 \beta_{Ti} \frac{\partial P_i}{\partial t} \quad . \tag{23}$$

$\phi_i = T_i - T_0$ is defined as in the RG model, Eq. (3), where T_0 is the
ambient temperature. The meaning of the suffix is: i = s : sample,
i = g : gas, i = b : backing. ρ_i is the density, α_i the thermal diffusivity, β_{Ti} the volume thermal expansion coefficient, $\varkappa_{Ti}$ the isothermal compressibility, and λ_i the thermal conductivity of medium i. The
term for heat sources in the sample; $Q_s = \frac{\beta J_0}{2} e^{\beta x}(1+e^{j\omega t})$, has the same
form as in the RG model, $Q_g = Q_b = 0$. β is the optical absorption coefficient to the sample. Readers, interested in the derivation of the
equations (22) and (23) should be referred to Sect. 6.4 in the book of
Morse and Ingard[7]. The left hand side of Eq. (22) differs from the equation of acoustic waves by the isothermal bulk modulus instead of the
adiabatic one. The left hand side of Eq. (23) is the equation for heat
conduction. Both expressions are completed by a term that takes into
account the thermal expansion. McDonald and Wetsel treated the problem
for plane geometry, Fig. 1, as Rosencwaig and Gersho. The time dependent

parts of the solutions of Eqs. (22) and (23) for pressure and temperature in medium labelled by i are:

$$P_i = (A_{Pi}e^{-jk_ix} + B_{Pi}e^{+jk_ix} + C_{Pi}e^{-\sigma_ix} + D_{Pi}e^{+\sigma_ix} + E_{Pi}e^{\beta x})e^{j\omega t}, \quad (24)$$

$$\phi_i = (A_{Ti}e^{-jk_ix} + B_{Ti}e^{jk_ix} + C_{Ti}e^{-\sigma_ix} + D_{Ti}e^{\sigma_ix} + E_{Ti}e^{\beta x})e^{j\omega t} \quad . \quad (25)$$

k_i is the wave number of the acoustic wave. There holds

$$k_i^2 = \rho_i \varkappa_{Si} \omega^2 \quad . \tag{26}$$

σ_i is the complex diffusion coefficient, Eq. (11), and the wave number
of the temperature wave. The solutions (24) and (25) describe generally
temperature and pressure in gas columns of arbitrary length. Therefore,
the window with thickness l_w has to be taken into account, Fig. 1.

Without the knowledge of the boundary conditions it is possible
to say something about certain ratios of the constants in Eqs. (24)
and (25). There hold

$$\frac{A_{Pi}}{A_{Ti}} = \frac{B_{Pi}}{B_{Ti}} = \frac{c_{Pi}\rho_i}{T\ \beta_{Ti}} \quad , \tag{27}$$

$$\frac{C_{Pi}}{C_{Ti}} = \frac{D_{Pi}}{D_{Ti}} = -\rho_i\ \beta_{Ti}\ \mu_i^2\ \omega^2 \quad , \tag{28}$$

$$\frac{E_{Pi}}{E_{Ti}} = \frac{\rho_i\ \beta_{Ti}^2\ \omega^2}{\beta^2 + k_i^2} \quad . \tag{29}$$

These ratios (27) to (29) show two facts: 1. The greater the thermal expansion coefficient and the smaller the optical absorption the greater
the pressure amplitude. 2. With increasing frequency the pressure amplitude increases also.

To determine the constants A_{Ti} to E_{Ti} etc. completely, the boundary
conditions for temperature, pressure and their derivatives should be
fullfilled.

The boundary condition for the pressure at $x = 0$ is

$$P_g = P_s \quad . \tag{30}$$

Further boundary conditions follow from the continuity of the velocity of the movement u_i at the boundary between different media,

$$u_i = \frac{j}{\omega \rho_i} \frac{d\rho_i}{dx} \; . \tag{31}$$

At the sample to gas boundary at $x = 0$ is

$$u_g = u_s \; . \tag{32}$$

At the remaining boundaries the media do not move. Therefore, at $x = l_g : u_g = 0$ and at $x = -l_s : u_s = 0$.

McDonald and Wetsel have given as an example the expression for the pressure in the gas for thermally thick samples (Eq. (41) in Ref.4) as

$$P_g \approx - \frac{j}{\omega} \frac{\gamma P_o}{T_g} \frac{J_o}{2\rho_s c_{ps}} \; [\frac{\beta}{\sigma_g T_o (g+1)(r+1)} + \beta_T (1-e^{-\beta l_s})] \tag{33}$$

$$= - \frac{j}{\omega} \frac{\gamma}{2l_g \rho_s c_{ps}} \frac{P_o \beta J_o}{\sigma_g T_o (g+1)(r+1)} \; [1 + \frac{\sigma_g \beta_{Ts} T_o (g+1)(r+1)}{\beta}(1-e^{-\beta l_s})] \; .$$

The first term is equal to that obtained by the RG theory. The second term describes the influence of the thermal expansion. It becomes more and more important with increasing thermal expansion coefficent of the sample β_{Ts} and sample length l_s and decreasing optical absorption co-efficient β. This may be seen in Fig. 4. In this figure is plotted the amplitude of P as a function of frequency for various values of β cal-culated by McDonald and Wetsel[4]. This feature can be explained in the following way. If in a thermally very thick sample the light absorption is weak, heat is created all over the volume. Therefore, all parts of the sample contribute to the change of volume that is transferred to the gas. But only light energy absorbed within an area in the order of a thermal diffusion length contributes to the temperature at the sample gas boundary independent on the optical absorption. When the frequency approaches the acoustic resonance at about 10^4 Hz the signal depends on the position of the detector on the x-axis, Fig. 4. The results accord-ing to the RG model are added. It becomes obvious that with increasing frequency or optical transparency the difference between the results of the RG model and the model of McDonald and Wetsel increases.

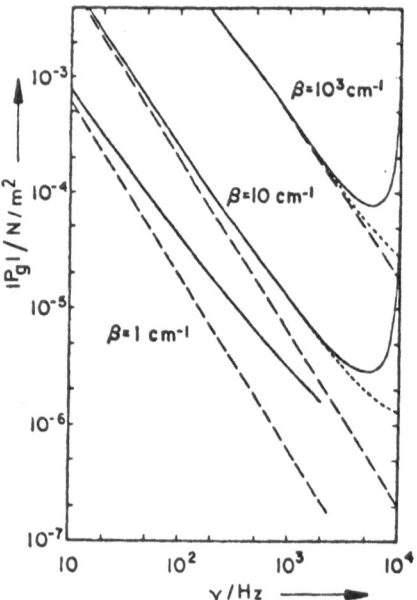

FIGURE 4   Pressure amplitude $|P_g|$ for a water sample as a function
of modulation frequency ν for various values of absorption
coefficient β. Calculated by McDonald and Wetsel[4] for
l_s = 0.48 cm, l_g = 1.74 cm, l_b = l_w = 0.2 cm. Position of
microphone at $x = l_g$: solid line, $x = l_g/2$: dotted line.
Results from RG model : dashed line.

For further results and details of the calculations the reader
should be referred to the original paper of McDonald and Wetsel[4].

3.2 Conclusion

McDonald and Wetsel start from an appropriate set of coupled temperature
and pressure dependent equations. It describes the pressure variation in
the gas for a condensed thermal and elastical isotropic sample more
correctly than the RG model. For an application to solid samples with
anisotropic elastic behaviour the model should be modified by using
the stress tensor instead of the pressure. McDonald and Wetsel assume
the temperature pressure relation to be adiabatic, Eq. (33). They begin
with the isothermal bulk modulus, the reciprocal of the compressibility,
in Eq. (22) [Eq. (1) in Ref. 4] and change-over to the adiabatic bulk
modulus in Eq. (26) [Eq. (19) in Ref. 4]. They apply their adiabatic
model even to frequencies down to 10^{-4} sec, corresponding to periodic

processes in the order of hours. This may not be correct. The problem
should be discussed in more detail.

REFERENCES

/1/   J.G. Parker, Appl.Opt. 12, 2974 (1973)

/2/   A. Rosencwaig, A. Gersho, J.Appl.Phys. 47, 64 (1976)

/3/   L.C. Aamodt, J.C. Murphy, J.G. Parker, J.Appl.Phys. 48, 927 (1977)

/4/   F.A. McDonald, G.C. Wetsel, Jr., J.Appl.Phys. 49, 2313 (1978)

/5/   P. Korpiun, R. Tilgner, J.Appl.Phys. 51, 6115 (1980);
      see also P. Korpiun: "The PAE at Phase Transitions", present volume

/6/   A.C. Tam, Y.H. Wong, Appl.Phys.Lett. 36, 471 (1980)

/7/   P.M. Morse, K.U. Ingard, Theoretical Acoustics, McGraw-Hill,
      New York, 1968

FREQUENCY DEPENDENT PHOTOACOUSTIC SPECTROSCOPY OF CONDENSED MATTER

J.Pelzl

Institut für Experimentalphysik VI, Ruhr-Universität,
4630 Bochum 1, FRG

ABSTRACT

This contribution deals with some specific aspects related to
photoacoustic spectroscopy (PAS) in the frequency domain. The
analysis is restricted to the photoacoustic effect in solids
or liquids which involves detection by a transducer gas. The
basic ideas and the fundamental differences of the frequency
dependent and time dependent measurement are briefly described.
Representative examples are presented to demonstrate the particular information obtained on optical, thermal and acoustical
properties from PA measurements at variable frequencies e.g.
on radiative and nonradiative energy transfer processes or on
optical absorption coefficients. New theoretical and experimental investigations of thin solids are reported. The results
yield evidence of non negligible thermoelastic effects.
Finally, the influence of the acoustic transfer function is
analyzed. The outlines of a generalized model of the Helmholtz
resonator are reviewed and experimental methods are proposed
which can be used to determine acoustic behaviour of PA cells.

INTRODUCTION

The photoacoustic (PA) effect involves different physical
processes in which the energy of electromagnetic radiation is
converted into heat and finally released as sound. The composite nature of the signal generation constitutes the essential
feature of the photoacoustic spectroscopy (PAS) in comparision
to other spectroscopy methods. The evolvement of heat foundates
the very characteristic information of PAS, that is to be sensitive only to the absorbed radiation, which is de-activated
by nonradiative processes.

PAS measurements may be carried out in the frequency domain or in the time domain. These two alternative methods are
well established in conventional spectroscopy such as nuclear
resonance (NMR,NQR), paramagnetic resonance (EPR) and ultrasonic and optical spectroscopy. The thorough investigation of

the time dependence of the spectroscopic signals are applied
e.g. to the study of relaxation phenomena. In this sense PAS
can be used to measure nonradiative relaxation of optically
excited electronic states. However, frequency or time domain
PAS bears a further potential. The different mechanism contributing to the PA signal involve characteristic times of interaction. The relative influence of the optical, thermal and
acoustic processes can be monitored in the time or frequency
domain and characteristic quantities involved in signal generation such as the optical absorption coefficient β or the
thermal diffusivity μ can be determined. The interplay of β
and μ , the latter exhibiting a time and a frequency dependence
constitutes the basis of the optical and thermal depth profiling which belongs to the most important potential application
of the PA effect in solids.

In the following the main features of the frequency and
time domain PAS are briefly described. The application of the
frequency dependent measurements to the study of optical, thermal and acoustic properties of matter are pointed out and new
experimental results are presented on thermal properties of
thin solids and on the acoustic behaviour of spatially extended PA cells. The representation is restricted to the PA effect
that makes use of a transducer gas. The application of the
piezoelectric detection is discussed in detail by Patel at this
conference.

FREQUENCY DOMAIN VERSUS TIME DOMAIN PAS

In the time domain PAS the thermoacoustic response following
the absorption of light is analyzed as a function of time.
Excitation by single pulses are preferable for optimal time
resolution. Frequency domain PAS is based on an optical absorption process that varies periodically in time. The amplitude and the phase angle of the resulting sound pressure in
the gas are analyzed as a function of the modulation frequency $\nu = \omega/2\pi$. PA effect in the time and frequency domain has
been treated theoretically by several authors. Starting from
the one dimensionel model developed by Rosencwaig and Gersho

/1/, Rosencwaig has studied theoretically the variation of the
amplitude and the phase angle as a function of the modulation
frequency /2/. A refined model, that uses the gas thermal transport equation, has been developed by Aamodt et al. who treated
both the frequency dependent /3/ as well as the time dependent
PA effect /4/. Mc Donald and Wetsel have extended the Rosencwaig-Gersho theory to include thermoelastic contributions /5/.
Extensive theoretical work has been done by Mandelis and Royce
on the applications of the frequency /6/ and time domain PAS
/7/.

     In Fig.1 the main features of the time and the frequency
domain PAS illustrated schematically for a semi-infinite one
dimensional solid (s) in contact with a gas (g). To simplify
the matter only the thermal response and the heat generated
by light intensity I in a slab at a characteristic distance
$x_0 = \mu_\beta$ are considered. The optical penetration depth $\mu_\beta = 1/\beta$
is defined by the reciprocal value of the optical absorption
coefficient of the solid. The heat flow in the solid and the
adjoining gas is governed by the thermal diffusion equations.
The characteristic quantities involved in the thermal processes are the thermal conductivity λ , the mass density ρ ,
the specific heat capacity C and the derived quantities: the
thermal diffusivity $\alpha_s = \lambda/\rho c$ and the thermal diffusion
length $\mu_s = (2\alpha_s/\omega)^{0.5}$. Heat excitation by a short pulse
of duration τ_p is assumed for the time domain study. The temperature distribution $\vartheta_s(x,t)$ at three different times following the excitation and the resulting time evolution of surface temperature $\vartheta_s(0,t)$ are sketched in the middle part of
Fig.1. A periodic heat source is required in the frequency
domain PAS. The resulting heat wave amplitude (wave length =
$2\pi \cdot \mu_s$) decays exponentially as a function of the distance from the source. It is obvious that the time delay τ_β
and the phase angle lag $\Delta\varphi$ of the surface temperature provide very similar informations. In this simplified exemple an
excited state life time would become evident in an increase
of τ_β or $\Delta\varphi$. Similarily the positions of isolated absorption
centres could also be determined from a measurement of the
time delay or of the phase lag. This ability constitutes the
basis of the depth sensitive PAS. However, in the real situ-

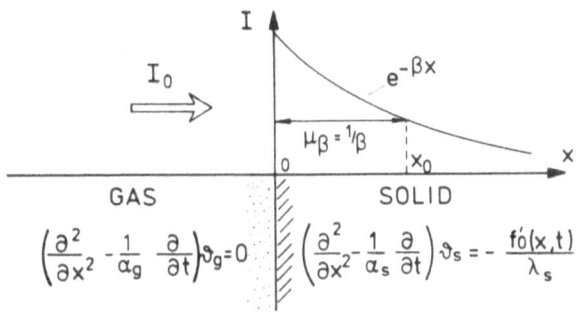

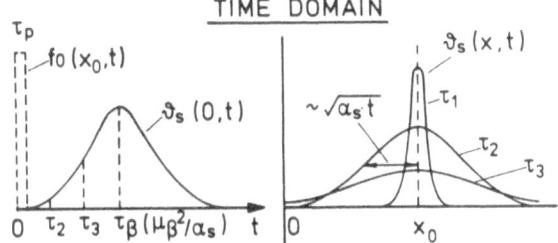

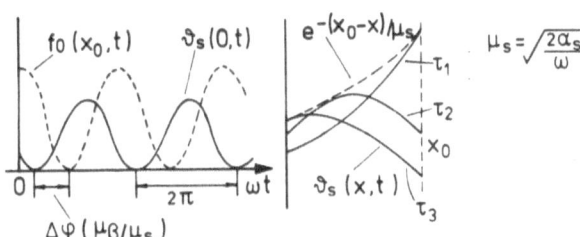

FIGURE 1 : Schematic representation of the photoacoustic effect in the time domain and in the frequency domain. Only thermal response resulting from optical absorption at $X_o = 1/\beta$ in a semi-infinite solid is considered by a one dimensional approximation. The symbols are explained in the text.

ation the signal includes heat generation from all parts of the sample and in addition the thermoacoustic process in the gas affects the signal shape. In the most cases it will be not possible to extract the desired quantities such as the lifetime in a straightforward way.

Basically the time domain and the frequency domain PAS deliver equivalent physical information , mutually correlated by a Fourier transformation. However, the experimental reali-

zation of both methods can lead to quite different statements
about the same physical content. These limitations are primarily caused by the finite time resolution and the restrained
frequency response of the detection system. The influence of
the microphone transfer function on the time domain PA signal
has thorougly been discussed by Mandelis and Royce /8/. The
graphs and signal traces in Fig.2 demonstrate the effect of the
low frequency limit of an electret microphone on the time domain (upper traces) and frequency domain (lower traces) PA signal obtained from graphon in a non resonant cell. The break

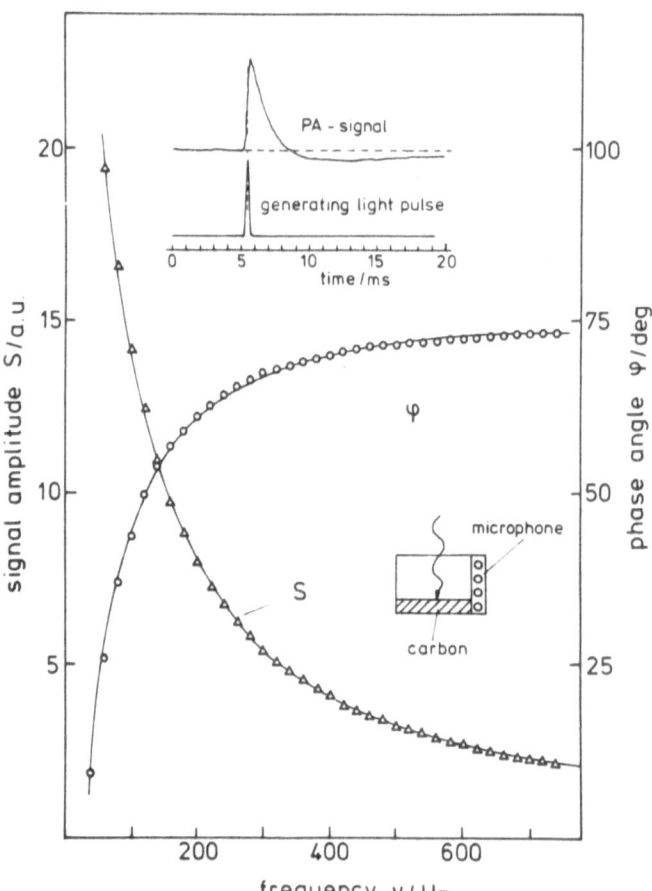

FIGURE 2 : PA effect from a graphon sample recorded in the time domain (upper traces) and in the frequency domain (lower
traces). Influence of microphone transfer function are evident in the long time limit and at low frequencies, respective
ly.

point (half power) frequency is 60 Hz. Besides the microphone
response two experimental factors are important: The acoustic
frequency behaviour of the cell arrangement - this point will
be discussed in the last paragraph - and in the time domain
PAS the shape of the exciting pulse as it has to be deconvoluted from the measured signal. The analysis of a time domain
experiment becomes relatively large scale because the signal
shape has to be fitted by complicated theoretical expressions,
which take into account all physical and experimental contributions to the PA signal. This may be one reason, why till
now most of the PA measurements have been carried out in the
frequency domain, as far as a transducer gas has been used.
If the PA effect is detected via the elastic waves following
a local thermal expansion of the solid or liquid the time domain spectroscopy provides many advantages. This has been demonstrated among others by Tam and Patel /9/.

Here we shall restrict us to frequency dependent studies
using gas-microphone detection.

DETERMINATION OF OPTICAL PROPERTIES

Frequency domain PAS can be used to deduce relaxation rates of
excited electronic states as well as absolute values of optical constants such as the quantum efficiency and the absorption coefficient.

The high frequency limit of the commercial condensor
microphones restricts relaxation studies of excited states to
lifetimes longer than 10 μsec. Assuming a two level system
and an opaque and thermally thick sample the nonradiative
lifetime τ_0 of the upper electronic level can be deduced directly from the frequency variation of the phase angle
with /10/

$$\varphi = \pi/4 + \arctan(\omega\tau_0)$$

Although, at present this method is applied successfully to
gases /11/, no reliable experimental results from solids are
available that could give quantitative evidence for the theoretical predictions in the case of condensed matter.

The situation is more favourable with regard to the determination of <u>absolute quantum efficiencies</u> . The quantum
efficiency QE is currently defined by the ratio of the rate
of the radiative transitions W_R and the sum of the radiative
(W_R) and the nonradiative (W_{NR}) transitions: $QE = W_R/(W_R+W_{NR})$.
Different techniques have been developed and applied to measure the QE by means of the PA effect. Most of them compare
the PA signals of the substance under investigation with a
standard which can be a reference sample or the same material
with a different concentration of absorbing ions or an electronic transition at another wave length. An alternative
method, that uses the frequency variation of the phase angle
was proposed by Quimby and Yen who applied it to the specific
case of ruby /12/. However, as the observed phase angle changes reach a few degrees at most, only a limited accuracy could
probably be achieved.

The potential of measuring <u>absolute optical absorption</u>
<u>coefficients</u> by PAS has been investigated by a great number
of research workers. An essential advantage provided by the
PAS is the quite large dynamic range for the absorption coefficient β. Changes in β from less than $10^{-2} cm^{-1}$ to more than
$10^4 cm^{-1}$ can principally measured within the frequency range
of a gas-contact microphone /13/. The lower limit is restricted by the background signals from the windows and the cell
walls. Large β-values are confined by optical saturation effects
which occur when the thermal diffusion length μ_s becomes much
smaller than the optical penetration depth $\mu_\beta = 1/\beta$. Insulators
possess a thermal diffusivity of the order of magnitude of
$\alpha = 10^{-7} m^2/s$ giving rise to a thermal penetration depth
$\mu_s = 2.5 \mu m$ at a modulation frequency $v = 5$ KHz. The accessible
range for the absorption coefficient can be considerably extended to smaller as well as to larger β-values if one makes
use of the thermoelastic response. Piezoelectric detection of
the elastic waves following the local heating after an optical absorption process may depress the detection limit of an
optical absorption coefficient below $\beta = 10^{-7} cm^{-1}$ /9/. In the
case of optically opaque thin solids thermoelastic effects
give rise to bending and buckling motions of the sample plane.
The forced mechanical vibration of such a thin platelet yields

a supplementary contribution to the PA signal which might be
used to determine ß-values of strongly absorbing solids.
A detailed description of the thermoelastic effect in thin
platelets is given in the next section. A similar procedure
which in contrast to the foregoing is based on the photosen-
sitivity of elastic constants is reported by Konstantinov
et al. who applied this method to semiconductors with absorption coefficients ranging up to values of $ß = 5 \cdot 10^5 cm^{-1}$ /14/.

In the field of the frequency domain PAS using gas-phase
microphone detection several alternative methods for the determination of absolute ß-values have been proposed and already tested with partial success. The first method makes use
of the frequency dependent interplay between the thermal diffusion length μ_s and theoptical penetration depth $\mu_ß = 1/ß$.
Investigating the amplitude S and the phase angle φ of the PA
signal at a fixed wavelength as a function of the chopping frequency ν , the absorption coefficient ß is obtained as a fit
parameter. This method was first applied by Wetsel and Mc Donald /15/ to absorbing liquids. Figure 3 shows theoretical results on the frequency dependence of the relative PA signal
amplitude for different absorption coefficients ß in cm^{-1}.
The two limiting case of a very strong and a weak absorption
with the PA amplitude varying as $\omega^{-3/2}$ and as ω^{-1} are indicated. Experimental data obtained by the authors from phenol
red sodium salt in destilled water are reproduced in the same
figure. In a subsequent publication of the same authors the
discrepancies between experiment and theory at low frequencies
have been attributed to thermoelastic contributions /5/. Including this acoustic term a satisfying agreement with optical
transmission data could be achieved.

The advantage of phase angle measurements especially in
those cases where the PA amplitude becomes saturated $(\mu_ß \ll \mu_s)$
was pointed out by Roark et al. /16/ who used an $Fe(bipy)_3 Br_2$
solution to test the ability of an photoacoustic phase angle
spectroscopy. A similar treatment has been reported by Poulet
et al. /17/ who considered both the amplitude and phase angle
as a functions of the ratio $\mu_s/\mu_ß$. Teng and Royce suggested
to combine amplitude and phase information to eliminate the
frequency response of the PA cell /18/. Alternatively, an elec-

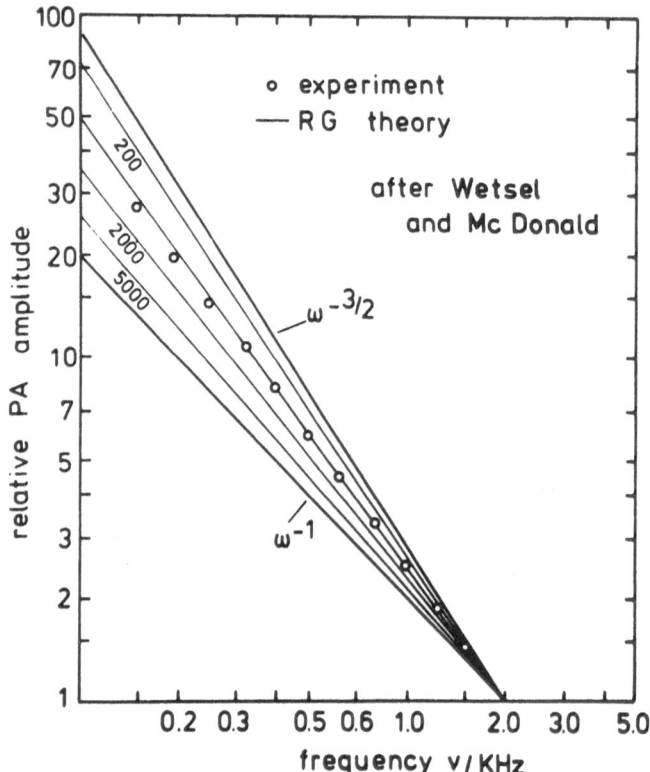

FIGURE 3 : Variation of the PA amplitude upon the chopping
frequency ν . The theoretical curves are determined on the basis of the Rosencwaig-Gersho (RG) theory for different values
of the optical absorption coefficient including the two limiting cases of an optically opaque (ω^{-1}) and optically transparent sample ($\omega^{-1,5}$). The experimental points are obtained
from phenol red sodium salt in destilled water with a concentration of 25 g/l. After Wetsel and Mc Donald /15/.

trical calibration procedure for an absolute measurement of the
cell transfer function was described among others by J.M.Mc
David and coworkers /19/.

Great efforts have been undertaken to determine by means
of the PA effect optical absorption coefficients in highly
transparent materials e.g. weakly absorbing windows. Extensive
experimental work is reported by Fernelius on IR antireflec-
tive coated ZnSe windows using a CO_2 laser /20/. The measured
photoacoustic signal variation upon the chopping frequency was
analysed using the Benett-Forman model and a modified Rosencwaig-Gersho theory with the attempt to determine both the

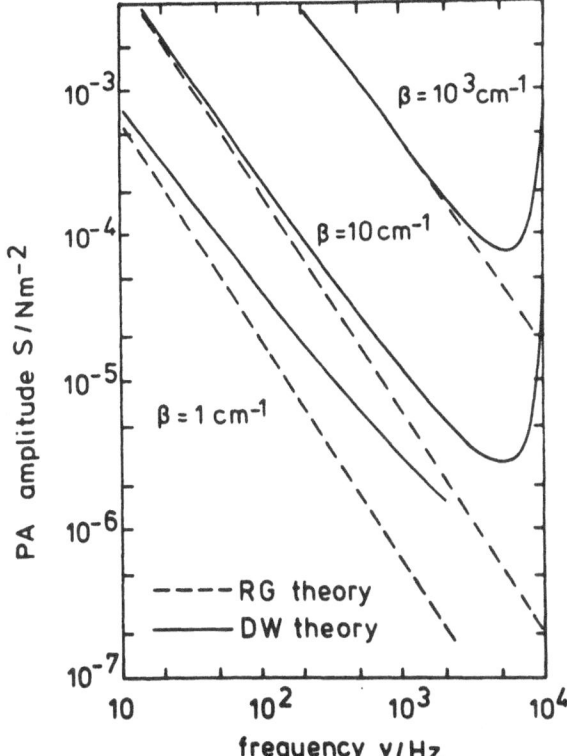

FIGURE 4 : Frequency dependence of the PA amplitude predicted by the Rosencwaig-Gersho (RG) theory (dashed lines) and by Mc Donald-Wetsel (DW) theory (solid lines) which includes the thermoelastic contribution. β is the optical absorption coefficient. After Mc Donald and Wetsel /5/.

bulk- and the surface absorption coefficient of the coated windows. The theoretical expression for the frequency dependence of the PA effect due to bulk- and surface absorption of windows in a cylindrical configuration derived by Benett and Forman are based on linearized hydrodynamic equations and include the heat flow from the sample to the surrounding gas /21/. The relative importance of direct thermoelastic contribution was pointed out by Mc Donald and Wetsel /5/. Particulary in weakly absorbing materials and at high chopping frequencies, the thermally generated mechanical motion of the sample can considerably modify the PA signal. Figure 4 shows theoretical results obtained by the authors in a one dimensio-

nel model calculation. In a subsequent paper Mc Donald extended the treatment to a cylindrical configuration /22/.

At the very beginning of PAS there was some hope to obtain quantitative results from powdered samples. Few attempts
have been made to investigate the influence of the grain size /23/ but the conclusions are still only qualitative.

The most important potential merit of the frequency dependent PAS is the depth profiling of the optical constants.
The depth sensitivity of the PA effect results from the interplay of the optical absorption length and of the thermal diffusion length, μ_s (see Fig.1). Whereas μ_β is independent of the
frequency μ_s decreases proportional to $\omega^{-0.5}$. The ability
of depth profiling was demonstrated convincingly with the
frequency dependent PA-spectrum of an apple peal /24/. A similar experiment has been performed on native and denatured lobster shell /25/. But till today no quantitative study has
been published in this field.

In summarizing this paragraph, frequency dependent PAS
may be useful to determine absolute values of optical absorption coefficient from weakly as well as from strongly absorbing materials. For highly transparent solids and liquids the
large thermoelastic contributions suggest to use piezoelectric detection as described by Tam and Patel /9/. However,
the most promising domain of the gas response PAS of condensed
matter constitutes the investigation of the depth profile and
the local variation of the optical absorption coefficient.

DETERMINATION OF THERMAL PROPERTIES

The basic requirement underlying the absolute measurement of
optical constants by means of PAS is the knowledge of the thermal diffusivity $\alpha = \lambda / \rho c$. Conversely, proceeding from known
optical parameters, the thermal properties can be determined
quantitatively by a frequency dependent PA experiment. Besides
those information which may be also obtained by conventional
techniques PA detection bears the great potential of local
measurements and of depth profiling of thermal characteristics.
In the last two years this field has developed rapidly. Semiquantitative frequency dependent studies of subsurface flaws

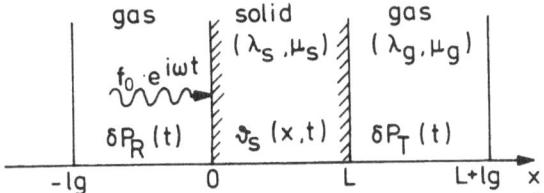

FIGURE 5 : One-dimensional model used for the theoretical treatment of the PA effect from thin solids.

in metals are reported by Thomas et al. /26/. The information provided by the amplitude and the phase angle in heat wave microscopy are demonstrated with several examples by Busse et al. /27,28/.

This section deals with the particular problem of quantitative studies of thermal properties of condensed matter by means of PA experiments in the frequency domain. The discussion is restricted to the special case of optically opaque thin solids with air backing.

A schematic description of the model used in the following theoretical and experimental treatment is illustrated in Figure 5. An infinite solid slab of thickness L is in thermal contact with a transducer gas at both sides. Periodically modulated light (modulation frequency $\nu = \omega/2\pi$) impinges from the left on the surface at x = 0. It is assumed that the light is absorbed in a thin surface layer giving rise to a periodic heat flux $f(0,t) = f_0 \exp(i\omega t)$. The temperature distribution in the solid $\vartheta_s(x,t)$ resulting from the planar heat source at x = 0 is obtained by a straightforward calculation:

$$\vartheta_s(x,t) = (f_0/\lambda_s \sigma_s)\{\text{Cosh}[\sigma_s(x-L)]/\text{Sinh}(\sigma_s L)\} \exp(i\omega t) \quad (1)$$

with $\sigma_s = (1+i)a_s$, $a_s = \mu_s^{-1} = (\omega/2\alpha_s)^{0,5}$
Defining an amplitude factor $\tilde{A} = A/q^2 = A/(a_s L)^2$ the oszillating sample temperature ϑ_R at x = o and ϑ_T at x = L can be expressed by

$$\vartheta_{R,T} = (f_0 a_s L^2/\lambda_s \sqrt{2}) \tilde{A}_{R,T} \exp[i(\omega t - \phi_{R,T})] \quad (2)$$

Following Rosencwaig and Gersho /1/ the corresponding acoustic pressures δP in a gas piston of the length l_g in front (X < 0) or at the back (X > 0) of the slab is directly proportional to the temperature ϑ_R or ϑ_T , respectively:

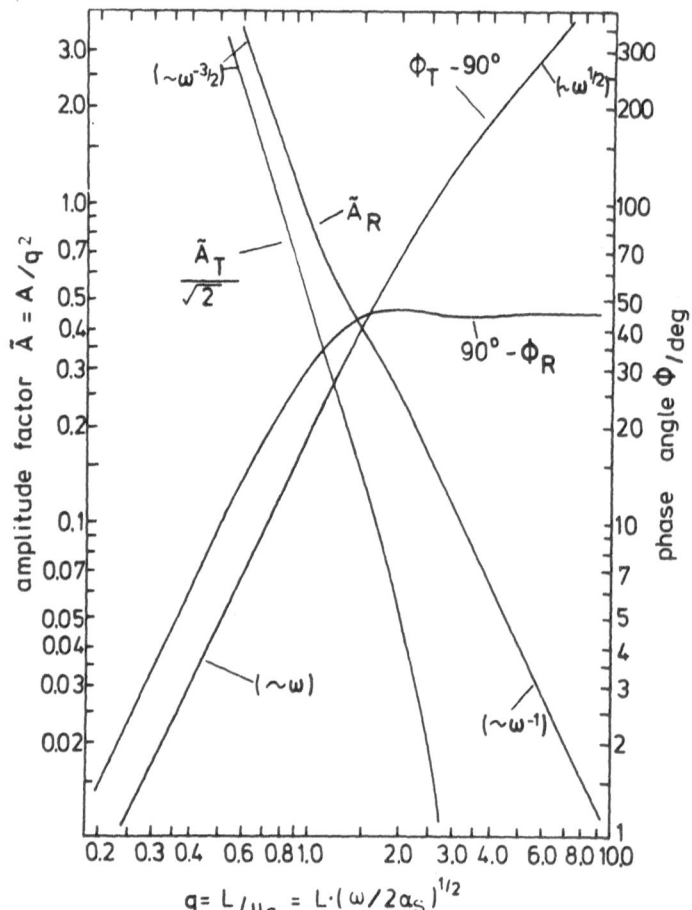

FIGURE 6 : Theoretical variation of the amplitude factor
$\tilde{A} = A/q^2$ and of the phase angle Φ upon the reduced sample
thickness $q = La_s$. The subscripts distinguish between the
thermal reflection (R) and the thermal transmission (T) case.

$$\delta P_{R,T} = (\varkappa P_0 f_0 a_s L^2 / 2T_0 l_g \lambda_s a_g) \tilde{A}_{R,T} \exp[i(\omega t - \Phi_{R,T} - \pi/4)] \quad (3)$$

P_0, T_0, $\varkappa$ and $\mu_g = 1/a_g$ are the gas characteristics: pressure,
temperature, the ratio of specific heats and the gas thermal
diffusion length, respectively.

    The only quantities of equation (3) which depend on the
chopping frequency are the amplitude factor $\tilde{A}$ and the phase
angle Φ . Their values have been calculated as a function of
the reduced sample thickness $q = L \cdot a_s = L/\mu_s$ and are
represented in Figure 6. The frequency behaviour of the PA

amplitude and phase angle are indicated for the limiting cases of small and large values of the reduced thickness q.

The predictions of the theory have been tested by two types of experiments. Crowley et al. measured the PA signal from
thin metal and ceramic samples which were fixed along one edge
in a nonresonant plexiglass cell /29/. Using different metals
with different thicknesses and varying the chopper frequency
between 250 Hz to 4500 Hz, a q-range of three orders of magnitude could be covered. Their experimental and theoretical
results are reproduced in Fig.7 in order to point out the very
good agreement between experiment and theory. With respect to
equations (1) to (3) the theoretical curve of Fig.7 would follow from a coherent superposition of δP_R and δP_T.

The PA signals δP_R as well as δP_T have been investigated
separately by the author and his coworkers /30,31,32/. Here
we want to report on some new results obtained from foils of
graphon and gadolinium. Cylindrical shaped foils were mounted

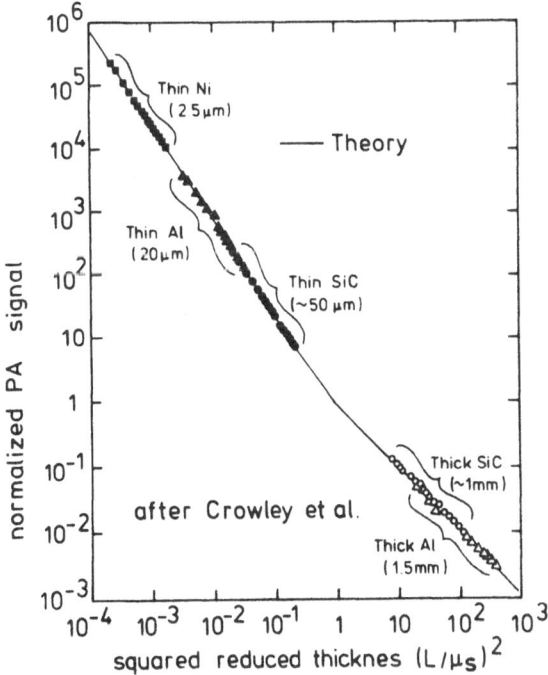

FIGURE 7 : Normalized photoacoustic signals recorded from various thin solids as a function of the square of the reduced
sample thickness (L/μ_s)^2. After Crowley, Faxvog and Roess-
ler /29/.

into a nonresonant aluminium cell dividing the cavity into two
equal chambers of 15 mm diameter and 8 mm length. One chamber
contained a miniature electret microphone (Knowles Electronics)
and was closed by a glass window. The end-off window of the
other cavity could be removed. Both chambers were supplied
with valves for gas exchange. The light of a 100 W halogen
lamp was focused with a 50 mm focal-length lense to a spot of
1 mm diameter at the sample surface. The light intensity was
modulated with a mechanical chopper and the amplitude S and
the phase angle φ of the PA signal were recorded simultanously
as a function of the chopper frequency by means of an Ithako
lock-in analyzer. To measure the PA signal in thermal reflexion
(S_R, φ_R) and in thermal transmission (S_T, φ_T) the sample was
first illuminated through the cavity which contained the micro-
phone and after that from the opposite side. A major objective
of this procedure was to eliminate the frequency dependent
acoustic transfer function of the cell. This goal is reached
by comparing the amplitude ratios deduced from experiment
(S_R/S_T) and theory (A_R/A_T) and the correlated experimental
and theoretical phase angle differences $\Delta\varphi = \varphi_T - \varphi_R$ and
    $\Delta\phi = \phi_T - \phi_R$, respectively.

Figure 8 shows the results obtained from three different
graphon foils. The full lines give the theoretical frequency
variation determined with the relations (1) to (3). $\varepsilon = q / v^{0.5}$
$= L a_s / v^{0.5}$ is a fit-parameter which has to be adjusted
independently to the amplitude ratio S_R/S_T and to the phase
angle lag $\Delta\varphi$. The compatibility condition provides a very sensitive proof of the theory. The observed agreement between
theory and experiment is most convincing in the case of 0.1 mm
graphon foil. For the thicker samples deviations are evident
in a frequency range where the thermal diffusion length μ_s
is of the order of magnitude or smaller than the sample thickness L ($\mu_s \leq L$). The observed discrepancies at frequencies
$\omega \geq 2\alpha / L^2$ arise from additional contributions to the acoustic pressure which are produced by small mechanical deflections of the foils.

These forced mechanical vibrations of the foil could have
four different origins : 1. The difference of the gas pressures in both cavities which are separated by the foil. 2. Li-

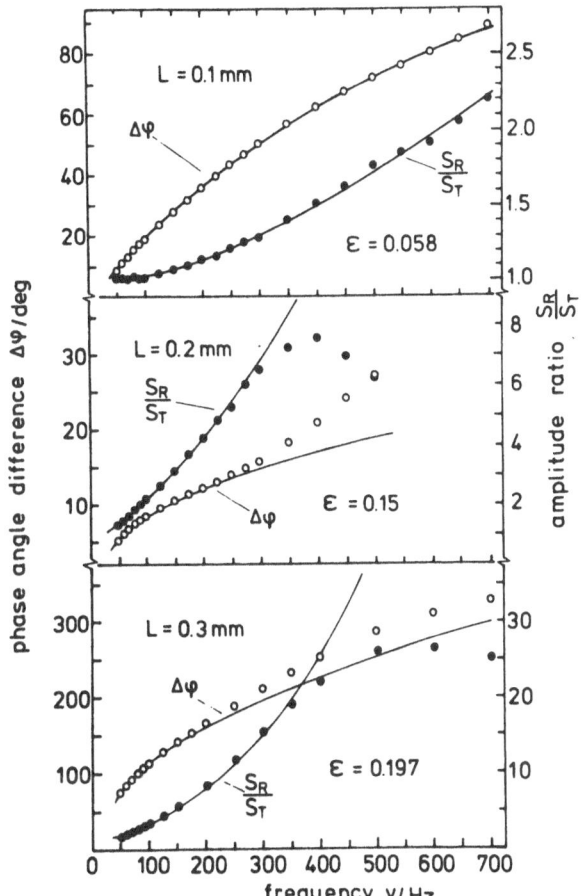

FIGURE 8 : Frequency dependence of the amplitude ratio S_R/S_T and of the phase angle difference $\Delta\varphi = \varphi_T - \varphi_R$ recorded from three graphon foils of different thickness L. Open and full circles are experimental data, the curves are theoretical results obtained with equations (3) and (1) adjusting the fit parameter $\varepsilon = q \cdot \nu^{-0.5} = L \cdot \pi^{0.5} \cdot \alpha_s^{-0.5}$

near thermal expansion of the sample. 3. Bending of the foil due to a temperature gradient through the sample and 4. foil buckling caused by thermoelastic stresses in the sample plane. Supplementary experiments and theoretical estimates indicate that the foil-bending yields the most important contribution /31,32/. The additional acoustic pressure due to the plane deflection can be represented by a change $\Delta\vartheta_{R,T}$ of the surface temperatures $\vartheta_{R,T}$. For a free plate or a simply suppor-

ted circular plate of the surface area $F = \pi R_0^2$ and the thickness L the calculation of the plate deflection becomes simple
provided that the temperature is independent of the position
in the plane /33/. Assuming a linear temperature gradient
$(\vartheta_R - \vartheta_T)/ L$ across the plate the effective temperature
change corresponding to the mechanical deflection of the plate
is given by

$$\Delta\vartheta_{R,T} = B \cdot (\vartheta_{R,T} - \vartheta_{T,R}) \qquad (4)$$

with $$B = -\beta_T \cdot (R_0^2 T_0 / 2L \cdot \mu_g) \exp(i\pi/4)$$
$$= -D \cdot v^{0,5} \exp(i\pi/4) \qquad (5)$$

β_T is the linear isothermal expansion coefficient of the material. Exactly the same result is obtained when determining
the curvature of an elastic band of thickness L and surface
area F with the assumption that the two surfaces experience
different thermal expansions proportional to ϑ_R at x = 0 and
to ϑ_T at x = L.

The thermoelastic contributions treated above become more
important at those frequencies which lead to large oszillating
temperature gradients across the sample thickness. For the
cases considered here the mechanical deflection of the mid-
plane ranges up to 1 nm. Taking into account the bending of
the graphon foils, the experimental data displayed in Fig.8
have been fitted again with D and ε as adjustable parameters.
Results obtained from 0,2 mm thick graphon foil are shown in
Figure 9. Although the extended treatment reproduces qualitatively the frequency behaviour of measured quantities, the
quantitative description suffers extremly from the fact that
small errors in the phase angle have a considerable influence
on the resulting amplitude.

Thermoelastic contributions to the photoacoustic effect
are governed by the magnitude of the thermal expansion coefficient β_T . It is well known, that β_T can diverge at structural
or magnetic phase transition, which then should lead to pro-
nounced effects on PA signals. Experimental results obtained
from Gd-foils confirm this supposition. Figure 10 shows temperature dependent recordings of the PA amplitude S_T and of
the phase angle φ_T at some selected frequencies. Only results

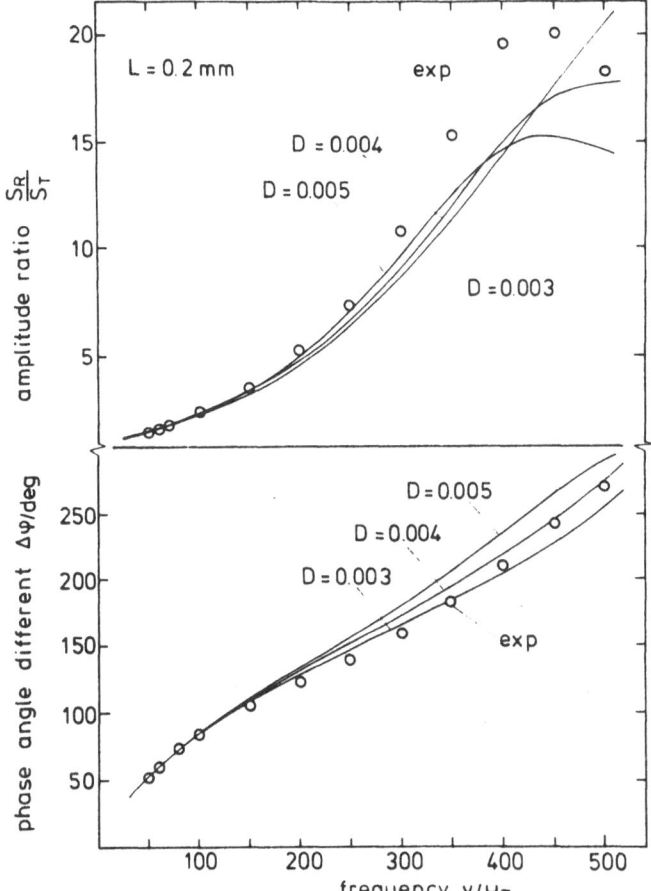

FIGURE 9 : Experimental results (open circles) obtained from
the 0.2 mm thick graphon foil and theoretical curves which include thermoelastically induced bending of the foil. D is the
adjustable bending parameter defined by equation (5).

of thermal transmission studies from a 0.1 mm thick Gd-plate
are presented. Gadolinium undergoes a magnetic transition to
a ferromagnetic order at T_c = 293,2 K /34/. This second order
transition is associated with anomalies in the specific heat
and in the lattice expansion the latter being a consequence
of magnetostriction effects. At low chopping frequencies
($v \le$ 60 Hz) the amplitude and the phase angle of the PA signal
reflect the specific heat anomaly which produces a phase angle
change of the order of magnitude of a few degrees. In the range

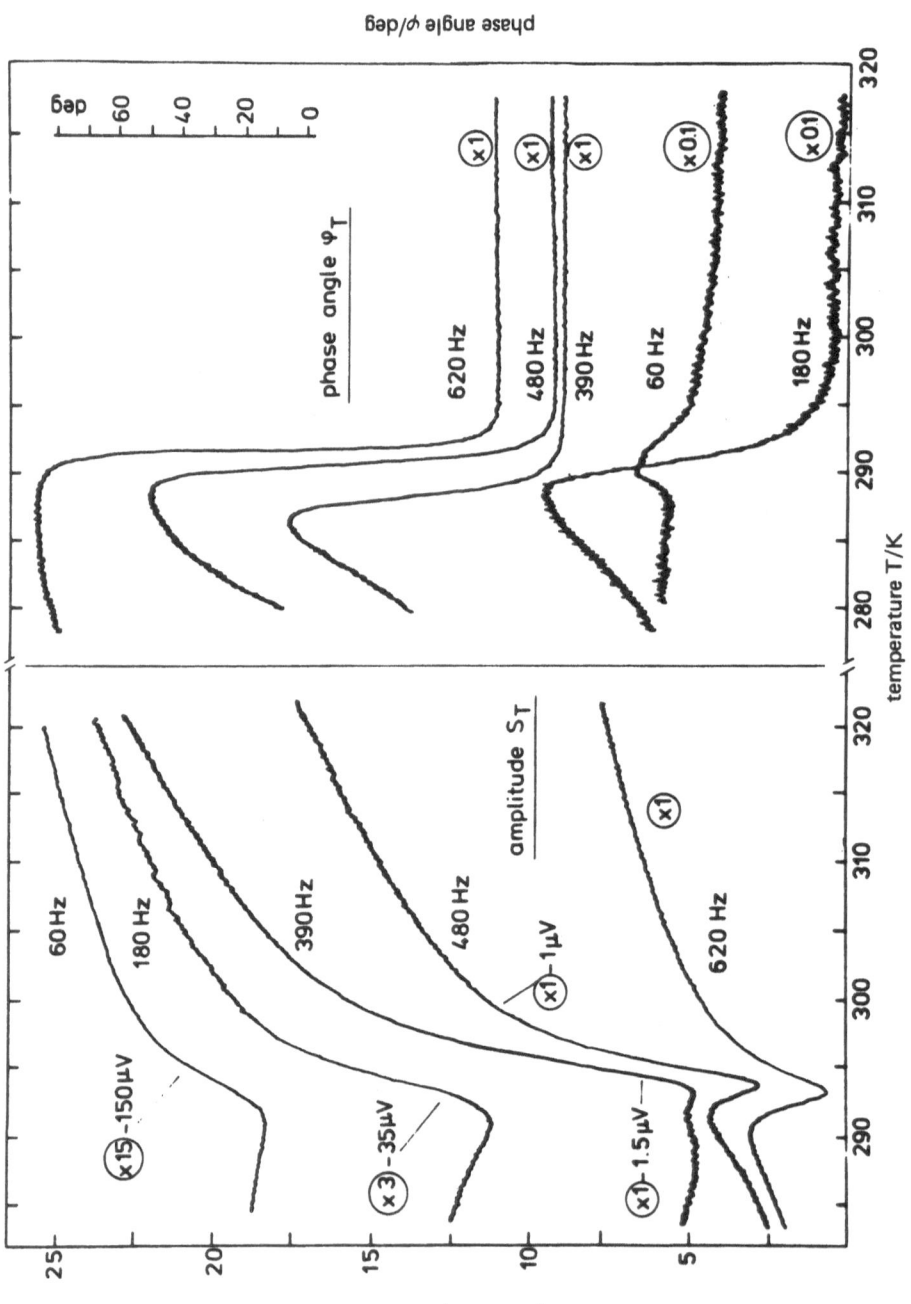

about 600 Hz ($\mu_s \approx L$) the thermally modulated magnetoelastic
contributions dominate. This becomes most evident in the dramatic phase angle change of about π at the transition temperature. The Gd results emphasize the relative importance of
thermoelastic effects in this kind of PA experiments.

With regard to the quantitative nature of the information
delivered by a frequency dependent PA measurement two basic
assumptions of the RG-theory have been illuminated critically
in the current literature. The RG theory proceeds from an intermediate thermal contact between the sample and the gas.
A thermal contact resistance taken into consideration by Cesar
et al. /35/ has been proved to be negligible by Quimby and Yen
/36/. The adequacy of the one dimensional heat flow underlying
the approach of Rosencwaig and Gersho has been discussed by
Mc Donald /37/, Quimby and Yen /38/ and Chow /39/ with essentially the same conclusion that the one-dimensional models
are fairly correct for the general case.

So far we have discussed the ability to study thermal
properties of a sample by means of the PA effect. As the basic
relations (1) to (3) underlying the quantitative analysis involve the product $q = a_s \cdot L$, the geometry may be also the
objective of a PA measurement. This potential application was
first pointed out by Adams and Kirkbright /40/.

ACOUSTIC TRANSFER FUNCTION

The discussion of the preceding chapters was devoted to the
various physical properties of condensed matter, which can be
studied with the frequency dependent photoacoustic effect.

FIGURE 10 : Temperature dependence of the photoacoustic amplitude (left) and phase angle (right) recorded at different chopping frequencies from a 0.1 mm thick gadolinium foil. The signals were detected in thermal transmission. Gd metal undergoes
a transition to a ferromagnetic order at Tc = 293.2 K.

However, the photoacoustic signal detected by a gas microphone contains additional experimental contributions which vary
with frequency : the acoustic response of the cell and that
of the detector. A flat response function of the amplitude as
well as of the phase angle is normally restricted to a narrow
frequency range and to nonresonant PA cells.

The experimental elimination of the acoustic transfer
function would be the most satisfactory way. An applicable
method described in the preceding section constitutes the
sequential measurement of PA signal in thermal transmission
and thermal reflection. But this procedure is restricted to
thin solids only and in addition at high frequencies thermoelastic effects impede a quantitative analysis.

Usually the acoustic transfer function of the detection
system has to be determined in an independent experiment and
deconvoluted from the PA signal of interest. The quantitative
treatment requires comprehensive knowledge of those two components which dominate the acoustic frequency response: the microphone and the acoustic properties of the cell cavity.

Signal detection by a condensor microphone has been thoroughly discussed e.g. by Kreuzer /41/. A detailed analysis
of microphone transfer function effects on time domain and
frequency domain PAS response reported by Mandelis and Royce
/8,10/ demonstrate the limitations imposed by the microphone
detector. The investigated condensor microphone was found to
reproduce accurately the cell pressure in the range 10 Hz to
$5 \cdot 10^4$ Hz whereas transfer function effects on the phase angle
were already present above $5 \cdot 10^3$ Hz.

The frequency range of a flat response is considerably
confined by acoustic resonances of the cell cavities. Cell designs exhibiting resonance behaviour have become increasingly
used in the last years. Basically,the resonance cells consist
of two cavities interconnected by a duct. They have to prevent
the microphone from light impact or they are used for temperature variable PAS. The frequency response of these arrangements
are commonly described by the Helmholtz resonator (HR) model,
the acoustic analogue of the electric RCL-circuit /42/. A generalized treatment which involves elements of the transmission
line theory has been reported recently /43/. This extended

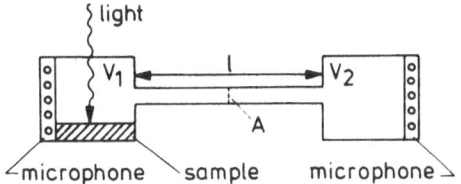

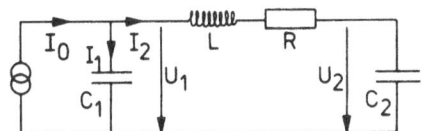

Helmholtz resonator model

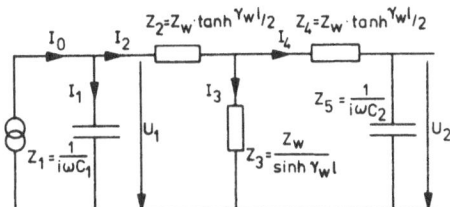

extended Helmholtz resonator model

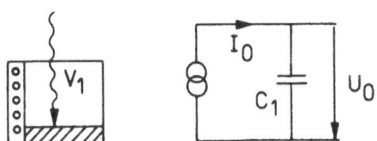

nonresonant reference cell

FIGURE 11 : Schematic representation of the two-cavity resonant cell (top). The analogous electrical equivalent circuit
of the corresponding Helmholtz resonator and of the extended
Helmholtz resonator (middle). Scheme and analogous electric-
circuit of the nonresonant reference cell (bottom).

Helmholtz resonator (EHR) model yields an improved description
of spatially extended cells such as low temperature PA cells
which make use of long ducts /44,45,46/. In the following the
outlines of the EHR model are briefly reviewed.

The resonance cell used for both the experimental and the
theoretical studies is sketched on the top of Figure 11. It
consists of two cavities which are interconnected by a narrow
tube. Typical dimensions are : $V_1 \cong V_2 \cong 700$ mm^3, A = 1,75 mm^2

and l = 200 - 400 mm. The analogous equivalent electrical circuits of both the conventional and the extended Helmholtz resonator are represented below. The capacitances C_1 and C_2 and
the current source are common to both approaches. C_1 and C_2
are proportional to the cavity volumes V_1 and V_2, respectively.
The signal generation due to the photoacoustic effect is described by a current source I_0. In the context of the Rosencwaig-Gersho theory /1/, the acoustic pressure in the gas is
produced by a moving gas piston at the solid gas interface and
the flow velocity is the acoustic analogue of the electrical
current. The duct is represented by discrete elements such as
the resistance R and the inductance L in the conventional
Helmholtz resonator approach. The extended Helmholtz resonator
model proceeds from the analogy to the electrical transmission
line, replacing R and L by line parameters the characteristic
impedance Z_w and the propagation constant γ_w. The retransfor-
mation of the transmission line into a T-network with discrete
elements leads to the equivalent circuit shown in Figure 11.
The circuit elements and the line parameters can be calculated
directly from system characteristics /47/. The EHR approach
also provides the possibility to determine the line parameters
semi-empirically by means of short circuit and open circuit
measurements /43/. Results of the experimental and theoretical
treatments are shown in Figure 12. The amplitudes A and the
phase angles φ of the resonant cells have been normalized with
respect to the signal (A_0, φ_0) of a nonresonant cell. The reference cell is sketched together with its electrical analogy
at the bottom of Figure 11. The good accordance between the
results of the EHR model and the experimental data is evident.
The HR model (dashed lines) yields a rough estimate of the
first resonance frequency, but the conventional model fails
to reproduce the Q values and the multiple resonance pattern
of the spatially extended cell system.

CONCLUSION

Photoacoustic spectroscopy in the frequency domain as well as
in the time domain provides a basis for quantitative measure-

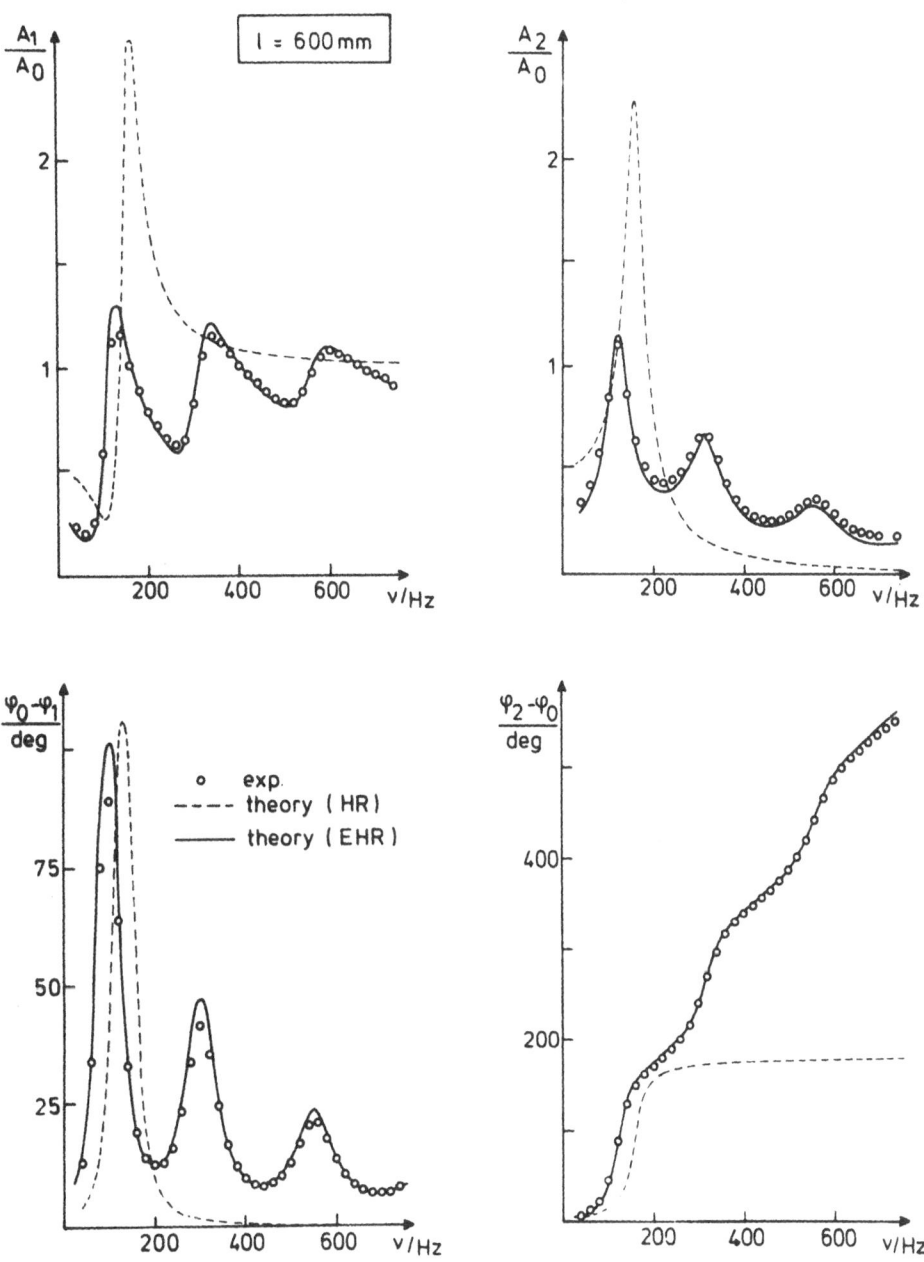

FIGURE 12 : Normalized amplitudes and relative phase angle of
the PA signal from the air filled cell arrangement of Fig. 11
versus the chopping frequency. The interconnecting tube length
is l = 600 mm. The circles are the experimental data. The theoretical curves obtained on the basis of the models of the
Helmholtz resonator (HR) and of the extended Helmholtz resonator (EHR) are represented by the dashed and solid lines,
respectively. After Nordhaus and Pelzl /43/.

ments. The frequency dependent investigation of the amplitude
or of the phase angle is capable to deliver information about
the nonradiative de-excitation processes. The ability of the
frequency domain PAS to measure optical and thermal properties
of condensed matter results from the interplay of the frequency
dependent thermal diffusion length and of the optical penetration depth. It constitutes the basis of optical and thermal
depth profiling which are the most important potential advantages of the frequency dependent measurement.

At present, gas-microphone detected PAS in the frequency
domain suffers mainly from the inability to deliver satisfactory quantitative results. Besides the determination of quantum efficiencies further attempts have been made to measure
absolute optical absorption coefficient. However, reliable experimental data are scarce and for a refined treatment one has
to take into account thermoelastic effects. Experimental results of thermal properties of thin solids which are optically
opaque are more convincing. Here the thermoelastic bending of
the sample can become important if the PA signal is excited in
thermal transmission. This is demonstratedclearly by PA studies performed on Gd foils in the vicinity of the Curie point.

The combination of thermal transmission and reflection
measurements provides the particular advantage that the acoustic frequency response can be eliminated. Conventionally, experiments in frequency domain require the knowledge of the
microphone transfer function and of the acoustic properties
of the cell cavities. As spatially extended cell arrangements
gain increasingly in significance in the PAS at variable temperature the acoustic behaviour of these kind of cells have
become also subject of experimental and theoretical studies.
The treatments proceed from the analogy with electrical resonance circuits. Experimental and theoretical results obtained
from a cell arrangement consisting of two cavities which are
separated by a long narrow duct demonstrate the adequacy of
the extended Helmholtz resonator model. The latter represents
a generalized treatment of the Helmholtz resonator which contains elements of transmission line theory.

ACKNOWLEDGEMENTS

The author would like to thank H.Fütterer, K.Klein, K.Junge,
P.Bechthold and M.Rosenberg for many useful comments.

REFERENCES

1.   A.Rosencwaig and A.Gersho, " Theory of the photoacoustic
      effect with solids ", J.Appl.Phys. 47, (1), 64 (1976)

2.   A.Rosencwaig, " Theoretical aspect of photoacoustic spec-
      troscopy ", J.Appl.Phys. 49, (5), 2905 (1978)

3.   L.C.Aamodt, J.C.Murphy and J.G.Parker, " Size considera-
      tions in the design of cells for photoacoustic spectros-
      copy ", J.Appl.Phys. 48, (3), 927 (1977)

4.   L.C.Aamodt and J.C.Murphy, " Size considerations in the
      design of cells for photoacoustic spectroscopy II:Pulsed
      excitation response ", J.Appl.Phys. 49, (6), 3036 (1978)

5.   F.A.McDonald and G.C.Wetsel,Jr. " Generalized theory of
      the photoacoustic effect ", J.Appl.Phys. 49, (4) 2313
      (1978)

6.   A.Mandelis, Y.C.Teng and B.S.H.Royce, " Phase measure-
      ments in the frequency domain photoacoustic spectroscopy
      of solids ", J.Appl.Phys. 50, (11) 7138 (1978)

7.   A.Mandelis and B.S.H.Royce, " Time-domain photoacoustic
      spectroscopy of solids ", J.Appl.Phys. 50, (6),4330 (1979)

8.   A.Mandelis and B.S.H.Royce, " Nonradiative lifetime mea-
      surements in Time-domain photoacoustic spectroscopy of
      condensed phases ", J.Appl.Phys. 51, (1) 610 (1980)

9.   A.C.Tam and C.K.N.Patel, " Optical absorption of light
      and heavy water by laser optoacoustic spectroscopy ",
      Appl.Optics 18, (19), 3348 (1979)

10.   A.Mandelis and B.S.H.Royce, " Relaxation time measurements
      in frequency and time-domain photoacoustic spectroscopy
      of condensed phases ", J.Opt.Soc.Am. 70, (5) 474 (1980)

11.   M.B.Robin, " Photoacoustic spectroscopy of gases in the
      visible and ultraviolet spectral region " in " Optoacous-
      tic spectroscopy and detection ", Ed.Y.H.Pao,Ac.Press
      N.Y. 1977, and references there in.

12.   R.S.Quimby and W.M.Yen, " Photoacoustic measurements of
      ruby quantum efficiency ", J.Appl.Phys. 51, (3)1780 (1980)

13.   A.Rosencwaig, " Photoacoustics and Photoacoustic Spectros-
      copy " in " Chemical Analysis " , Vol.57, Ed.P.J.Elving,
      J.D.Winefordner, J.Wiley and Sons, N.Y., 1980

14.   L.L.Konstantinov, V.P.Hinkov, J.I.Burov and M.I.Borissov,
      " A new aspect in the photoacoustic spectroscopy of semi-
      conductors ", Inst.Phys.Conf.Ser. 43, 219 (1979)

15. G.C.Wetsel,Jr. and F.A.McDonald " Photoacoustic deter-
    mination of absolute optical absorption coefficient ",
    Appl.Phys.Lett. 30, (5) 252 (1977)

16. J.C.Roark, R.A.Palmer and J.S.Hutchison, " Quantitative
    absorption spectra via photoacoustic phase angle spectros-
    copy ", Chem.Phys.Lett. 60, (1), 112 (1978)

17. P.Poulet, J.Chambron and R.Untereiner, " Quantitative
    photoacoustic spectroscopy applied to thermally thick
    samples ", J.Appl.Phys. 51, (3), 1738 (1980)

18. Y.C.Teng and B.S.H.Royce, " Absolute optical absorption
    coefficient measurements using photoacoustic spectrosco-
    py amplitude and phase information ", J.Opt.Soc.Am. 70,
    (5) 557 (1980)

19. J.M.McDavid, K.L.Lee, S.S.Yee and M.A.Afromowitz, " Photo-
    acoustic determination of optical absorptance of highly
    transparent solids ", J.Appl.Phys. 49, (12), 6112 (1978)

20. N.C.Fernelius, " Photoacoustic signal variations with
    chopping frequency for ZnSe laser windows ", J.Appl.Phys.
    51, (3) 1756 (1980)

21. H.S.Benett and R.A.Forman, " Frequency dependence of pho-
    toacoustic spectroscopy: surface-and bulk-absorption coef-
    ficients ", J.Appl.Phys. 48, (4), 1432 (1977)

22. F.A.McDonald, " Photoacoustic determination of small ab-
    sorption coefficients: extended theory ", Appl.Optics
    18, (9), 1363 (1979)

23. E.M.Monaham,Jr. and A.W.Nolle, " Quantitative study of a
    photoacoustic system for powdered samples ", J.Appl.Phys.
    48, (8), 3519 (1977)

24. A.Rosencwaig, " Photoacoustic spectroscopy ", Adv.Elec -
    tron. Electron Phys. 46, 207 (1978)

25. M.L.Mackenthum, R.D.Tom and T.A.Moore, " Lobster shell
    carotonoprotein organisation insity studied by photoacou-
    stic spectroscopy ", Nature 279, 265 (1979)

26. R.L.Thomas, J.J.Ponch, J.H.Wong, L.D.Favro, P.K.Kuo and
    A.Rosencwaig, " Subsurface flaw detection in metals by
    photoacoustic microscopy ", J.Appl.Phys. 51,(2),1152
    (1980)

27. G.Busse, " Optoacoustic and photothermal imaging and mi-
    croscopy ", this conference

28. A.Rosencwaig and G.Busse, " High resolution thermal-wa-
    ve microscopy ", Appl.Phys.Lett. 36, (9) 725 (1980)

29. T.P.Crowley, F.R.Faxvog and D.M.Roessler, " Photoacoustic
    effect with thermally thin solids ", Appl.Phys.Lett. 36
    (8) 641 (1980)

30. J.Pelzl, H.Fütterer, D.Krüger, K.Junge and H.Mang,
    " Investigation of phase transitions with the photoacou-
    stic effect ", 1st Topical Meeting on Photoacoustic Spec-
    troscopy, Technical Digest, FA 7/1-4, Ed.Optical Soc.Am.
    Washington 1979

31. H.Fütterer, " Untersuchung thermischer Eigenschaften von
     Festkörpern mit Hilfe des Photoakustischen Effekts ",
     Diplomarbeit 1980, Abteilung XII, Ruhr-Universität,
     Bochum

32. K.Junge, " Untersuchung von Phasenübergängen mit Hilfe
     des Photoakustischen Effekts ", Diplomarbeit 1981,
     Abteilung XII, Ruhr-Universität, Bochum

33. H.Parkus, " Thermoelasticity ", Blaisdell Publ.Comp.,
     Waltham 1968

34. K.N.R.Taylor and M.I.Darby, " Physics of Rare Earth Solids"
     Chapman and Hall LTD, London, 1972

35. C.L.Cesar, H.Vargas, J.A.Meyer and L.C.M.Miranda " Photo-
     acoustic effect in solids ", Phys.Rev.Lett. $\underline{42}$, 1570(1979)

36. R.S.Quimby and W.M.Yen, " The effect of thermal contact
     resistance on the photoacoustic signal " Topical Meeting
     on Photoacoustic Spectroscopy, Ames, Iowa, 1979, post
     deadline paper

37. F.A.McDonald, " Three-dimensional heat flow in photoacou-
     stic effect ", Appl.Phys.Lett. $\underline{36}$, (2), 123 (1980)

38. R.S.Quimby and W.M.Yen, " On the adequacy of one-dimensi-
     onal treatments of photoacoustic effect ", J.Appl.Phys.
     $\underline{51}$, (2), 1252 (1980)

39. H.C.Chow, " Theory of three-dimensional photoacoustic
     effect with solids ",J.Appl.Phys. $\underline{51}$, (8) 4053 (1980)

40. M.J.Adams and G.F.Kirkbright, " Thermal diffusivity and
     thickness measurements for solid samples utilizing the
     optoacoustic effect ", Analyst $\underline{102}$, 678 (1977)

41. L.B.Kreuzer, " The physics of signal generation and de-
     tection " in " Optoacoustic spectroscopy and Detection ",
     Ed. Y.H.Pao, Acad.Press, N.Y. 1977

42. N.C.Fernelius, " Helmholtz resonance effect in photoacou-
     stic cells ", Appl.Optics $\underline{18}$, (11), 1784 (1979)

43. U.Nordhaus and J.Pelzl, " Frequency dependence of reso-
     nant photoacoustic cells : The extended Helmholtz reso-
     nator ", Appl.Physics, in press

44. J.C.Murphy and L.C.Aamodt, " Photoacoustic spectroscopy
     of luminiscent solids : Ruby ", J.Appl. Phys. $\underline{48}$,(8)
     3502 (1977)

45. P.S.Bechthold, M.Campagna and J.Chalzipetros, " Variable
     temperature photoacoustic spectroscopy ", Opt.Comm. $\underline{36}$,
     (5), 369 (1981) and this conference

46. K.Klein, J.Pelzl and H.Fütterer, " Low temperature photo-
     acoustic cell ", this conference

47. P.M.Morse, K.U.Ingard, " Theoretical Acoustics " , MC
     Graw Hill, N.Y. 1968

PHOTOACOUSTIC SPECTROSCOPIES

D. FOURNIER and A.C. BOCCARA

Laboratoire d'Optique Physique

10, rue Vauquelin

75231 PARIS CEDEX 05 - FRANCE

ABSTRACT

Photoacoustic spectroscopy of solids is now well-known as a method which allows the measurement of the absorption spectroscopy of non conventional samples. However, besides the classical photoacoustic spectroscopy (sample - gas - microphone) other methods have recently been developed. Thus, new kinds of signal production (e.g. multiplex excitation), of signal detection (e.g. mirage effect, photothermal radiometry) associated with more complete theoretical works, have recently led to studies of more "exotic" problems (e.g. highly transparent or inhomogeneous samples...).

During the same time, polarized light photoacoustic studies have allowed (by using dichroism techniques), the measurement of natural or induced anisotropies.

Low temperature measurements, which have been so important for studies of condensed matter, have been successfully introduced in order to enlarge greatly the scope of application of photoacoustic spectroscopy.

At last, the use of photoacoustic for the detection of non linear effects will be discussed.

INTRODUCTION

Upon the absorption of electromagnetic radiation by a sample, a more or less important fraction of the incident energy will be converted into heat. Recently, this non-radiative mechanism has provided the physical basis for a new class of photothermal spectroscopies such as photoacoustic spectroscopy (PAS). When a sample is irradiated by a periodic monochromatic light beam, the induced periodic heating which is related to the absorption coefficient α, is classically detected by measuring either the correlated pressure variations within an enclosed

cell with a microphone, or the thermally induced vibrations with a
piezoelectric transducer glued on the sample. Thus, PAS allows absorption
spectra recording for solids and semi-solids whatever their physical
aspect is, because only the absorbed light generates a signal.

In a first section of this paper, we compare the classical photoacoustic scheme (monochromatic excitation - gas microphone detection)
with other kinds of excitation and detection, mainly in the case of the
spectroscopy of solids or semi solids. Then, in Section II, we introduce
some specific examples for which PA or photothermic spectroscopies are
particularily powerful. The third Section is devoted to dichroism and the
fourth one to low temperature measurements. Finally, some future trends
are outlined in the non-linear effects domain.

I - EXCITATION AND DETECTION OF PHOTOACOUSTIC AND PHOTOTHERMIC SIGNALS

Rather than going through the entire analytical expressions of
Rosencwaig and Gersho theory (1), we shall outline the significant parameters through a simplified example.

Fig. 1a shows the attenuation of an intensity-modulated monochromatic
flux ϕ_0 irradiating a sample and Fig. 1b represents the temperature distribution inside a material whose surface temperature is ΔT_o cos ωt.

We can define two main parameters :

- α, the absorption coefficient of the sample, and $1/\alpha$ which is the
  length for which the sample has absorbed about 2/3 of incident flux ϕ_0
- μ_s the thermal diffusion length which is the length for which thermal
  wave amplitude decreases by a factor of 2/3. This coefficient is
  related to various thermal parameters of the sample :

$$\mu_s = \sqrt{\frac{2k}{\rho C \omega}}$$

with k = thermal conductivity
C = specific heat
ρ = density
ω = pulsation of the incident light.

Let us point out that this parameter depends strongly on the modulation frequency of the light.

In the extreme cases $\mu_s \ll 1/\alpha$ (Fig. 1c) and $\mu_s \gg 1/\alpha$ (Fig. 1d),
the surface amplitude ΔT of the modulated part of the temperature of the
sample can be estimated by calculating the detailed balance within the

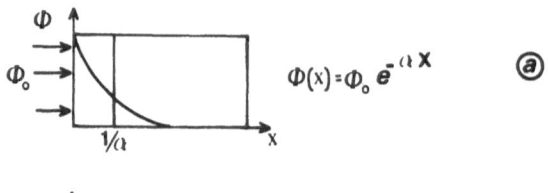

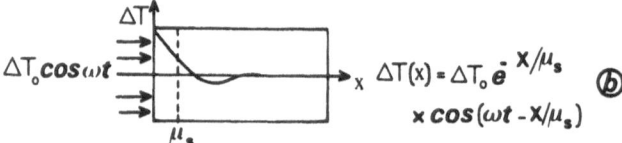

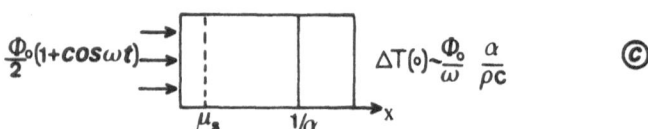

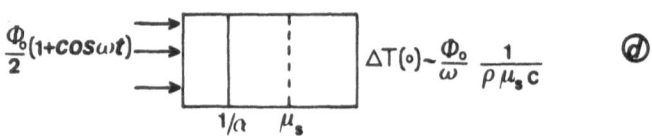

FIGURE 1 - a) and b) light and temperature distribution in a solid,
c) and d) modulated part of the surface temperature.

length μ_s adjacent to the surface in contact with a transparent gas.
whose heat capacity is negligible.

Therefore, the measurement of the surface temperature ΔT has to
lead to the knowledge of the optical absorption coefficient α (One must
choose the modulation frequency so that $\mu_s \ll 1/\alpha$ in order to avoid saturation effect). Indeed, it is well-known that the pressure variations in
a PA cell are proportional to ΔT (1). But there are other ways to detect
the surface temperature by using either infra-red radiometry, or the
deflection of a probe laser beam propagating along this surface ("mirage
effect"). The former, based on the detection of the infra-red emission
of the surface supposed to act as a grey body, will be detailed in the
"Photothermal radiometry paper"(2). In the latter case, the deflection
is induced by the periodic temperature gradient at the interface between
the solid and the adjacent fluid (3). In these methods, the sample does
not have to be enclosed in a cell and thus one avoids stray coherent
signals, generated by the walls or the windows of the cell.

In a good PA cell, temperature fluctuation measurements as small as a few 10^{-5} degrees (time constant of about 1s) can be achieved. For photo-thermal radiometry, the temperature signal is also correlated to the IR emissivity of the sample and a sensitivity of about 10^{-2} – 10^{-4} degrees can be reached depending on the detection noise equivalent power (2). It can be shown that in the mirage effect (Fig. 2), the thermally induced deflection is :

$$\Theta = \frac{\ell}{n} \cdot \frac{dn}{dT} \cdot \frac{dT}{dx} \sim \frac{\ell}{n} \cdot \frac{dn}{dT} \cdot \frac{\Delta T}{\mu_f}$$

where n is the refraction index and μ_f the thermal diffusion length of the fluid. With a commercial position sensor, it is easy to measure angular variations as small as 10^{-9} radians, so the corresponding ΔT is about 10^{-4} degrees when the surrounding fluid is air. We have achieved much better sensitivity (10^{-7} degrees) by immersing the sample in an organic liquid.

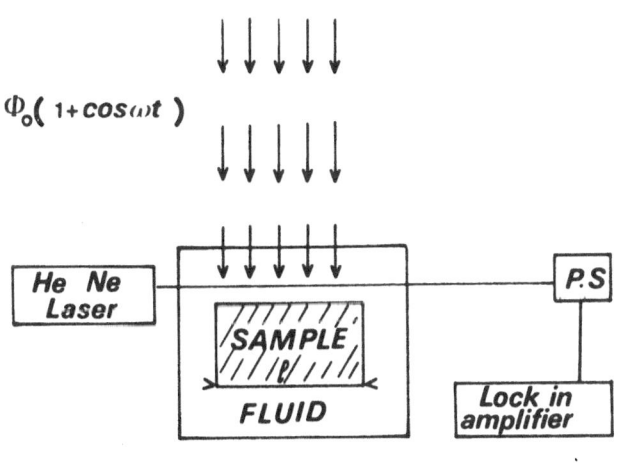

FLUID

$10 H_z$:
- air = $n=1$ $\frac{dn}{dT} = 10^{-6}$ $\mu_f = 1.2\,mm$
- CCl_4 = $n=1.45$ $\frac{dn}{dT} = 5.8\ 10^{-4}$ $\mu_f = 44\mu$

$\ell = 1\,cm$
$\theta\,\tilde{min}\,10^{-9}\,rad$
- air $\Delta T\,min = 10^{-3}\,deg.$
- CCl_4 $\Delta T\,min = 10^{-7}\,deg.$

FIGURE 2 – "Mirage effect" experimental set-up and sensitivities.

Even better sensitivities have been claimed with interferometric
detections, (4) but a much more difficult optical alignment is needed.

Moreover, let us point out that the PA signal production does not
necessarily need the properly intensity modulated monochromatic flux of
a laser or a monochromator. Indeed, incoherent sources may be coupled
with very efficient spectrometers, like e.g. a Michelson interferometer
(5) or grid-spectrometer (6). In these kinds of systems, which exhibit
a high luminosity and a good resolution, the light is inherent modulated
and can be directly used for PA signal generation (7). Finally, for
anisotropic samples, polarization modulation can be used to create a PA
signal ; this point will be the topic of the third section.

II - SOME EXAMPLES OF PA ABSORPTION STUDIES

We would like to outline a few specific applications of PA absorption spectroscopy taken from the solid state literature.

1. Opaque Samples.

PA spectroscopy is particularly suitable for this kind of sample,
which cannot be studied by classical photoelectric methods. Absorbance
as large as a few hundred cm^{-1} may be measured. If saturation effects are
to be avoided (1st section) several ways are possible by :

a) increasing the modulation frequency in order to reduce μ_s (9) (Fig. 3).
b) mechanically dispersing the absorbing sample in a transparent material
   (for instance by cogrinding (10)).
c) carefully analysing the phase of the photoacoustic signal (11) which
   is given by $\varphi = -\pi + \text{arctg} (1 + \alpha \mu_s)$ (Fig.4 (12).

In this later case, the phase study is particularly fruitful,
because it allows also an absolute calibration (if μ_s is known).

2. Highly Transparent Samples :

Since the PA signal is proportional to the absorbed energy, if the
incident light power is strong enough (laser), one can achieve absorption
measurements much smaller (10^{-5} cm^{-1} for bulk absorption and losses of
10^{-7} of the incident power in the case of surface or coating absorption)
than the reflexion or diffusion losses. Detection has been accomplished
using either PZT detector (13) or a thermal deflection method (14).

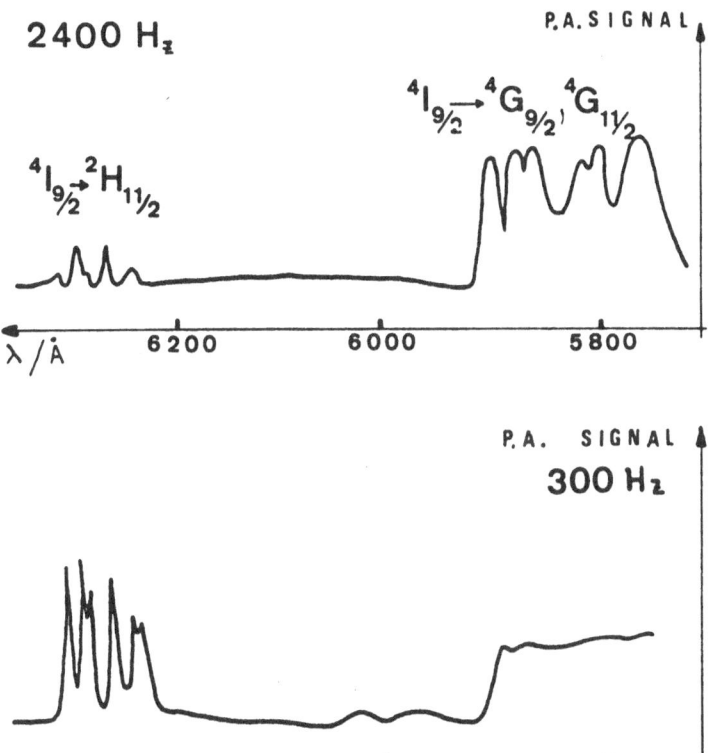

FIGURE 3 - Example ofde-saturation by increasing the modulation frequency (9).

Finally, immersed mirage effect detection (1st section), associated with a conventional incoherent source has been found to be very powerful in the study of thin film sample spectroscopy. Fig. 5 shows the result obtained on an amorphous silicon film deposit on a fused silica substrate.

3. Inhomogeneous Samples.

PA detection offers the unique feature of distinguishing various chromophores as a function of their distance from the surface. This has been accomplished by using the phase lag associated to the thermal wave generated by each local source (15). Fig. 6 illustrates this point with the spectroscopy of a single $CaWO_4/Nd^{+++}$ crystal : the narrow lines are the $4f^3 \rightarrow 4f^3$ transitions while the background is due to adsorbed parti-cles introduced by the polishing procedure.

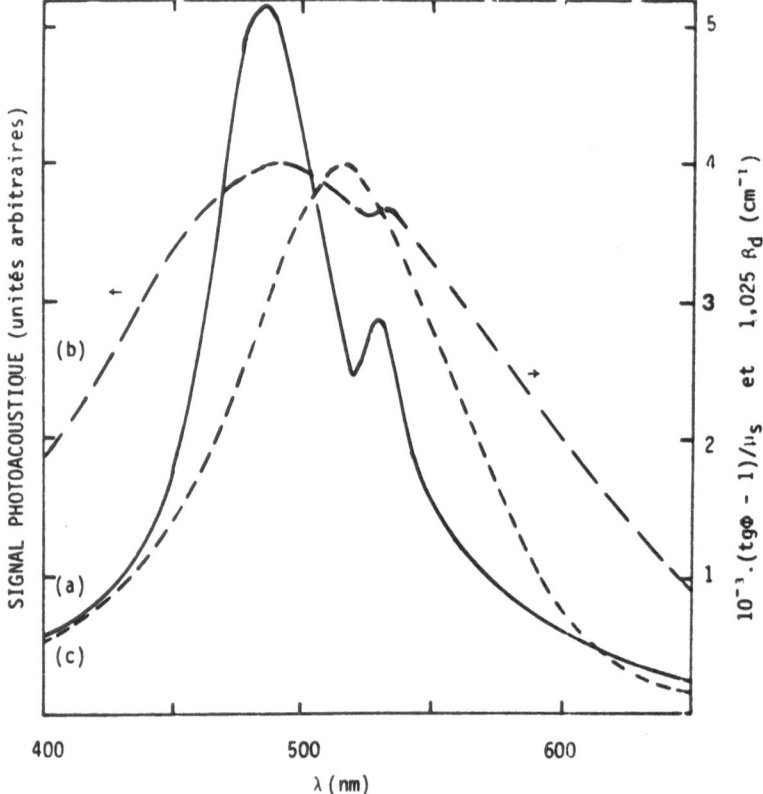

<u>FIGURE 4</u> - Recovery of the absorption signal by using the phase of the
PA signal (from ref. 12).

Finally one can consider the case of absorbing immersed samples such
as living cells (e.g. algae in sea water) or chemical products inter-
facing with a specific liquid.

The thermal wave damping is very noticeable in the liquid, so the
liquid surface temperature is strongly reduced compared to that of the
sample. One takes advantage of this fact to probe directly the surface
temperature of the sample within the liquid with the mirage effect.

4. <u>Radiative and non Radiative Transitions</u>

Only the non-radiative part of the de-excitation generates the PA
signal. So by carefully analysing the different de-excitation channels
it is possible to determine the absolute quantum efficiency (16) (17).
Moreover, time dependent PA signals (or phase lag measurements) have been
used for this kind of analysis (18).

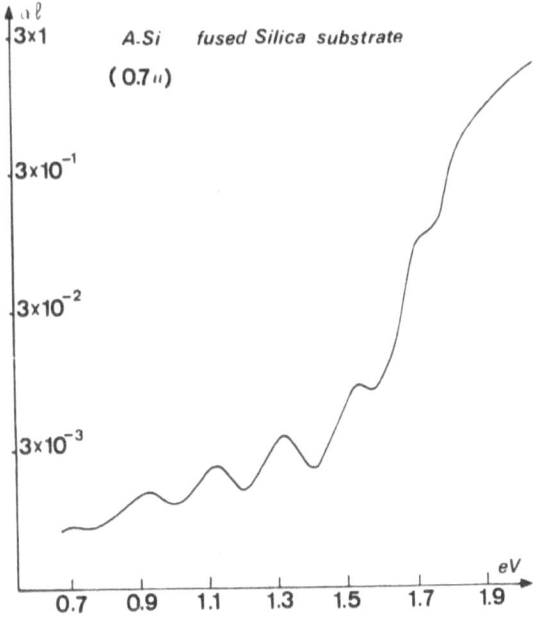

FIGURE 5

PA spectrum of an amorphous
silicon thin film deposited
on a fused silica substrate
(Mirage effect detection in
CCl_4).

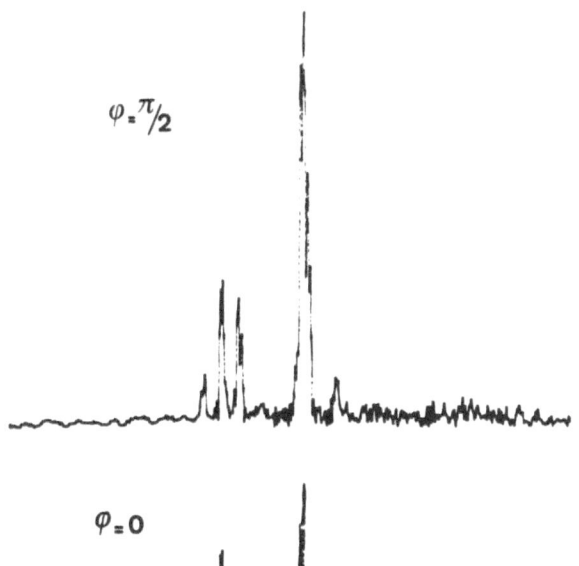

FIGURE 6

PA spectra of $CaWO_4/Nd^{3+}$
(1 %) single crystal in
phase (a), and in
quadrature (b) on the
surface signal.

III - <u>PHOTOACOUSTIC DICHROISMS</u>

The difference between a PA absorption experiment and a PA dichroism one is that in the latter case the light is polarization modulated rather than intensity modulated. One can define two kinds of dichroisms :
- circular dichroism, which is the difference between absorption coefficients associated with left and right circularly polarized light,
- linear dichroism, which is the difference between absorption coefficients associated with two orthogonal linear polarizations.

Both effects can be natural or induced. In the first case they are related to the anisotropic properties of the sample, and it is well known that such chiral properties are of paramount importance in chemistry and biology (19).

In the second case, the presence of an external field (magnetic, .stress or electric field) induces an anisotropy, the determination of which leads to important spectroscopic parameters (20) (e.g., Landé factors of the ground and excited levels).

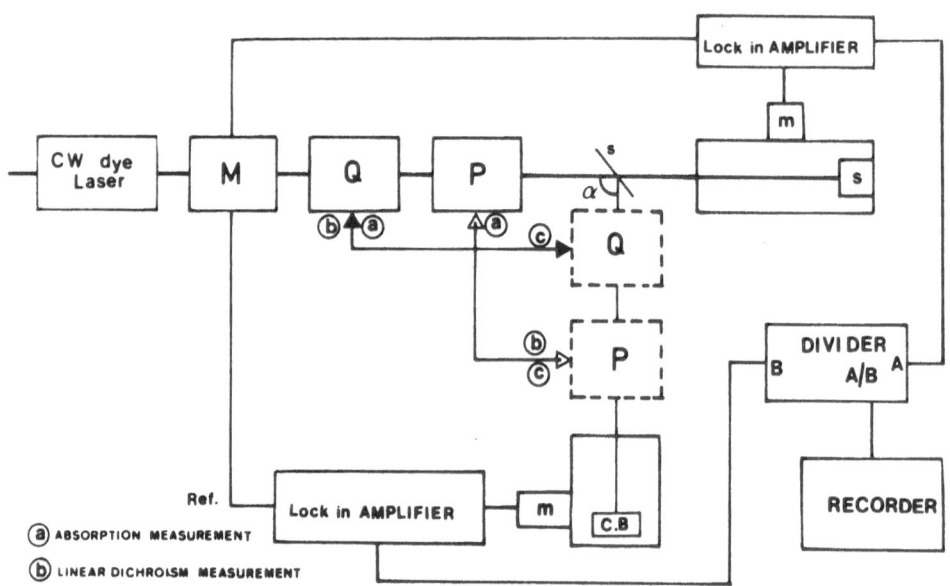

FIGURE 7 - PA dichroism and absorption experimental set-up : m electret microphon National Panasonic WMo64 ; S sample ; CB carbon black ; bs beam splitter ; (α is set very small to avoid polarization effects)(21).

Up to now, two experimental set-ups have been described in the literature for measuring dichroism with photoacoustic detection. Fournier et
al (21) have used the same apparatus in order to record either circular
dichroism (c), linear dichroism (b) or absorption (a) PA spectra (Fig.7).
Palmer et al (22) describe another method which allows measuring during
the same experiment both circular dichroism and absorption PA spectra
(Fig. 8).

Figure 9 is an example of magnetic circular dichroism $(Nd_2(MoO_4)_3$
single crystal) at room temperature.

Let us point out a particular feature of PA dichroism : the signal
decreases drastically when obtained from PA absorptions starting to be
saturate   (Fig. 10).

In conclusion, PA dichroism measurements are very fruitful for
anisotropic opaque samples and for powders. It is noticeable, that, in
the latter case, one does not loose the circular anisotropy of the
system whereas the linear one is completely cancelled. In principle,
this allows the measurement of the CD of low symmetry systems which
cannot be studied by another way (23).

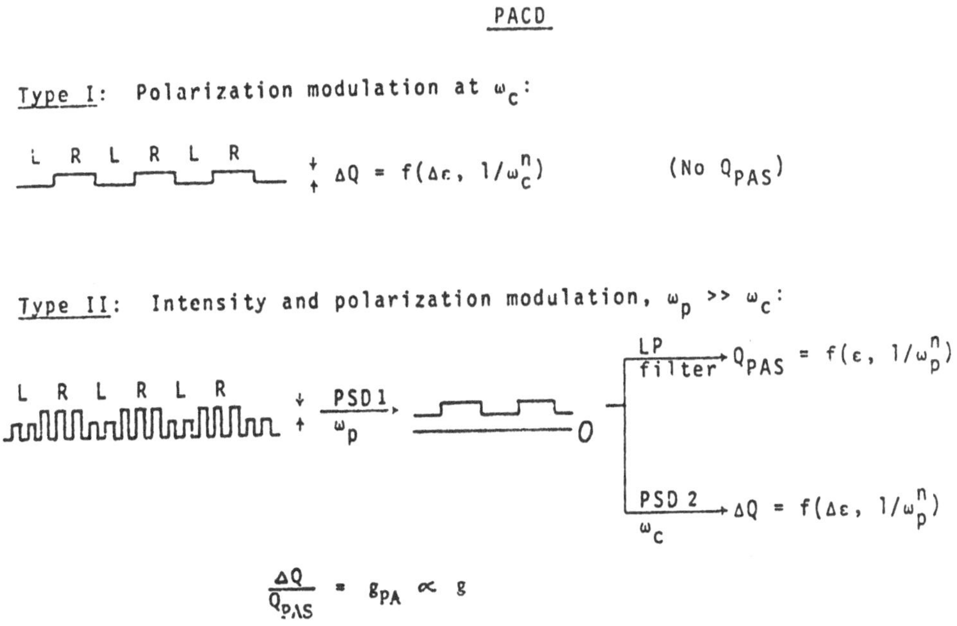

<u>PACD</u>

<u>Type I</u>: Polarization modulation at ω_c:

$$\Delta Q = f(\Delta\epsilon, 1/\omega_c^n) \qquad (No\ Q_{PAS})$$

<u>Type II</u>: Intensity and polarization modulation, $\omega_p \gg \omega_c$:

$$\xrightarrow{\text{PSD1}} \omega_p$$

$$\xrightarrow{\text{LP filter}} Q_{PAS} = f(\epsilon, 1/\omega_p^n)$$

$$\xrightarrow{\text{PSD 2}} \Delta Q = f(\Delta\epsilon, 1/\omega_p^n)$$
$$\omega_c$$

$$\frac{\Delta Q}{Q_{PAS}} = g_{PA} \propto g$$

Two possible methods for photoacoustic detection of circular dichroism (PACD).

<u>FIGURE 8</u> - Schematic diagram for achieving dichroism measurement
(from ref. 22).

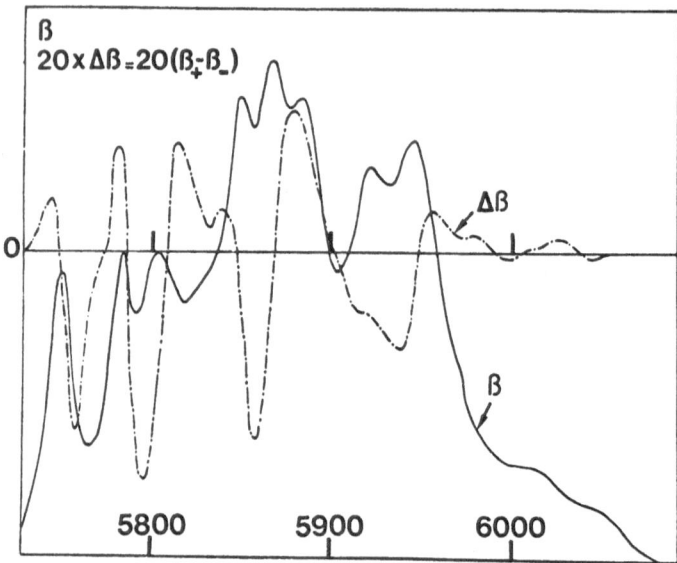

FIGURE 9 - Magnetic circular dichroism of a single crystal of $Nd_2(MoO_4)_3$ at room temperature (21).

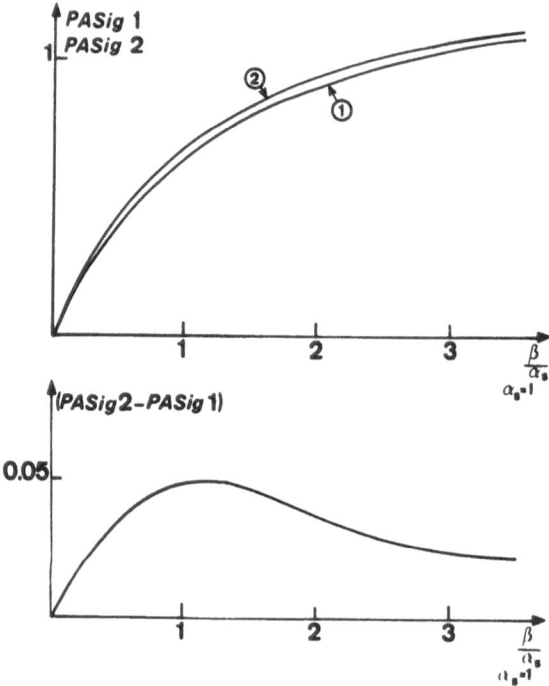

FIGURE 10 - Saturation effect in PA dichroism measurements.

IV - LOW TEMPERATURE PA MEASUREMENTS

When studying electronic levels of solid state materials, there is often a strong need to operate at low temperature. Here again, we can find in the literature two kinds of PA apparatus. In the first one only the sample cell is cooled (at liquid N_2 temperature), and the microphone, coupled by a narrow stainless steel tube, is maintened at room temperature (24). In contrast, in the second one (25) the whole apparatus (specially designed microphone, sample and cell) is immersed in a flow of cooled helium gas (the temperature varying continuously from 300 K to 5 K). These two set-ups can be applied either to spectroscopic studies (absorption, dichroism or de-excitation studies) or to thermal measurements such as phase transition determination.

V - CONCLUSION

In this paper we have briefly outlined some salient features of PA spectroscopies. The growing interest in this field allows us to speculate on new and more elaborate developments of the techniques. We believe that in the near future non linear effect studies (such as saturated absorption spectroscopy, stimulated Raman effect, excited state absorption, etc...), which have been applied successfully to gases and liquids (26) (27), will be extended to solid state spectroscopy, using PA detection.

ACKNOWLEDGMENTS

We want to thank R. PALMER for his helpful comments.

REFERENCES

(1) A. ROSENCWAIG and A. GERSHO.
Theory of the photoacoustic effect with solids.
J. Appl. Phys. _47_, 64 (1976).

(2) P.E. NORDAL and S.O. KANSTAD.
Photothermal Radiometry.
Physica Scripta _20_, 659 (1979).

(3) A.C. BOCCARA, D. FOURNIER and J. BADOZ
Thermo-optical spectroscopy : detection by the "mirage effect".
Appl. Phys. Lett. _36_, 130 (1980).

(4) C.C. DAVIS and S.J. PETUCHOWSKI
    Studies of energy absorption in gases and liquids by phase fluctua-
    tion optical heterodyne spectroscopy.
    Topical meeting on photoacoustic spectroscopy, August 1-3, 1979,
    AMES, Iowa.

(5) Aspen International Conference on Fourier Spectroscopy.
    G.A. VANASSE, A.T. STAIR, D. BAKER Editors.
    Special Reports n° 114, US Air Force Systems, Command 1971

(6) A. GIRARD and P. JACQUINOT
    "Principles of instrumental methods in spectroscopy" in Advanced
    Optical Techniques, edited by ACS van Heel, North Holland,
    Amsterdam 1967.

(7) D. DEBARRE, Thèse de 3ème Cycle, Paris 1981.

(8) D. DEBARRE, A.C. BOCCARA and D. FOURNIER
    High luminosity, visible and near infrared Fourier transform spec-
    trometer.
    (To be published).

(9) D. FOURNIER, A.C. BOCCARA and J. BADOZ.
    "Variable temperature absorption dichroism and thermal measurements
    in photoacoustic spectroscopy". Topical meeting on photoacoustic
    spectroscopy. August 1-3 1979, AMES (Iowa).

(10) J.W. LIN and L.P. DUDECK.
     Signal saturation effect and analytical techniques in photoacoustic
     spectroscopy of solids.
     Anal. Chem. $\underline{51}$, 1627 (1979).

(11) J.C. ROARK, R.A. PALMER and J.S. HUTCHINSON.
     Quantitative absorption spectra via photoacoustic phase angle
     spectroscopy.
     Chem. Phys. Lett. $\underline{60}$, 112 (1978).

(12) P. POULET, J. CHAMBRON and R. UNTERREINER.
     Quantitative photoacoustic spectroscopy applied to thermal thick
     samples.
     J. Appl. Phys.$\underline{51}$(3),1738 (1980).

(13) A. HORDVIK and H. SCHLOSSBERG.
     Photoacoustic technique for determining optical absorption coeffi-
     cients in solids.
     Appl. Opt. $\underline{16}$, 101 (1977).

(14) A.C. BOCCARA, D. FOURNIER, W. JACKSON and N. AMER.
     Sensitive photothermal deflection technique for measuring absorp-
     tion in optically thin media.
     Opt. Lett. $\underline{5}$ (9), 377 (1980).

(15) M. ADAMS and G. KIRKBRIGHT
     Part III : "The optoacoustic effect and thermal diffusivity.
     Analyst $\underline{102}$, 281 (1977).

(16) F. AUZEL, D. MEICHENIN and J.C. MICHEL.
     Determination of quantum yield of self actived mini-laser materials
     by photoacoustic spectroscopy.
     International Conference on Luminescence 17-21 July 1978.
     Paris, D. CURIE, J. MATTLER, R. PARROT Editors North Holland,
     Publishing Company, Amsterdam.

(17) R.S. QUIMBY and W.H. YEN,
     Photoacoustic measurement of absolute quantum efficiencies in solids.
     Opt. Lett. 3, 181 (1978).

(18) L.D. MERKLE and R.C. POWELL
     Photoacoustic spectroscopy investigation of radiationless transi-
     tions in Eu^{++} ions in KCl crystals.
     Chem. Phys. Lett. 46, 303 (1977).

(19) T.M. LOWRY
     Optical Rotatory Power, Dover
     New-York, 1964.

(20) J. BADOZ, M. BILLARDON, A.C. BOCCARA and B. BRIAT.
     Measurements and interpretation of magnetic circular dichroism and
     magnetic linear dichroism spectra. Symposium of the Faraday Society,
     3, 27 (1969).

(21) D. FOURNIER, A.C. BOCCARA and J. BADOZ.
     Dichroism measurements in photoacoustic spectroscopy.
     Appl. Phys. Lett. 32, 640 (1978).

(22) R. PALMER, J. ROARK, J. ROBINSON and J. HOWELL.
     Photoacoustic detection of natural circular dichroism in solids.
     Topical meeting on photoacoustic spectroscopy. Aug. 1-3 1979,
     AMES (Iowa).

(23) R. PALMER (Private communication).

(24) J.C. MURPHY and L.D. AAMODT
     Photoacoustic spectroscopy of luminescent solids : Ruby.
     J. Appl. Phys. 48, 3502 (1977).

(25) C. PICHON, M. LE LIBOUX, D. FOURNIER and A.C. BOCCARA
     Variable temperature photoacoustic effect : application to phase
     transition.
     Appl. Phys. Lett. 35(6) 435 (1979).

(26) J. BARRET, G. WEST and D. SIEBERT.
     Photoacoustic Raman Spectroscopy (PARS) of gases
     Topical meeting on photoacoustic spectroscopy. Aug. 1-3 1979,
     AMES (Iowa).

(27) C.K. PATEL.
     Opto-acoustic spectroscopy of excited molecular states.
     Topical meeting on photoacoustic spectroscopy. Aug. 1-3 1979,
     AMES (Iowa).

COMPARISON BETWEEN PHOTOACOUSTIC
AND OTHER SPECTROSCOPIES

R. TILGNER
Siemens AG,
Zentrale Fertigungsaufgaben
D-8 MÜNCHEN 70

ABSTRACT

Well established techniques as transmission-, reflec-
tance- and diffuse reflectance spectroscopy (TS, RS and DRS,
resp.) are compared with photoacoustic spectroscopy (PAS).

For weakly absorbing samples showing no light scattering, PAS gives precise absolute values of absorption coefficients, providing the unique possibility of depth profiling, which is inaccessible to all the other techniques.
With strongly absorbing and light scattering samples,
however, one may not dispense with the other methods. Especially with powders PAS will not give spectra in the sense
of the Rosencwaig-Gersho theory, but rather DRS spectra,
which may be more adequately understood by following the
theory of Kubelka and Munk. Thus many famous "PAS"-spectra
as the ones of rare earth oxides turn out to be in fact DRS-
spectra monitored by acoustical means.

INTRODUCTION

Doing optical spectroscopy in principle one is interested to gain informations on a microscopic scale, that is,
about the molar/atomic coefficient of absorption ε, which is
to be considered as the true fingerprint of the very mole-
cule/atom under interest. The established methods to measure
this quantity include transmission spectroscopy, which provides the most direct way to ε. While the application of this
method remains restricted to not too highly concentrated
molar systems, reflectance spectroscopy as well as diffuse
reflectance spectroscopy are used with bulk and powdered
materials, respectively, where the exact nature of the

measured quantity is not equally well accessible for interpretation as the transmittance of diluted systems showing no
light scattering.

Photoacoustic spectroscopy in the past has been demonstrated to be applicable within a very large field of questions. Up to now a considerable amount of its results has remained qualitative, however. In the following it will be
tried to compare applications of these different spectroscopic methods and to destillate the ones which are unique for
photoacoustics.

TRANSMISSION SPECTROSCOPY (TS)

With optically transparent or only weakly absorbing samples
showing neglegible scattering the way to obtain the molar absorption coefficient is straight forward, if one uses the
Lambert-Beer law:

$$I = I_o \cdot \exp(-\beta x) \quad , \tag{1}$$

where I_o denotes the incident, I the transmitted light intensity. It has been shown experimentally, that for low
concentrations c and neglegible scattering there holds:

$$\beta = 2,303 \cdot \varepsilon \cdot c \quad . \tag{2}$$

Even for considerable concentrations, in some cases the deviations from this law may be small /1/.

REFLECTANCE SPECTROSCOPY (RS)

This technique is applied to flat surfaces, where the
incident light behaves specularly. In this case the reflectance may be described by Fresnel's formula in terms of the
real and imaginary part of the samples' refractive index. In
case of perpendicular incidence, we have for the surface reflectance $R_{\perp reg}$

$$R_{\perp reg} = \frac{(n_1 - n_o)^2 + K_1^2}{(n_1 + n_o)^2 + K_1^2} \quad , \tag{3}$$

where n_1 and K_1 are the real and the imaginary part of the
samples refractive index, respectively, while the subscript
O refers to the adjacent gas, normally air. From a measurement of the reflectance and the phase shift of the reflected
wave one obtains n_1 and K_1 /2/. Therefrom:

$$\beta(\lambda_o) = \frac{4\pi K_1(\lambda_o)}{\lambda_o} \quad , \tag{4}$$

λ_o is the wavelength at which the measurement is made.

DIFFUSE REFLECTANCE SPECTROSCOPY (DRS)

In case of diffusely scattering samples as most pro-
minently realized by powders, the situation becomes even more
complicated /3/. The reflectance of such a sample depends not
only from the molar absorption coefficient ε, but also from
parameters as particle size, packing density, crystal form
and the refractive index of the powder /2/.

The situation is treated theoretically by phenomeno-
logical approaches like that one of Kubelka and Munk /4/.
This theory uses only two coefficients to characterize the
experimental situation: an absorption coefficient K and a
scattering coefficient S. The theory relates these two coefficients to quantities accessible in experiment as:

  R_b, the reflectance of the backing material
  R_d, the reflectance of the sample of thickness d
  R_{d_o}, the reflectance of the sample of thickness d with
       a black backing material and
  R_∞, the reflectance of a sample of infinite thickness.

One obtains

$$S = \frac{R_\infty}{d(1 - R_\infty^2)} \ln \frac{R_\infty(1 - R_{d_o} \cdot R_\infty)}{R_\infty - R_{d_o}} \tag{5}$$

$$K = \frac{1 - R_\infty}{2d(1 + R_\infty)} \ln \frac{R_\infty(1 - R_{d_o} \cdot R_\infty)}{R_\infty - R_{d_o}} \tag{6}$$

From this one finds

$$\frac{K}{S} = \frac{(1 - R_\infty)^2}{2R} \equiv F(R_\infty) . \tag{7}$$

F is called "Kubelka-Munk function". The way from K to β is an experimental one: Diluting the powder one approaches the true absorption spectrum /2/, while the powder's other characteristics as mentioned above are eliminated by re-fering the diluted sample to the pure diluent (mostly $BaSO_4$, MgO and other white standards).

PHOTOACOUSTIC SPECTROSCOPY (PAS)

The widely adopted theory of Rosencwaig and Gersho /5/ assumes the incident radiation to become absorbed in a Lambert-Beer way, that is, proportional to $\exp(-\beta x)$, as mentioned above. For optically transparent samples, the PAS-signal is proportional to β. As the thermal diffusion length μ_p may be shifted by changing the modulation fre-quency of the light, the PAS-signal depends also from the thermal properties of the backing material or of the sample itself, respectively. In case of $\mu_\beta = \frac{1}{\beta} > \mu_p$ PAS offers one unique possibility: depth profiling.

For optically opaque samples the condition $\mu_\beta > \mu_p$ may be met, too, to gain optical information on them. In case of $\mu_p < \mu_\beta$ the PAS-signal is saturated, however, and depends only from the sample's and the backing material's, resp., thermal properties.

EXPERIMENTAL RESULTS

"PAS"- and DRS-spectra of opaque powders as Nd_2O_3, Ho_2O_3 and Cr_2O_3

Up to now PAS is claimed for to be insensitive to light scattering. This has misled several authors to consider the

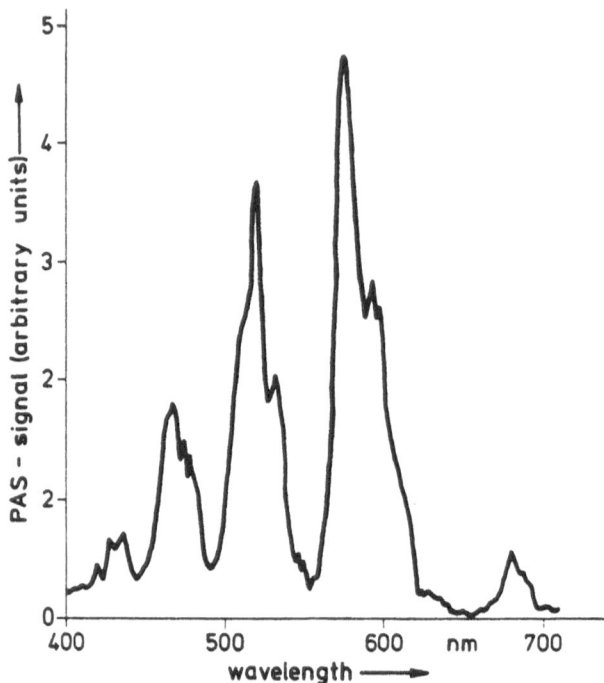

Fig. 1:   "PAS"-spectrum of Nd_2O_3-powder.
          Modulation frequency: 190 s^{-1} (after /6/)

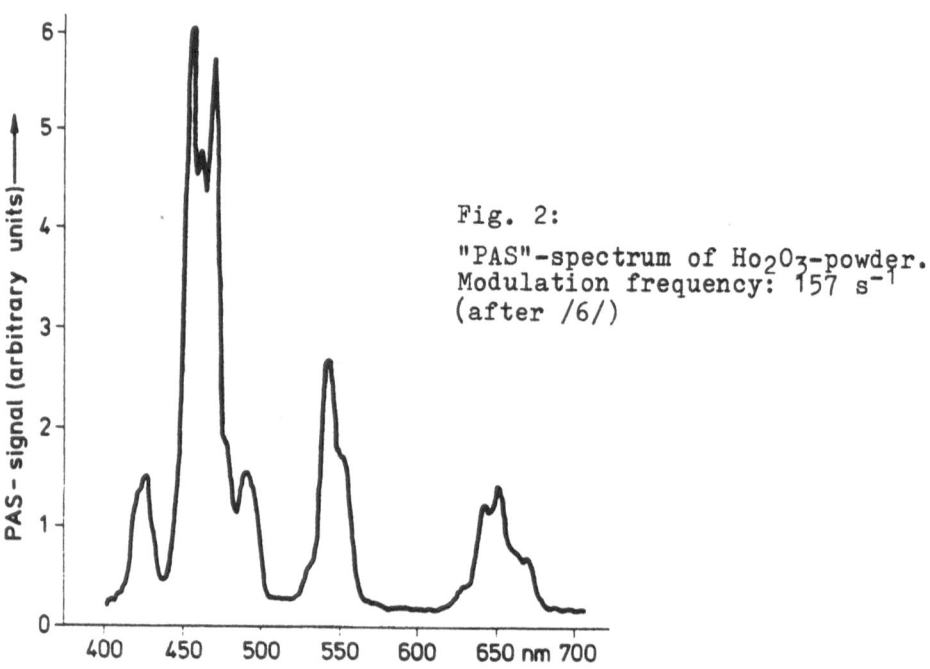

Fig. 2:

"PAS"-spectrum of Ho_2O_3-powder.
Modulation frequency: 157 s^{-1}
(after /6/)

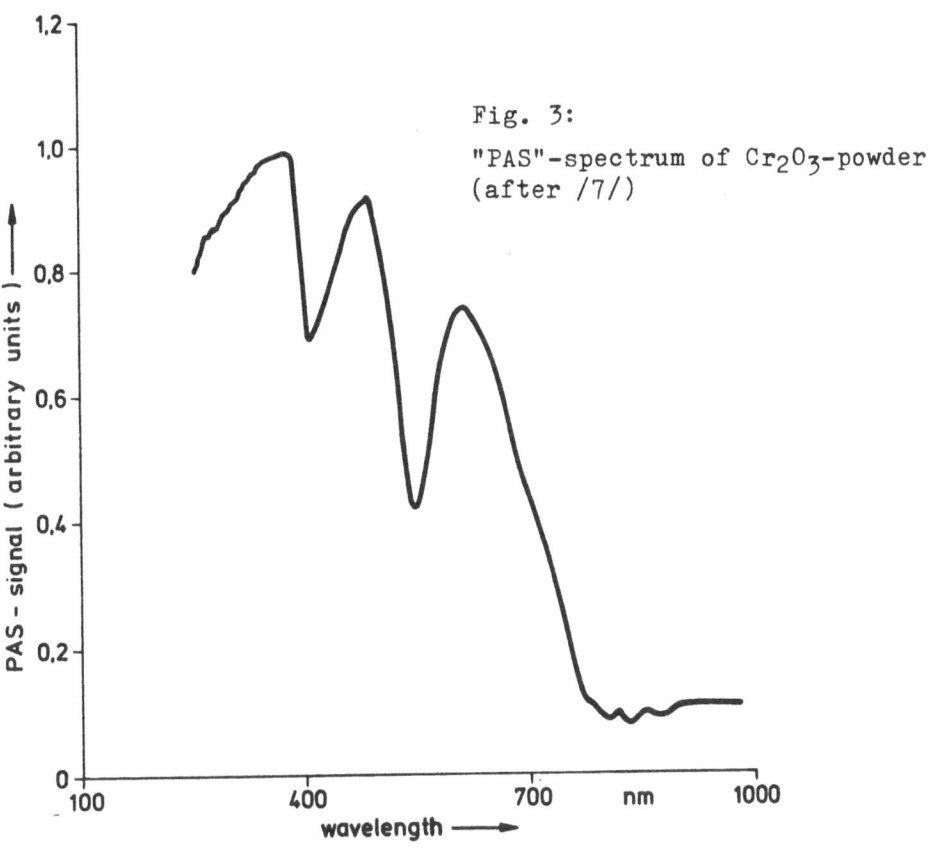

Fig. 3:

"PAS"-spectrum of Cr_2O_3-powder (after /7/)

TABLE 1

Sample	wavelength λ (nm)	absorption-coefficient β (μm^{-1})	absorption-length $\mu_\beta = \frac{1}{\beta}$ (μm)	thermal-diffusivity α ($\frac{cm^2}{s}$)	thermal diffusion length at 50 s^{-1} μ_p^{50} (μm)	500 s^{-1} μ_p^{500} (μm)	modulation frequency for $\mu_\beta = \mu_p$ v_g (s^{-1})
Nd_2O_3	500	3,77 [17]	0,27	$\sim 10^{-2}$ [20]	~ 80	~ 25	$4,5 \times 10^6$
Cr_2O_3	450	$\sim 1,7$ [18]	$\sim 0,6$	$\sim 6 \times 10^{-3}$ [20],[22]	~ 61	~ 19	$5,5 \times 10^5$

Optical and thermal properties of opaque substances.
In /22/ the thermal conductivity is only given for a con-
glomerate of different oxides as MgO, Fe_2O_3, Al_2O_3 and
others. Cr_2O_3 represents the majority, however, making up
some 40 % of the mixture. In /20/ the thermal diffusivity of
Nd_2O_3 is not given. From the value for Sm_2O_3 and other rare
earth oxides the value of α was estimated.

spectra obtained from opaque powders (conf. fig. 1 to 3)
photoacoustic. Actually for these substances the condition
$\mu_p < \mu_\beta$ is not met up to modulation frequencies as high as
500 kHz (see table 1). The resemblance to DRS-spectra is
striking, however (conf. fig. 4 to 6) and in fact these
spectra may be better understood as DRS-spectra recorded by
acoustic means. This idea was recently verified in experiment by Freeman, Friedman and Reichard /8/.

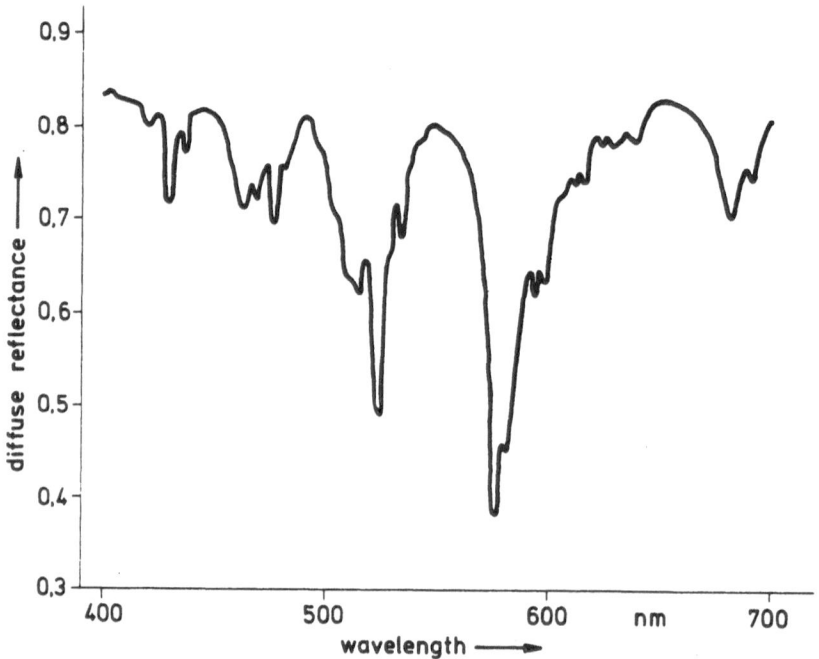

Fig. 4:   DRS-spectrum of the same Nd_2O_3-powder as in fig. 1.
           Measured with geometry $0°$/diffuse against a $BaSO_4$
           white standard (after /6/)

Substances showing sharp edges as

Semiconductors

The figures 7 and 8 show photoacoustic spectra of CdS
and Si powders. The interpretation of these spectra is considerable complicated by the fact, that the condition
$\mu_p < \mu_\beta$ is not met for the whole spectrum in both cases.

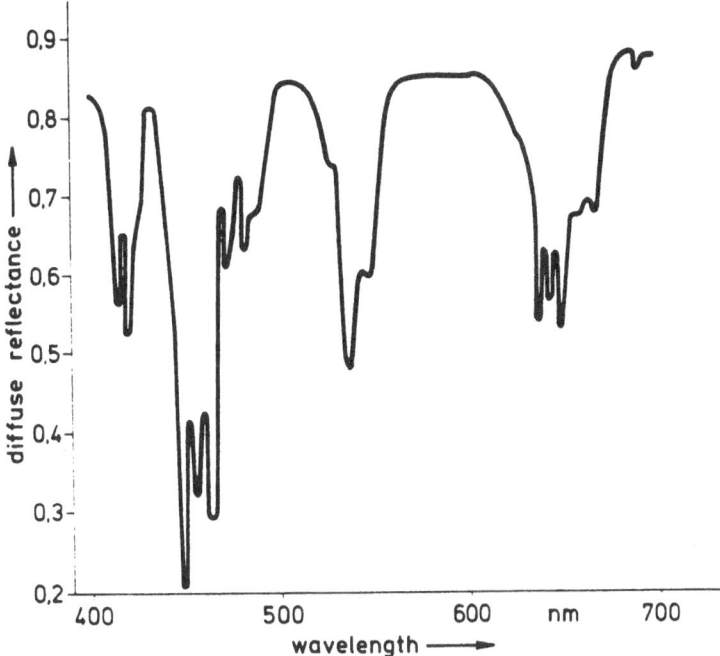

Fig. 5: DRS-spectrum of the same Ho_2O_3-powder as in fig. 2.
Experimental parameters as in fig. 4

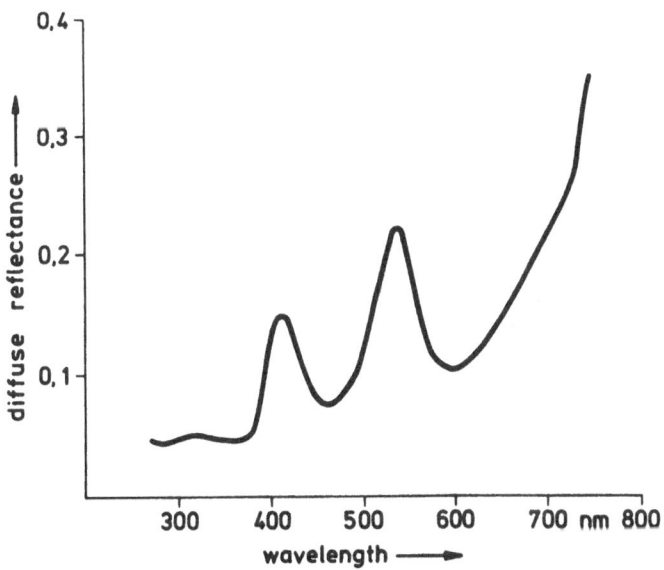

Fig. 6: DRS-spectrum of Cr_2O_3-powder. Experimental parameters as in fig. 4

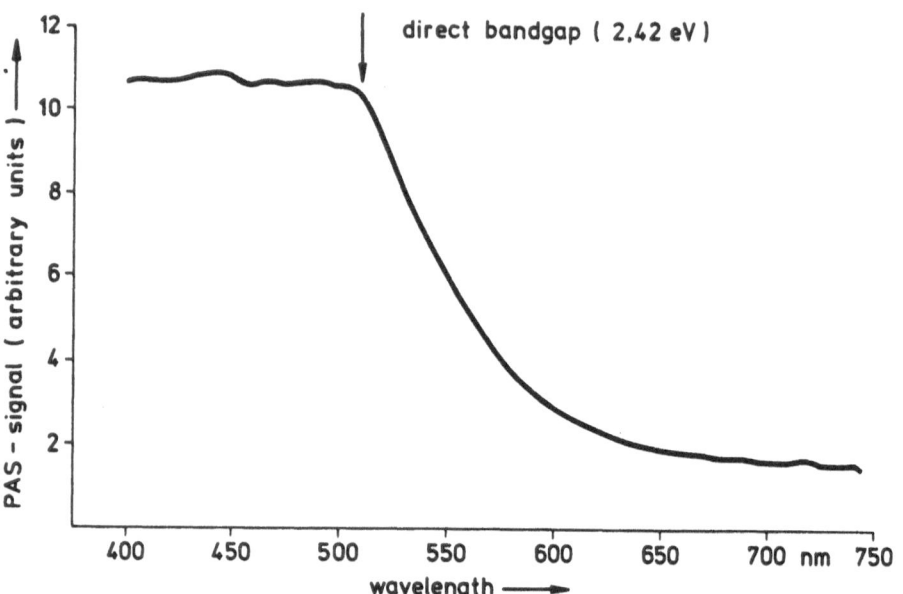

Fig. 7: Photoacoustic spectrum of CdS-powder (after /15/).
Modulation frequency: 190 s^{-1}. The arrow indicates
the true bandgap /16/

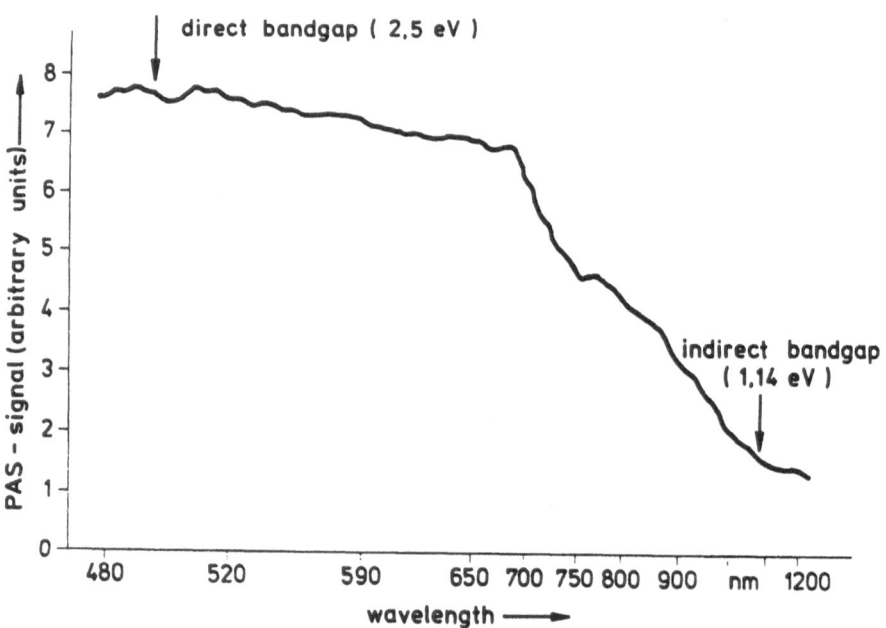

Fig. 8: Photoacoustic spectrum of Si-powder (after /6/).
Modulation frequency: 190 s^{-1}. The arrows indicate
the true values of the direct and the indirect
bandgap /16/

Approaching the edge from the long wave side the condition
is met. At some wavelength one reaches $\mu_\beta = \mu_p$, however, and
the PAS-signal becomes saturated, that is, independent from
the incident light's wavelength. Thus for the powdered samples again we have "PAS"-spectra for the wavelength, were
$\mu_p < \mu_\beta$ actually is not met. This may be taken as a hint to
consider these spectra DRS like, too (see also table 2).

TABLE 2

Sample	wavelength λ (nm)	absorption-coefficient β (μm^{-1})	absorption-length $\mu_\beta = \frac{1}{\beta}$ (μm)	thermal-diffusivity α ($\frac{cm^2}{s}$)	thermal diffusion length at 50 s^{-1} μ_p^{50} (μm)	500 s^{-1} μ_p^{500} (μm)	modulation frequency for $\mu_\beta=\mu_p$ v_g (s^{-1})
Si	700 750	0,3 0,21 [6]	3,3 4,7	0,88[23]	749	237	$2,5 \times 10^6$ $1,2 \times 10^6$
CdS	510 517	$9,7 \times 10^{-2}$ $4,5 \times 10^{-2}$[19]	10,4 22,1	015[21]	311	98	$4,5 \times 10^4$ $9,7 \times 10^3$

Optical and thermal properties of semiconductors

Spectroscopic characterization of a
layer close to surface

Fig. 9 shows a TS-spectrum of a 3 mm thick coloured
Polymethylmethacrylate sheet, which is strongly absorbing in
the red region. Therefore direct spectroscopic examination
by TS is not possible. After thinning the plate down to
290 µm one may obtain a TS-spectrum as also shown in fig. 9.
This laborious procedure can be avoided by producing the
spectrum of the thick sheet photoacoustically with a properly choosen modulation frequency (conf. fig. 10). From RS no
reasonable spectrum may be expected as one can derive from
equation 3 using K_1 as taken from fig. 9. Assuming $n_1 \sim 1.5$
the absorption contributes only in the order of 10^{-4} to the
total reflection of the sample. DRS will not produce any
spectrum of the flat sheet. After grinding it one obtains
the DRS-spectrum shown in fig. 11.

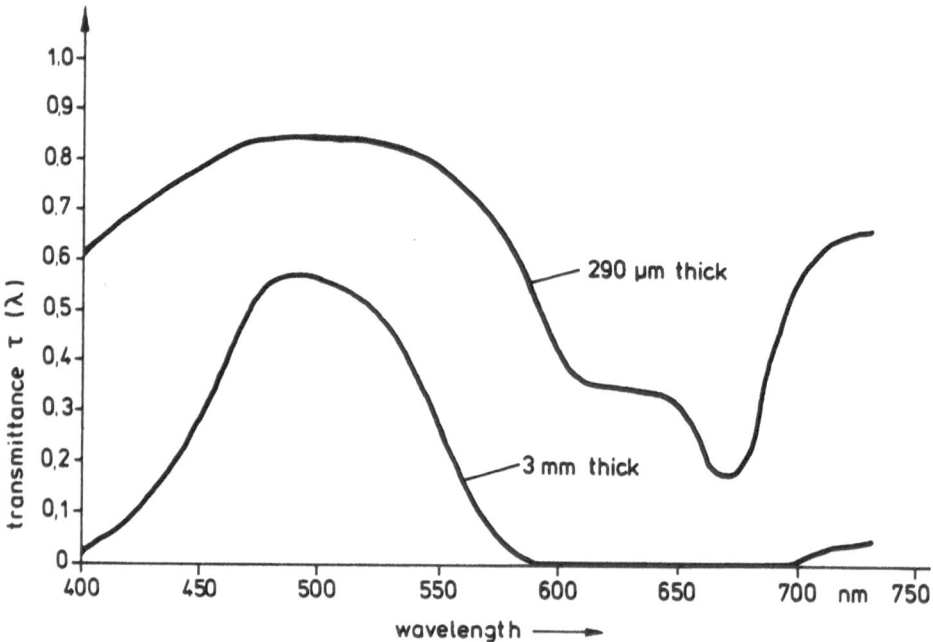

Fig. 9:   TS-spectra of coloured Polymethylmethacrylate
           sheets of 3 mm and 290 μm thickness (after /6/)

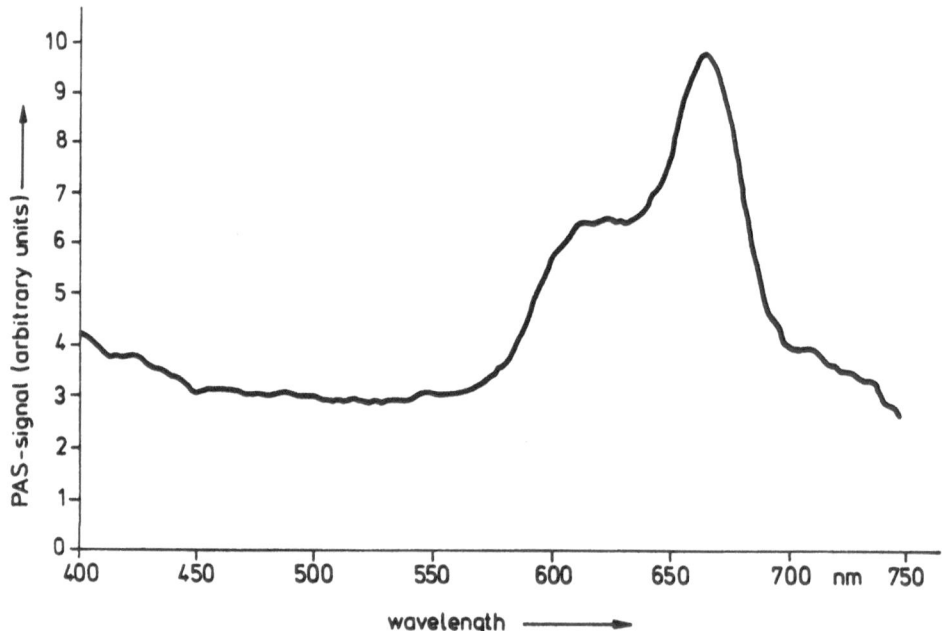

Fig. 10:   PAS-spectrum of a 3 mm thick coloured Polymethyl-
           methacrylate sheet. Modulation frequency: 63 s^{-1},
           $\mu_p \sim 150$ μm (after /6/)

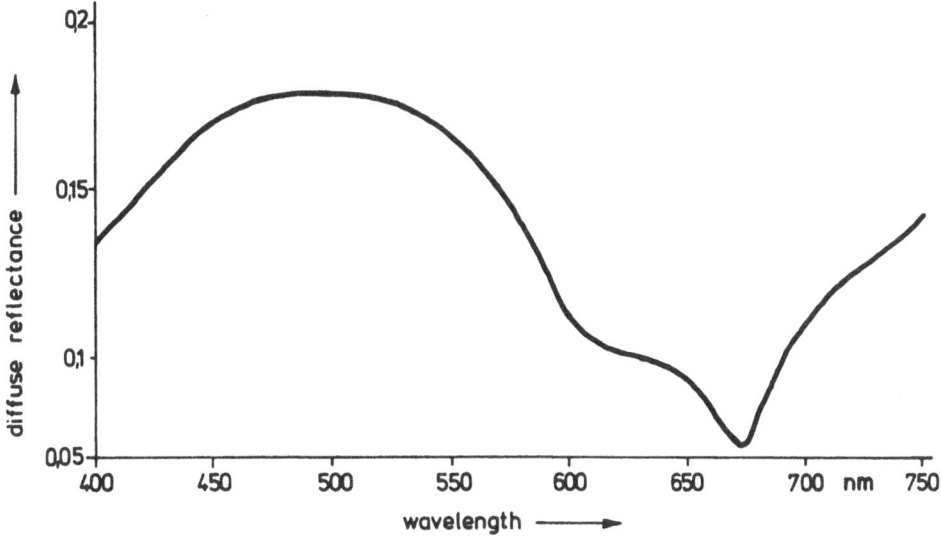

Fig. 11:   DRS-spectrum of coloured Polymethylmethacrylate
           after grinding. Geometry: 0°/diffuse

RÉSUMÉ

The hope PAS in case of light scattering samples to be
a convenient means to avoid the more laborious way using DRS
seems not to be satisfied, especially for strongly absorbing
substances. Actually the observed "PAS"-spectra of Nd_2O_3,
Ho_2O_3 and others are no PAS-spectra in the sense of the
Rosencwaig-Gersho theory. They are much more DRS-spectra,
monitored acoustically, diffuse scattering beeing a conditio
sine qua non to obtain any wavelength dependent PAS-signal
at all.

For weakly absorbing samples with low scattering PAS
gives highly correct values of β as it was shown for gases
/9, 10/, liquids /11/ and solids /12, 13/. Samples of medium
absorptance as for instance coloured plastics /6/ are most
easily analyzed by TS. If the condition $\mu_{\beta} > \mu_p$ may be met,
however, PAS enables one to gain additional information, not
accessible to TS, as depth profiling.

Light scattering also for PAS presents an inevitable
problem, which may be adequately treated only by performing
model experiments /14/ or applying a Kubelka-Munk treatment,
as for DRS, if possible.

ACKNOWLEDGEMENT

The cooperation with the group of Prof. E. Lüscher,
Technische Universität München, at whose institute the photoacoustic experiments were carried out, is gratefully acknowledged.

REFERENCES

/1/ H.Ulich; W.Jost, "Kurzes Lehrbuch der physikalischen
                     Chemie", Steinkopff, Darmstadt 1963

/2/ G.Kortüm, "Reflexionsspektroskopie", Springer,
              Heidelberg 1969

/3/ R.Tilgner, Photoacoustic spectroscopy with light
                scattering samples, Appl.Optics to be published

/4/ P.Kubelka; F.Munk, Ein Beitrag zur Optik der Farb-
                       anstriche, Z.Techn.Phys. $\underline{12}$ (1931) 593

/5/ A.Rosencwaig; A.Gersho, Theory of the photoacoustic
                            effect with solids, J.Appl.Phys. $\underline{47}$ (1976) 64

/6/ R.Tilgner, E.Lüscher, Photoakustische Spektroskopie im
                          Vergleich mit Transmissions- und Reflexions-
                          spektroskopie an einigen Testsubstanzen,
                          Z.Phys.Chem. NF $\underline{111}$ (1978) 19

/7/ A.Rosencwaig, Photoacoustic spectroscopy,
                   Adv. Electronics Electron Phys. $\underline{46}$ (1978) 207

/8/ J.J.Freeman; R.M.Friedman; H.S.Reichard, Photoacoustic
                                             spectroscopy of particulate solids,
                                             J.Phys.Chem. $\underline{84}$ (1980) 315

/9/ L.B.Kreuzer; C.K.N.Patel, Nitric oxide air pollution:
                               detection by optoacoustic spectroscopy,
                               Science $\underline{173}$ (1971) 45

/10/ L.B.Kreuzer, Laser optoacoustic spectroscopy for GC
                  detection, Anal. Chem. $\underline{50}$ (1978) 597 A

/11/ A.C.Tam; C.K.N.Patel, Optical absorptions of light and
                          heavy water by laser optoacoustic spectro-
                          scopy, Appl. Optics $\underline{18}$ (1979) 3348

/12/ A.Hordvik; L.Skolnik, Photoacoustic measurements of
        surface and bulk absorption in HF/DF laser
        window materials, Appl. Optics 16 (1977) 2919

/13/ J.M.Mc David; K.L.Lee; S.S.Yee; M.A. Afromowitz, Photo-
        acoustic determination of the optical
        absorptance of highly transparent solids,
        J. Appl. Phys. 49 (1978) 6112

/14/ P.Helander; I.Lundström; D.Mc Queen, Light scattering
        effects in photoacoustic spectroscopy,
        J. Appl. Phys. 51 (1980) 3841

/15/ U.Möller, Diplomarbeit TU München 1978

/16/ C.Kittel, "Einführung in die Festkörperphysik", 3.
        Aufl., Oldenbourg, München 1973

/17/ G.Haas; J.B.Ramsey; R.Thun, Optical properties of va-
        rious evaporated rare earth oxides and
        fluorides, J.O.S.A. 49 (1959) 116

/18/ D.S.Mc Clure, Comparison of the crystal fields and op-
        tical spectra of Cr_2O_3 and ruby,
        J. Chem. Phys. 38 (1963) 2289

/19/ C.C.Klick, Luminescence and photoconductivity in Cad-
        mium Sulfide at the absorption edge,
        Phys. Rev. 89 (1953) 274

/20/ L.Gmelin, "Handbuch der Anorganischen Chemie", 8. Aufl.

/21/ Cleveland crystals, data sheet, 1975

/22/ Y.Touloukian ed. "Thermophysical properties of matter",
        TPRC data series, Vol. 2, IFI/Plenum,
        New York 1973

/23/ dito, Vol. 10

GENERATION OF ACOUSTIC WAVES IN
LIQUIDS BY PULSED LASERS

M.W. Sigrist
Solid State Physics Laboratory
ETH Hönggerberg
CH-8093 ZURICH, SWITZERLAND

ABSTRACT

The generation of acoustic transients in liquids by laser impact is
described both experimentally and theoretically. The experiments
were performed with a hybrid-CO_2-laser whose beam was focused perpendicular onto the free surface of various liquids. Special piezoelectric transducers developed in our laboratory were used for the
detection of the acoustic waves. Three different regimes could be
distinguished with respect to the time dependence of the acoustic
signal. One of them corresponds to the pure photoacoustic effect as
the wave generating mechanism. The possibility of inducing acoustic
waves with tunable high frequencies in the MHz range in this case
indicates the transient character of this process. On the basis of
this result a new spherical model on the photoacoustic effect is
proposed in which the laser heating process is represented by the
three-dimensional heat pole. Our analytical model provides excellent agreement with the experimental results.

INTRODUCTION

There are four interaction mechanisms responsible for the genera-

tion of acoustic waves in liquids by laser impact. These are

dielectric breakdown, electrostriction, vaporization and thermo-

elastic or photoacoustic process. The first two mechanisms are

responsible for the stress generation at high laser intensities

($\gtrsim 10$ GW cm^{-2}) and in transparent liquids. On the other hand,

vaporization and the photoacoustic effect dominate the generation

of acoustic waves in absorbing liquids and at laser intensities

below the breakdown threshold. This study concentrates on the last

two mechanisms.

EXPERIMENTS

Our experiments were performed with a hybrid-CO_2-laser operating in the TEM_{oo} mode. It delivers pulses of 350 ns halfwidth and approximately 100 kW peak power. The laser beam was focused perpendicular onto the free surface of different liquids where an intensity of up to 10^8 W cm^{-2} was achieved. The liquids investigated were chosen on the basis of their absorption coefficient at the laser wavelength and their thermal properties. The laser-generated acoustic waves in the liquid were investigated with both high speed shadow- or Schlieren photography[1] and piezoelectric transducers. These gauges were developed in our laboratory. They exhibit a response time of a few ns and a sensitivity of the order of 100 mV/bar.[2] These values were achieved with transducers operating with thin z-cut $LiNbO_3$ disks.

RESULTS AND DISCUSSION

High speed shadow- and Schlieren photography provides information on the absorption and vaporization process of the liquid. Interesting statements can be made on the geometrical structure of the shock waves. Whereas in strongly absorbing liquids like water only hemispherical waves are generated, additional cylindrical waves appear in liquids with weak absorption like carbon tetrachloride. The measured propagation speed of the acoustic waves is equal to the sound speed in the liquid. The pressure amplitude and the time dependence of the acoustic signals which can both not be deduced from the shadowgraphs are provided by our fast transducers.

The striking influence of the transverse mode structure of the laser beam on the pressure amplitude[3] demonstrates the need for a well defined and controlled laser beam for this type of study.

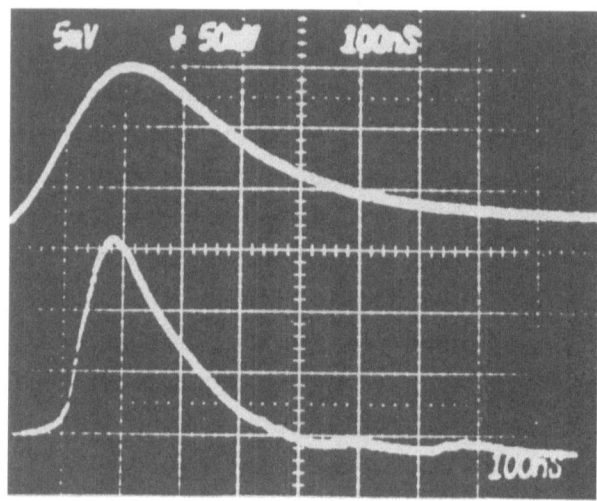

FIGURE 1a   Upper trace: Pulse of the hybrid-CO_2-laser.

                   Lower trace: Acoustic transient for absorbed laser
                                energy densities well above the vapo-
                                rization threshold.

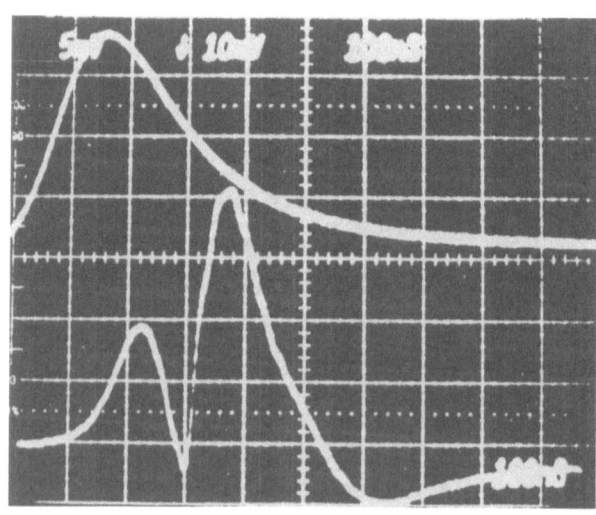

FIGURE 1b   Upper trace: Pulse of the hybrid-CO_2-laser.

                   Lower trace: Acoustic transient for absorbed laser
                                energy densities only slightly above
                                the vaporization threshold.

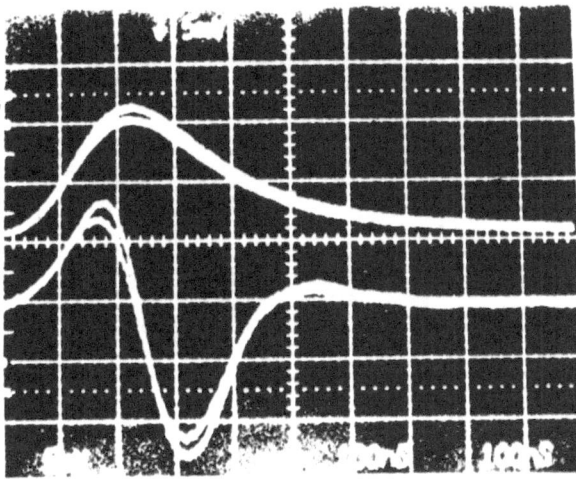

FIGURE 1c   Upper trace: Pulse of the hybrid-CO_2-laser.
                Lower trace: Acoustic transient for absorbed
                             laser energy densities below the
                             vaporization threshold.

Three different cases can be distinguished with respect to the
time dependence of the generated acoustic signal. This is shown
for water in Figures 1a to 1c. For absorbed laser energy densities well above the vaporization threshold of the liquid
(Figure 1a), a shock wave is induced with high amplitude and
steep front. If the absorbed laser energy density is just
above the vaporization threshold (Figure 1b) a double pulse
results. This type of signal is due to the individual contributions of both the vaporization and the photoacoustic effect.
The two processes could be observed separately for the first
time. The weak prepulse is attributed to the photoacoustic effect
whereas the main pulse is caused by the delayed onset of vaporization. Figure 2 shows the dependence of the time delay Δt
between the two pulses on the laser pulse peak power P for two
liquids. In both cases Δt decreases with increasing laser power.
This tendency is explained by taking into account that the vaporization threshold of the liquid is attained earlier with laser
pulses of increased power. The larger Δt for n-heptane compared
to water is due to its lower absorptivity.

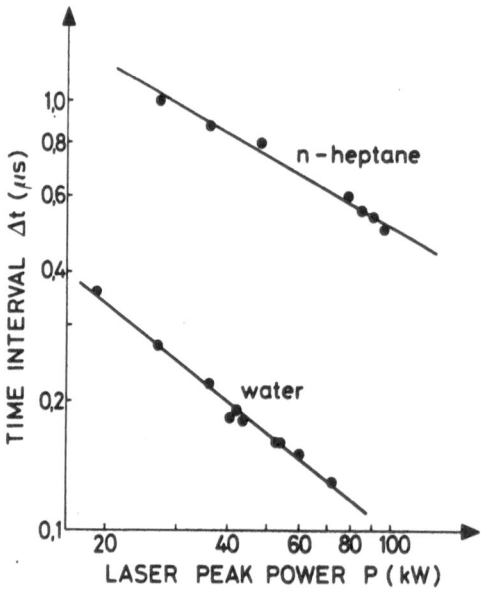

FIGURE 2   Time interval Δt between the photoacoustic pre-
                  pulse and the vaporization pulse as a function
                  of the laser peak power P for water and n-heptane
                  at a constant distance r from the impact. (Double
                  logarithmic scale).

If the absorbed laser energy density is not enough to induce
vaporization (Figure 1c) a symmetric acoustic signal with
respect to the time axis is observed. This is characteristic
for the photoacoustic process. Apart from the behaviour of the
acoustic amplitude at low laser power[2] there is another phenomenon which demonstrates the transient character of the photoacoustic process. This is the generation of acoustic signals
in liquids with tunable high frequencies in the MHz range[2,4].
With the cw section of the hybrid-CO_2-laser operated just at
the laser threshold, the amplitude of the output pulses is
modulated due to longitudinal mode beating. Therefore the
modulation frequency can be tuned simply by changing the
resonator length. The same amplitude modulation also appears
in the generated acoustic signal. With this method we were
able to generate and record for example 60 MHz waves in water.
In other liquids these high frequency waves may not be observed
due to the large sound attenuation at these frequencies.

THEORY

As has been discussed in a previous publication[5], the only
existing analytical model in three dimensions on the thermoelastic generation of acoustic transients[6] does not yield a
satisfactory explanation of our experimental results. Therefore we developed a new spherical model in which the laser
heating process is represented by the three-dimensional heat
pole[5]. This description takes the transient character of the
photoacoustic effect into account which is manifested in
the experiment. Furthermore the spatial distribution of our
heat source corresponds to the TEM_{oo} mode of the incident
laser beam. An analytical solution for this case was presented
by us for the first time. The predicted dependence of the peak
pressures on the laser pulse energy, on the physical properties
of the liquid and on the radius of the spherical wave is in
agreement with the experiment. The essential result however is
the time dependence of the acoustic wave. Our theoretical model
predicts a signal consisting of a compressive part followed by
a symmetric rarefaction part. The two peak amplitudes are separated by a time T which is proportional to the initial radius
of the heat source. This radius is determined by both the laser
focus diameter and the absorptivity of the liquid at the laser
wavelength. Hence, T can be adjusted to each individual liquid.
The excellent agreement with experimental results obtained
under various conditions has been demonstrated[5]. Our model is
applicable not only to the photoacoustic process in liquids
but generally to acoustic wave generation by transient heating
processes in liquids and solids as long as no phase change is
involved.

REFERENCES

[1] D.C. Emmony, M.W. Sigrist and F.K. Kneubühl
"Laser-induced shock waves in liquids"
Appl. Phys. Lett. 29, 547 (1976)

[2] M.W. Sigrist and F.K. Kneubühl
"Laser-generated stress waves in liquids"
J. Acoust. Soc. Am. 64, 1652 (1978)

[3] M.W. Sigrist, F. Fuchs and F.K. Kneubühl
"Influence of the transverse mode structure on laser-induced
acoustic waves"
Helv. Phys. Acta 52, 53 (1979)

[4] M.W. Sigrist, F. Fuchs and F.K. Kneubühl
"Laser-generated 60 MHz acoustic waves in liquids"
Digest 1st European Conference on Optical Systems and Applica-
tions (ECOSA 1), Brighton, p. 33 (1978)

[5] M.W. Sigrist and F.K. Kneubühl
"Spherical Model on Thermoelastic Generation of Acoustic Waves"
J. Appl. Math. and Phys. (ZAMP) 29, 353 (1978)

[6] C.L. Hu
"Spherical Model of an Acoustical Wave Generated by Rapid
Laser Heating in a Liquid"
J. Acoust. Soc. Am. 46, 728 (1969)

PIEZOELECTRIC DETECTION IN PHOTOACOUSTIC SPECTROSCOPY. THEORY AND APPLICATION

.H. D. BREUER

FR 13.2 Physikalische Chemie,

Universität des Saarlandes

D-6600 Saarbrücken, West Germany

Among the various detection techniques in PAS the use of piezoelectric ceramics is becoming more and more important. This is due mainly to the relative simplicity which is thought to be in this technique. If one uses a piezoelectric transducer in a photoacoustic or photothermal device it is quite obvious that there are some advantages over conventional electret or condenser microphones. There is a great temptation just to attach the transducer to the sample for instance with epoxy. In most cases it will work and the signal is sufficient for further processing in a lock-in amplifier.

Ambient noise is very much less troublesome than with microphones and everything works beautifully. Performing the same measurements with another sample makes things usually somewhat disappointing. The results seem to be unreproducible. In most events, however, this is not due to the properties of the transducer but mostly to the tempting easyness of its use. For an efficient employment of piezoelectric transducers it is necessary to know for what purpose they are made and how they have to be installed. Therefore a brief description of the piezoelectric effect will be given followed by some principal modes of operation and a discussion of the electronic requirements which are necessary for optimum signal processing.

THE PIEZOELECTRIC EFFECT

Piezoelectricity is the ability of certain materials to generate electric charges, when stressed mechanically. The reverse procedure, namely that of creating strain by applying electric fields is known as the inverse piezo-effect. In the low signal range, which is the usual case in PAS, the effect is linear, i. e. the electrical field strength and the deformation are proportional to each other. The direct piezo-effect is described as the conversion of mechanical energy into electrical energy and the inverse piezo-effect as the conversion of electrical energy into mechanical.

The piezoelectric effect was described first by Pierre and Jacques Curie in 1880, the same year as A. G. Bell discovered the principles of PAS. The natural piezoelectric crystals include Rochelle salt (sodium potassium tartrate) tourmaline, which had been regarded as "lapis electricus" long before an explanation of the effect was given, and the most well-known representative, quartz.

Since these materials are piezoelectric only as single crystals they have to be cut along certain defined axes. Today more and more polycrystalline materials are used whereby the range of technical applications is considerably broadened.

The ceramics which can be used in PAS are mostly based on $BaTiO_3$ or mixed crystals of lead zirconate-titanate (PZT). Their chemical composition can be expressed by the general formula

$$Pb \ [Ti_{1-x}Zr_x] \ O_3, \quad x \approx 0.5$$

The physical characteristics can be modified by varying the ratio of titanium to zirconium and by a limited addition of other elements. On a microscopic scale these compounds are ferroelectric, i. e. they possess an inherent polarisation.

In a sintered ceramic, however, no macroscopic dipole moment is observable. Therefore in the manufacturing process a remanent polarisation is produced by a applying a high d. c. vol-

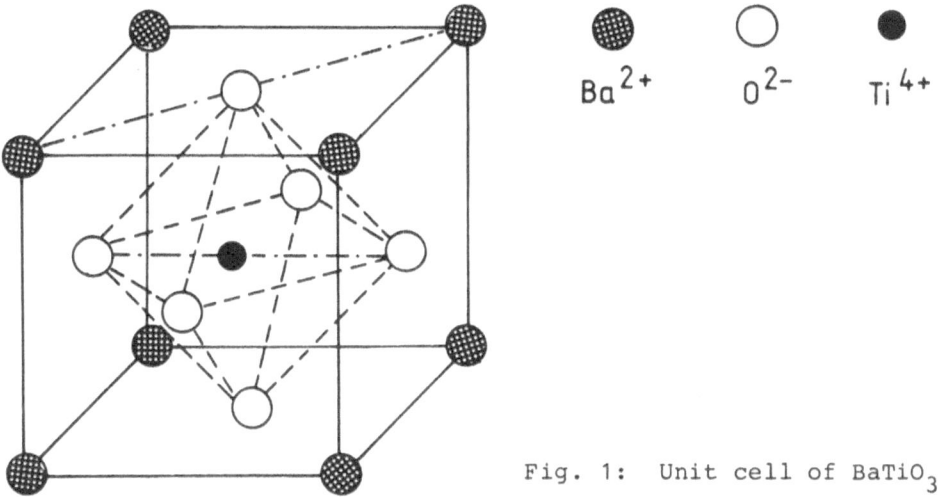

Ba^{2+} O^{2-} Ti^{4+}

Fig. 1:   Unit cell of $BaTiO_3$

tage at an elevated temperature. The piezoelectric materials
possess a perovskite structure whose unit cell is illustrated
in fig. 1 for barium titanate. Above the Curie temperature
the unit cells are cubic, while below this temperature they
become distorted tetragonally in the direction of the C-axis,
whereby the space between the positive and negative ions also
changes. This displacement causes a separation of the centers
of gravity of the charges within the cells and thus creating
an electrical dipol moment. Analogous to ferromagnetism, domains are formed whose polarisation direction is distributed
statistically in the polycrystalline material. By applying
an electric field they can be oriented in a preferred direction. If now an alternating mechanical load is applied to the
transducer an alternating electric field is produced. On the
other hand mechanical deformation can be induced by applying
an electric field.

SOME DEFINITIONS

Piezoelectric elements are anisotropic. Their piezoelectric coefficients are therefore dependent on the direction
of polarisation and on the direction of mechanical force. The
direction of polarisation is always the Z-axis in a right-ang-

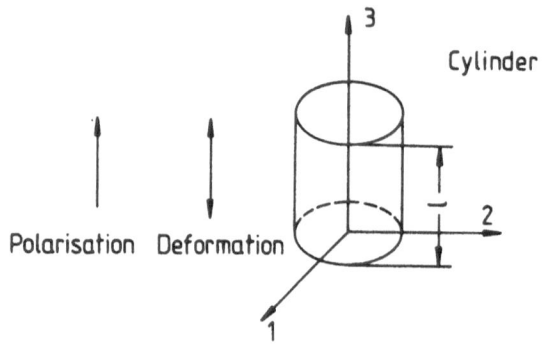

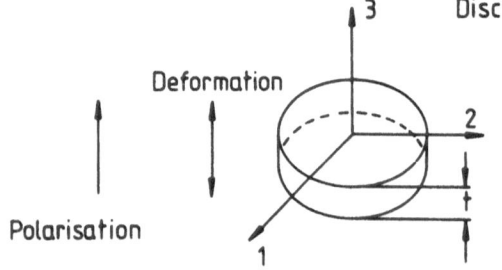

Fig. 2:

Schematics of piezoelectric transducers

led system of coordinates (X, Y, Z). According to international terminology the numbers 1, 2, 3, will be used for these
coordinates. Electrical and mechanical magnitudes that act
in any of these directions are indicated by these numbers.

Since discs and cylinders are mostly used for photoacoustic detection these will be discussed in some more detail,
fig. 2. The electrical and mechanical boundary conditions are
indicated by the following superscripts:

> E   constant electric field
> D   constant dielectric displacement
> T   constant stress
> S   constant strain

1: Relative Permittivities

$\varepsilon^T_{33}/\varepsilon_o$ = Relative permittivity with constant stress
(ε_o = 8.85 · 10^{-12} F m^{-1})

$\varepsilon^S_{33}/\varepsilon_o$ = Relative permittivity with constant strain

2: Coupling Factors

The electro-mechanical coupling factors are a measure of the ratio of convertible to absorbed energy.

k_{33} = Coupling factor of a fundamental vibration of a longitudinally excited bar.

k_t = Coupling factor of a fundamental thickness vibration of a thin plate.

3: Piezoelectric Charge Coefficients

d_{33} = Ratio of the dielectric displacement D to the mechanical stress T at constant E, or, ratio of strain S to the electrical field E at constant T:

$$d_{ij} = \left(\frac{\partial D}{\partial T}\right)_E = \left(\frac{\partial S}{\partial E}\right)_T$$

4: Piezoelectric Voltage Coefficients

g_{33} = Ratio of the electric field strength E to the mechanical stress T at constant D, or, ratio of the strain S to the dielectric displacement D at constant T:

$$g_{ij} = \left(\frac{\partial E}{\partial T}\right)_D = \left(\frac{\partial S}{\partial D}\right)_T$$

5: Frequency constants

These are the product of the fundamental resonance frequency, f_s, and the dimension of the vibrating direction in kHz $\cdot$ mm.

$N_3 = f_s \cdot 1$ = Frequency constant of the longitudinal vibration of a longitudinally polarised bar of length 1.

$$N_t = f_s \cdot t = \text{Frequency constant of the thickness}$$

vibration of a thin plate of thickness t.

6: Capacitance

The capacitance C of a transducer can be calculated using the capacitance formula. For a disc we thus have:

$$C = \varepsilon_r \varepsilon_o \cdot \frac{A}{t}$$

C = Capacitance
A = Area of the metal electrodes
t = thickness.

7: Mechanical Quality Factor Q_m

The quality factor is a dimensionless quantity representing the resonance step-up. For broadband operation Q_m should have a rather low value to give a flat response. On the other hand, working on the resonance frequency of the detector and choosing a high Q_m can enhance the signal considerably.

MOUNTING OF PIEZOELECTRIC TRANSDUCER

A "nude" piezoelectric element cannot be used for detection of mechanical stress or strain. Although to date many papers have appeared in which the authors simply have glued the transducer to the sample, this is not the optimum way to convert a mechanical signal into an electric signal. In any mode of operation the ceramic has to be placed between two force-transducing elements, one of which may be the sample itself. The way in which these transducing elements act on the piezoelectric ceramic depends on the direction of polarisation and the physical shape. In the case of discs or cylinders with a polarisation as indicated in fig. 2 the transducing elements may be two metal plates with radius r, mounted on top and

bottom of the ceramic. The only stress component in the ceramic is than

$$T_3 = \frac{F}{\pi r^2}$$

where F is mechanical force acting on the element. The dielectric displacement is calculated to be

$$D_3 = d_{33}T_3 = \frac{d_{33}F}{\pi r^2}$$

The charge q at the electrodes is obtained by multiplying D_3
by the area of the electrodes

$$q = d_{33} \cdot F$$

If the force F is not acting normally to the surface of the
piezoelectric element only the normal component F_n is contributing to the electric signal. Care has to be taken that no
unwanted force components can reach the transducer. The main
source for these interferences may be structure-borne noise.
Any material in contact with the transducer acts as a seismic
mass. Therefore the whole detector unit should be carefully
isolated from its surrounding.

Since piezoelectric ceramics are also pyroelectric, they
have to be shielded from the probing light. This will be easy
when working with optically opaque samples and having the transducer attached to the rear side of the sample. If the transducer is mounted at the front side of the sample and exposed to
the modulated light it will produce a signal much higher than
the photoacoustic signal.

SIGNAL PROCESSING

In all piezoelectric elements the electric signal is produced by a dielectric displacement in the transducer. If a force is acting on the transducer an electric charge q is available at the electrodes. The corresponding voltage is U = q/C,

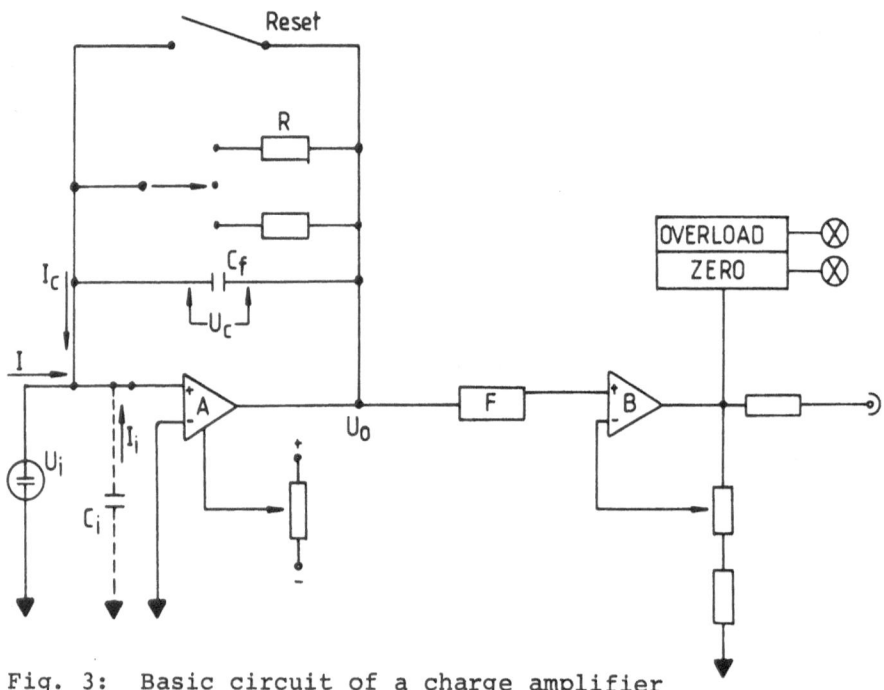

Fig. 3:   Basic circuit of a charge amplifier

where C is the capacitance of the transducer. Considering a
transducer with a capacitance of 100 pF and a sensitivity of
10 pCN^{-1} and a force of 1 N acting on it, a voltage of 0.1 V
is produced. Since the internal resistance of piezoelectric
transducers is in the GΩ or TΩ range and the physical property to be measured is an electrical charge, charge amplifiers
are superiour to electrometer amplifiers.

The basic circuit of a charge amplifier is shown in fig. 3.
It consists mainly of an operational amplifier (A) with high
input impedance and high open loop gain. Input and output
are connected by a highly insulating capacitor C_f providing
a capacitive feedback. By this the input impedance is virtual-
ly kept close to zero. The high insulation resistance at the
input is, however, retained if the amplifier is equipped with
a MOSFET input stage. The amplifier should be selected to have
input drift values as low as possible to prevent it from running into saturation if C_f is not shunted by a resistor.

The principle of operation is as follows: The relation between input voltage U_i and output voltage U_o is

$$U_o = - v_i U_i \qquad (1)$$

where v_i is the internal gain of the amplifier. The voltage measured across the feedback capacitor C_f is

$$U_c = U_o - U_i = U_o + \frac{U_o}{v_i} = (1 + \frac{1}{v_i}) U_o \qquad (2)$$

According to Kirchhoff's law and assuming an ideal amplifier the sum of all currents at the summing point is zero

$$I + I_c + I_i = 0 \qquad (3)$$

Since

$$I = \frac{dq}{dt} \qquad (4)$$

$$I_c = C_f \frac{dU_c}{dt} = (1 + \frac{1}{v_i}) C_f \frac{dU_o}{dt} \qquad (5)$$

and

$$I_i = C_i \frac{dU_i}{dt} = \frac{1}{v_i} C_i \frac{dU_o}{dt} \qquad (6)$$

C_i is the total capacitance of the transducer, the cable and the input of the amplifier. Inserting equations 4, 5 and 6 into 3 leads to

$$\frac{dq}{dt} = - (1 + \frac{1}{v_i}) C_f \frac{dU_o}{dt} - \frac{1}{v_i} C_i \frac{dU_o}{dt} \qquad (7)$$

The output voltage U_o is obtained by integrating and neglecting the constant term

$$U_o = \frac{q}{(1 + \frac{1}{v_i}) C_f + \frac{1}{v_i} C_i} \qquad (8)$$

Since in an ideal amplifier v_i is infinity equation 8 reduces to

$$U_o = \frac{q}{C_f} \qquad\qquad (9)$$

From equation 9 it follows that the output voltage of the amplifier is proportional to the charge q delivered by the transducer. There is no influence of transducer capacity and cable capacity. Since the charge at the electrodes of the transducer is directly proportional to the mechanical force the output voltage is directly proportional to the force acting on the transducer.

By selecting appropriate resistors R in the feedback the low frequency limit is

$$f_1 = \frac{1}{2\pi RC_f} \qquad\qquad (10)$$

The upper frequency limit is mainly given by the frequency response of the amplifier. Care has to be taken, however, if the modulation frequency is close to a resonance frequency of the transducer. In this case errors in magnitude and phase of the signal may be introduced.

The operational amplifier may be followed by a Butterworth filter (F) providing a very flat frequency response and a steep roll-off of 12 dB/octave above the cutoff frequency. The use of filters, however, depends on the frequency range one wants to cover and on the noise which has to be eliminated. The rest of the circuit is used to indicate overload condition and a deviation from zero in the Reset position.

Since MOSFET input stages are very sensitive to high voltages (>50 V) special care has to be taken when connecting the transducer via a cable to the amplifier. Even starting a xenon lamp can produce a transient high voltage puls which may cause damage in the input stages.

For a more detailed description of the principles and ap-
plications of piezoelectric devices the reader is refered to:
J. Tichý, G. Gautschi, Piezoelektrische Meßtechnik, Springer-
Verlag, Berlin, Heidelberg, New York, 1980.

A PRACTICAL EXAMPLE

How piezoelectric detection can be employed in PAS will be
shown by describing a spectrometer which has been set up in
the author's laboratory.

As light source we use a 450 W Xenon arc (Osram XBO 450/
W1). This lamp can be operated in a horizontal position. This
is very important since the lamp housing we are using has an
elliptical reflector. A very high collection efficiency can
be achieved by positioning the arc of the lamp in one of the
foci and having the monochromator entrance slit at the other.
The aperture of the housing is f/2.5. These housings are made
by PRA (Canada) which also made the regulated power supply.
The output of the lamp can, of course, be chopped mechanically.
However, in our spectrometer we make use of the possibility
to modulate the lamp current. Current modulation has some
influence on the life time of the lamp but on the other hand
it is free of noise and vibrations. The modulator supplied
by PRA may be used but we decided to have a more flexible one
built in our electronic shop. In the manual mode any frequency between 10 Hz and 5 kHz can be selected, the degree of modulation can be set between 0 % and 90 % and the lamp current
can be off-set from its value determined by the power supply.
In the automatic mode the frequency increases from 10 Hz to
a preselected upper limit at selectable rates. In this mode
a frequency proportional analog voltage is available to control a xy-recorder. In both modes a 1 V pp reference voltage
is delivered to synchronise a lock-in amplifier.

As monochromator we use either a 0.25 m Ebert (Jarrel-Ash)
or a 0.32 m Czerny-Turner (ISA). Both are equipped with a
stepping motor which is under microcomputer control. The output of the monochromator is focussed by an achromatic fused

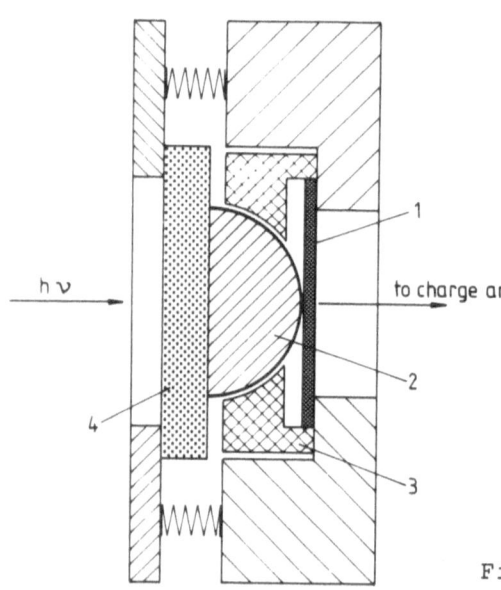

1 Piezoceramic
2 Hemisphere
3 Teflonring
4 Sample

h ν →

to charge amp. →

Fig. 4: Piezoelectric detector

silica lens onto the sample which is placed in the detector
cell. A cut-away view of the detector is shown in fig. 4.
The "heart" of the detector is the piezoelectric ceramic which
in our case is a thin round wafer which is glued to a metal
disc (Sonox, Rosenthal). These devices are originally manufactured as piezo sound-sources for instance in electronic wat-
ches. However, we make use of the direct piezo-effect to detect vibrations induced by the absorption of light.

Of some importance is the mounting of this membrane. Both
impedance and resonance frequency are influenced by this. A
rather flexible fixation at the nodal circle results in a very
low resonance resistance. In our detector the disc is supported at the nodal circle and held down by a teflon ring. In the
center this teflon ring keeps a stainless steel hemisphere in
position which acts as a pressure transmitting device between
the sample and the ceramic. Because the hemisphere aligns to
the surface of the sample the latter has not to be flat. For
any shape of the sample surface an optimum contact is established. The sample itself is spring loaded against the hemisphere. The signal to noise ratio can be optimized by varying
the spring tension. The detector is mounted on a table which

is mounted on four adjustable shock-absorbing elements (Physik
Instrumente) to eliminate any structure-borne noise.

In the detector housing we have a charge amplifier of the
type shown in fig. 3. The operational amplifiers are Teledyne
Philbrick 142101. The feedback capacitor C_f is 10 nF and R
can be chosen between 10^6 and 10^{12} Ω. The output of the charge
amplifier is fed into a Model 5204 lock-in analyzer (PAR).
This instrument provides a simultaneous readout of magnitude
and phase of the photoacoustic signal.

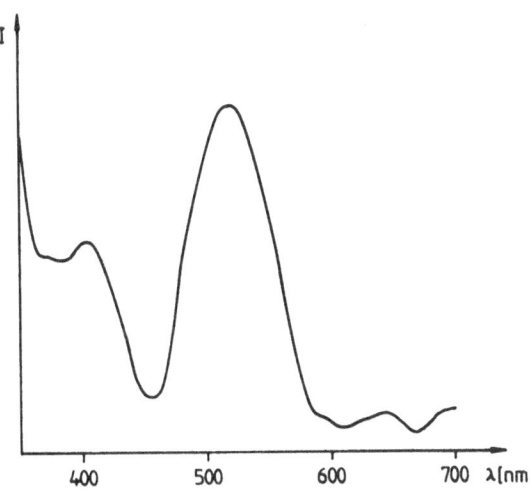

Fig. 5: Spectrum obtained with piezoelectric detection

As an example of the performance of our spectrometer fig.
5 shows the spectrum of an ancient Roman brick. The discussion of the spectrum is beyond the scope of this article but
some technical details may be interesting:

    slit width: 1 mm
    scanning speed: 100 nm/min
    time constant: 30 msec
    sensitivity (lock-in): 100 mV
    modulation frequency: 72 Hz.

From this short description of our spectrometer it follows
that a PA spectrometer with a microphone detector can easily
be converted into one using piezoelectric detection. All of
the optical and most of the electrical equipment is the same
for both modes of detection. However, there is much more flexi-
bility in the design of the detector than with microphones
as long as the basic principles pointed out are obeyed.

ACKNOWLEDGMENT

The author wants to thank Dr. K. Stärk, Rosenthal Technik
A. G. for providing him with piezoelectric transducers and
for helpfull and stimulating discussions. Financial support
by the Fonds der Chemischen Industrie is greatfully acknowledged.

PHOTOTHERMAL RADIOMETRY

Per-Erik Nordal and Svein Otto Kanstad

Laser and applied optics laboratory
P.O. Box 303 Blindern
Oslo 3, Norway

ABSTRACT

A specimen that is subjected to pulsating illumination may experience oscillating temperatures due to absorbed radiation. Depending on the spatial, spectral and temporal properties of the illuminating radiation, information on the chemical composition and the material qualities of the specimen may be obtained by observing pulsations in the thermal re-radiation from the sample. The basic principles of the technique are reviewed, and examples are presented on infrared and visible spectral studies of powders and biological material and on scanning/imaging applications. Distinguishing features of the technique are discussed, with a view to possible future developments.

INTRODUCTION

The recent years have seen a proliferation of analytic techniques that exploit sample heating effects in conjunction with time-dependent irradiation. Many of these differ markedly from the air-microphone-based photoacoustic method, both as regards irradiation (e.g., electron beam[1]) and detection (e.g., based on elastic waves[1], photo-refraction[2], electrical resistance[3]). Below, a brief introduction is given to another technique, photothermal radiometry (PTR), that relies on the detection of *thermal re-radiation from irradiated objects*[4].

BASIC PRINCIPLES

Consider the system shown in Fig. 1, where the sample is illuminated by pulsed, quasi-monochromatic radiation. Absorption of the incident radiation (followed by conversion to heat) causes

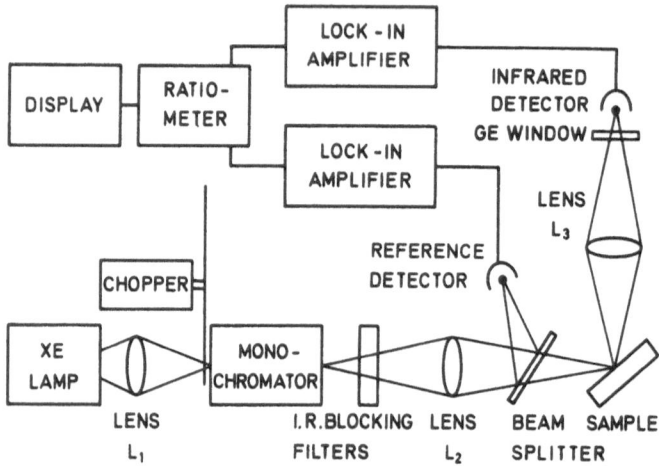

Fig. 1: Experimental arrangement for visible-light PTR spectroscopy.

surface temperature oscillations with amplitudes δT, which again
modulate the total radiant emittance of the sample by amounts

$$\delta W = 4\varepsilon\sigma T^3 \delta T. \qquad (1)$$

Here, σ is the Stefan–Boltzmann constant and $\varepsilon \leq 1$ is the emissivity of the specimen, assumed to be an ideal greybody at tempera-
ture T. *Pulsewise sample irradiation thus gives rise to a component in the radiant emittance that varies at the pulse frequency*,
superimposed on the (generally much larger) CW radiant emittance
$W = \varepsilon\sigma T^4$. The amplitude and phase of the former (PTR) component
contain information on optical as well as thermal properties of
the illuminated specimen, as will be illustrated in the next
section. Indeed, to a first approximation, the same functional
dependence on the object's optical and thermal parameters is
found for the PTR signal[5] as is known from photoacoustic spectroscopy[6]. In particular, under photoacoustically transparent condi-
tions $\delta W \propto \alpha(\lambda)$ (where $\alpha(\lambda)$ is the spectral absorption coefficient
of the sample material), whereas photoacoustic opacity results in
δW being independent of $\alpha(\lambda)$.

EXPERIMENT

Optical absorption spectra in the visible-NIR regions have
been recorded from a variety of samples, using the set-up in Fig.1.
A Xe lamp/monochromator combination was used, with spectral resolution of ∿ 12 nm. The net time averaged (50 % duty cycle) irradiation power was less than 0.5 mW across the visible spectrum.
Thermal re-radiation from the illuminated spot was collected by a
lens and detected as shown, the lock-in amplifier extracting only
that part of the detector signal which is phase coherent with the
pulsewise irradiation of the sample.

PTR spectra from several different specimens have been
obtained[7]; Fig. 2 shows the absorption spectrum from a smear of
whole blood on a glass slide. The spectrum in Fig. 2 was recorded
in visible-light irradiation, with detection in the middle infra-

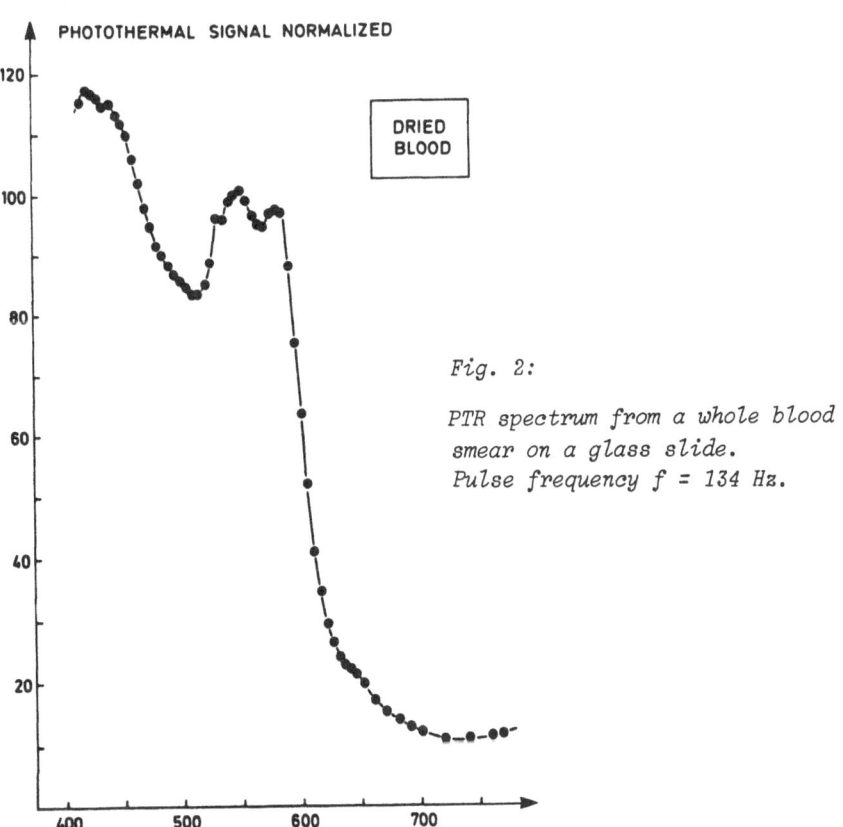

Fig. 2:

*PTR spectrum from a whole blood
smear on a glass slide.
Pulse frequency f = 134 Hz.*

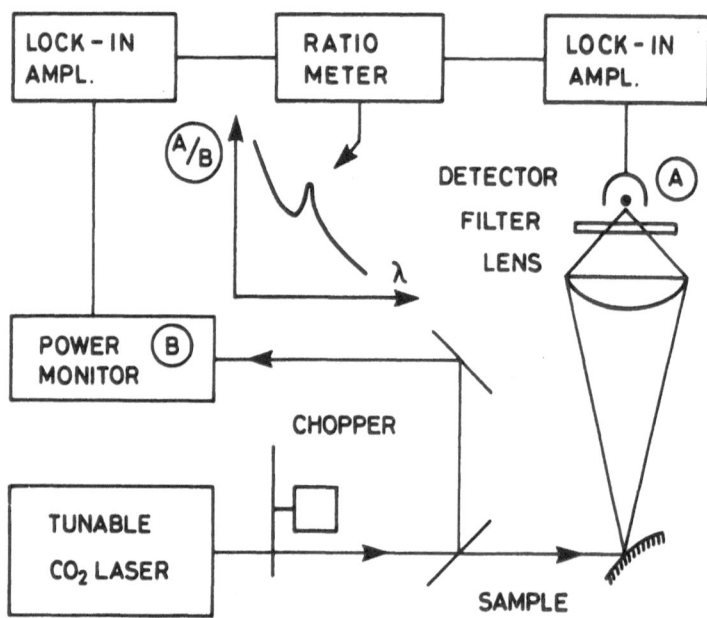

Fig. 3: *Experimental arrangement for PTR spectroscopy with CO_2 laser irradiation near 10 μm wavelength.*

red using a PbSnTe detector whose peak sensitivity was near 10 μm wavelength. Experiments[4] have also been carried out with the set-up shown in Fig. 3, where a wavelength-tunable CO_2 laser was employed to record absorption spectra from powdered K_2SO_4 in the 10 μm region, at temperatures from 295 K to 942 K, as shown in Fig. 4. Special care was taken to ensure that scattered illuminating radiation did not interfere with the detection of thermal re-radiation, by using a sapphire filter in front of the (InSb) detector. It appears that the PTR technique should be particularly well suited for studies of objects at high temperatures: In addition to the absence of physical contact with the sample, one notes from Eq. (1) that the PTR signal (as well as the signal-to-noise ratio at the detector) increases as the temperature T is raised.

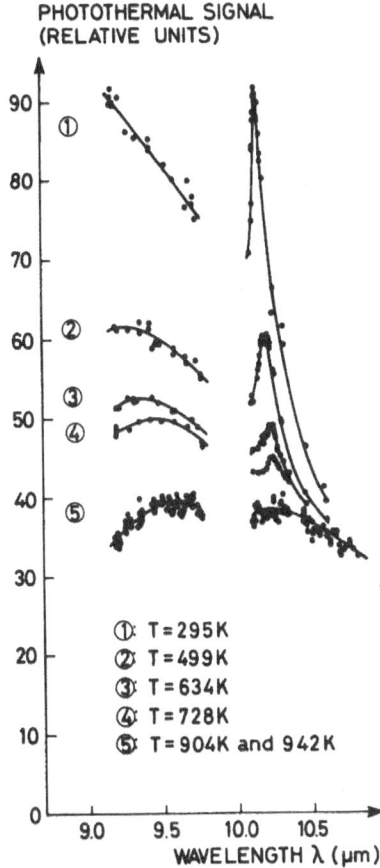

Fig. 4:

PTR spectra of powdered K_2SO_4 at several temperatures. Pulse frequency f = 77 Hz.

The PTR technique also appears well suited to the scanning and/or imaging of surfaces, with respect to spectral characteristics as well as thermal/material properties[8]. Fig. 5 shows the results from measurements made by scanning a bluish-pink chromatogram stain past the focal point of a 47.5 mW He-Ne laser beam at 632.8 nm wavelength. The estimated detection limit in the experiment corresponds to approximately one monolayer of material in the chromatogram stain. An interesting extension of this work would be to map out locations of different substances in a chromatogram by proper choice of irradiating wavelengths.

Since as a dynamic technique PTR relies on the thermal

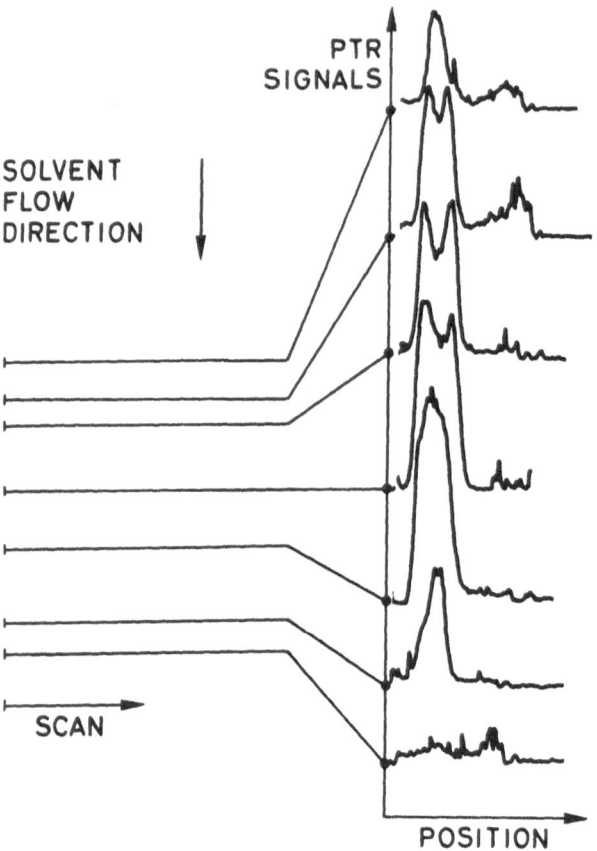

Fig. 5: PTR signals from a chromatogram stain (10 μg aspartic acid developed by ninhydrin). Pulse frequency f = 80 Hz.

transport properties of the sample, it allows one to probe surface and subsurface qualities such as surface film thickness, delaminations, cracks, voids, density variations etc., and at the same time enables the probing depths to be controlled within limits set by the thermal diffusion length through adjustments of the irradiation pulse frequency. Thus by scanning the He-Ne laser beam across a metal surface carrying a strip of adhesive tape underneath a uniform coating of mat black paint, the results presented in Fig. 6 were obtained. The presence of the tape underneath the paint is strongly evident in the scans at low pulse frequencies, where the thermal diffusion length extends through

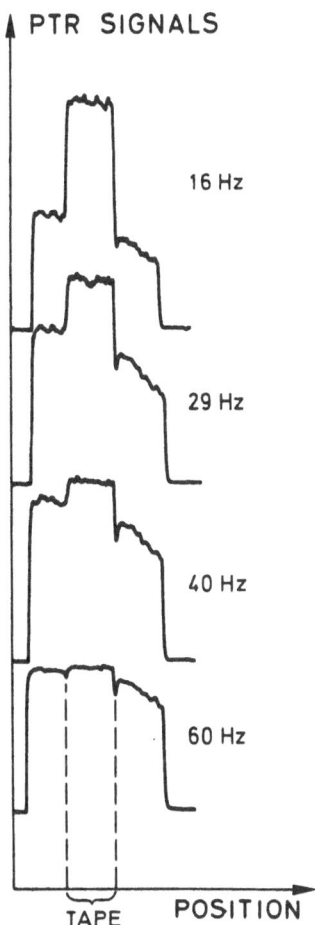

Fig. 6: PTR signals at different pulse frequencies, as recorded
from an aluminium sheet partially covered by a strip of
adhesive tape, with black paint on top. All scans followed
the same path across the sample. Zero signal levels before
and after each scan were obtained by blocking the laser.

the paint layer but not through the additional thickness of the
tape.

Busse[9] has demonstrated that probing depths of several mm
can be realized in metals by monitoring *phase* changes in PTR
signals from the *rear surface* of irradiated samples. That same
author[10] also obtained a PTR image of an integrated circuit, and
Luukkala[11] has used PTR scanning to observe surface and subsurface
flaws in metals.

FUTURE DEVELOPMENTS

Although it is too early to make firm predictions regarding
the future of the PTR technique, some distinguishing features
may be identified that point to possible fruitful areas of
technical development and scientific applications.

No physical contact

- Studies of samples *in vacuo* or in controlled atmospheres.
This would be of interest in connection with chemical surface
processes (catalysis, corrosion) and in biological applications
(cultures, aseptic environments).
- Studies of samples in hostile environments or at high
temperatures.
- Inspection or study of material streams.
- Remote spectroscopy/scanning/imaging.

Optical transmission.

The thermal re-radiation as well as the illuminating light
may penetrate a wide range of materials, which may allow
- Studies of interfaces between transparent media
- Probing of difficult-to-reach areas by means of fiber
optic techniques.

Rapid response.

Although thermal diffusion processes play an important part
in most PTR applications, the practical limit to attainable time
resolution may in many instances be set by the duration of the
illuminating pulse or by the detector time constant, both of which
can be made shorter than 1 μs. This facilitates
- Studies of energy decay pathways.
- Transient measurements on rapid thermal events.

Emissivity dependence.

In contrast to the other thermal detection techniques referred to above, PTR methods are explicitly dependent on emissivity, which contains additional information on material composition as well as on surface micro-structure.

CONCLUSION

Photothermal radiometry has been found to yield optical absorption spectra in the visible and in the infrared of a wide variety of specimens in their natural state. Studies have also been made that demonstrate the potential of the technique in contact-free scanning and imaging. The sensitivity appears to be comparable to that of photoacoustic spectroscopy, to which PTR methods bear strong similarities in principle as well as in its applications. The PTR technique, however, enables completely contact-free measurements to be made at some distance from the object. In particular, the latter property permits simple PTR analysis of high temperature samples. It might reasonably be expected, therefore, that PTR techniques would present themselves as future alternatives to photoacoustic methods in several instances, in addition to suggesting exciting new possibilities of their own.

REFERENCES

(1) R.M. White, "Generation of elastic waves by transient surface heating", *J. Appl. Phys.* <u>34</u> *3559 (1963)*.

(2) A.J. Twarowski and D.S. Kliger, "Multiphoton absorption spectra using thermal blooming", *Chem. Phys.* <u>20</u> *253,259 (1977)*.

(3) G.H. Brilmyer, A. Fujishima, K.S.V. Santhanam and A.J. Bard, "Photothermal spectroscopy", *Anal. Chem.* <u>49</u> *2057 (1977)*.

(4) P.-E. Nordal and S.O. Kanstad, "Photothermal radiometry", *Physica Scripta* <u>20</u> *659 (1979)*.

(5)   S.O. Kanstad and P.-E. Nordal, "Photoacoustic and photother-
      mal techniques for powder and surface spectroscopy", *Appl.
      Surface Sci. (in press).*

(6)   A. Rosencwaig and A. Gersho, "Theory of the photoacoustic
      effect with solids", *J. Appl. Phys. 47 64 (1976).*

(7)   P.-E. Nordal and S.O. Kanstad, "Visible-light spectroscopy
      by photothermal radiometry using an incoherent source",
      Appl. Phys. Lett. (in press).

(8)   P.-E. Nordal and S.O. Kanstad, "Photothermal radiometry for
      spatial mapping of spectral and material properties". *In*
      "Scanned Image Microscopy" (E.A. Ash, Editor), *p. 331,*
      Academic Press. London, 1980.

(9)   G. Busse, "Photothermal transmission probing of a metal",
      Infrared Physics 20 419 (1980).

(10) G. Busse, "The optoacoustic and photothermal microscope: The
      instrument and its applications". *In* "Scanned Image Micro-
      scopy (E.A. Ash, Editor), *p. 341.* Academic Press, London 1980.

(11) M. Luukkala, "Photoacoustic microscopy at low modulation
      frequencies". *In* "Scanned Image Microscopy" (E.A. Ash, Editor),
      p. 273. Academic Press, London 1980.

2. Spectroscopic Applications

APPLICATIONS OF RESONANT PHOTOACOUSTIC SPECTROSCOPY

J. Röper, K. Frank, and P. Hess

Institut für Physikalische Chemie
der Universität Heidelberg
6900 Heidelberg 1, F.R.G.

ABSTRACT

An apparatus is described which allows an accurate determination of the acoustic resonance frequencies of a
cylindrical cell. Three different applications of such
a device are discussed: 1. Static gas analysis. The
resonance frequency of the first radial mode is determined in the mixtures 1 % CH_4 + 99 % N_2 and 1 % CH_4 +
99 % CO to test the accuracy of the method and to obtain an accurate value of the cell radius. 2. Dynamic
gas analysis. The shift of the resonance frequency
during the mixing of CH_4 and N_2 is detected by repeated
recording of the acoustic resonance curves. 3. Determination of vibrational relaxation times. The dispersion
of the resonance frequency is measured in the pressure
region where relaxation processes occur.

INTRODUCTION

The resonant photoacoustic effect was discovered in 1881
by Bell[1]. He observed that the sound produced by the
photoacoustic effect was much louder when the chopping
frequency corresponded to an acoustic resonance frequency of the photoacoustic cell. A detailed discussion of
resonant photoacoustic spectroscopy and its application
to detect atmospheric pollutants was given 92 years la-

ter by Dewey, Kamm, and Hackett[2]. In the following years
resonant photoacoustic spectroscopy found increasing interest. Not only experimental work, for example, on trace
gas analysis, but also theoretical investigations of dis-
sipative processes in a resonator were published[3,4]

In this paper three applications of resonant photoacoustic spectroscopy are discussed. The resonance frequencies depend on the properties of the resonator and
on gas properties such as molecular weight and heat capacity. If the two components of a binary mixture possess
different properties, the measurement of the resonance
frequency allows an accurate determination of the composition of the mixture. With the apparatus described in
the next section about 30 s are needed for such an analysis. Thus, dynamic processes such as chemical reactions or mixing processes can be monitored by a repeated
measurement of a resonance frequency. A third possibility
of resonant photoacoustic spectroscopy is the investigation of energy transfer processes. In the frequency-to-
pressure region where relaxation occurs, the dispersion
of sound velocity causes a dispersion of the resonance
frequency. Therefore, the measurement of the resonance
frequency as a function of pressure allows an accurate
determination of the corresponding relaxation time.

METHOD

A schematic diagram of the experimental arrangement is
shown in Fig. 1. The $3,39\,\mu$m beam of a He-Ne laser was
intensity modulated by a chopper. The chopper frequency
was continuously varied in an interval of about 100 Hz
around the selected acoustic resonance. In all measurements reported here, the first radial mode of a cylindrical sample cell was excited. The standing acoustic wave
in the resonator was produced by V-R,T transfer from vibrationally excited methane, which was the optically absorbing species in all experiments. The pressure change

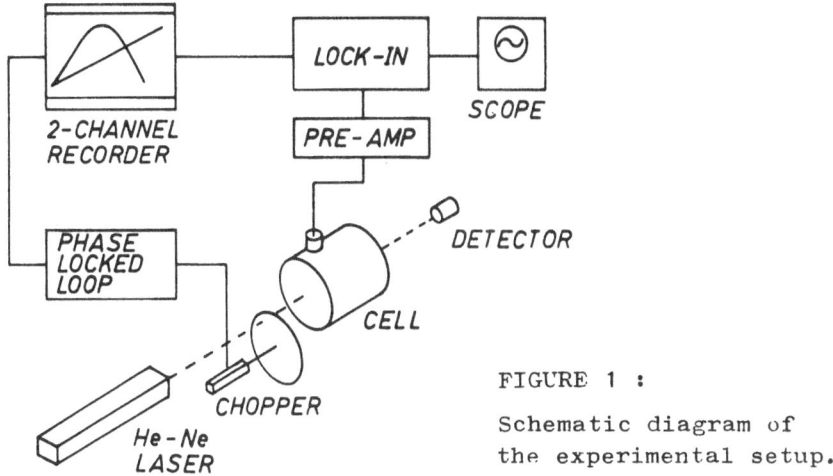

FIGURE 1 :

Schematic diagram of
the experimental setup.

in the sample cell was detected with a small electret
microphone. The microphone signals were preamplified and
then a lock-in analyzer was employed for further signal
processing. Both the linear frequency change of the chopper and the corresponding microphone signal were recorded
simultaneously with a two channel recorder. It was possible to make several successive runs through the selected frequency interval to detect a frequency shift. From
the plots of the chopper frequency variation and the
acoustic resonance curves, the frequency of the excited
mode was determined. In the dynamic mixing experiments,
the shift of the resonance frequency was recorded as a
function of time; in the energy transfer studies, the variation of the resonance frequency as a function of pressure was observed. Further experimental details will be
published elsewhere [5].

RESULTS AND DISCUSSION

To test the accuracy of the technique, two equilibrated
mixtures were studied. Fig. 2 shows the resonance curves
measured for a mixture of 1 % CH_4 + 99 % N_2 and a mixture
of 1 % CH_4 + 99 % CO. Both mixtures were at 293 K and a

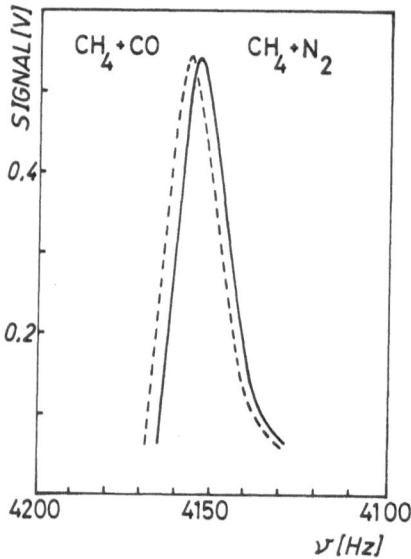

FIGURE 2 : Resonance curves of the first radial mode for the mixtures 1 % CH₄ + 99 % CO (4158 Hz) and 1 % CH₄ + 99 % N₂ (4155 Hz).

pressure of 700 torr. The experimental values obtained from these curves for the resonance frequencies are 4155 Hz and 4158 Hz for the CH_4-N_2 and CH_4-CO mixtures, respectively. A similar frequency is expected for the two mixtures, because the molecular weights and heat capacities of N_2 and CO are nearly identical (N_2: M = 28,01 g/Mol, c_p^o = 29,12 J/K,Mol; CO: M = 28,01 g/Mol, c_p^o = 29,15 J/K,Mol.)

Assuming that the errors due to the temperature and composition of the mixtures are small, the resonance frequencies of the first radial mode in the two mixtures can be used to determine the radius of the photoacoustic cell. Applying the formula for the resonant frequencies in a cylindrical cell[6], a mean value of 5,128 $\pm$ 0,003 cm is obtained for the cell radius. This result illustrates the high accuracy obtainable with resonant photoacoustic spectroscopy. In fact, it was shown that the sound velocity and thermophysical properties of gases can be determined with high precision using this technique [7]. Recently the velocity of sound was measured with an accuracy of 0,02 % in a spherical acoustic resonator[8].

As mentioned above, about 30 s are needed to record one resonance curve. Thus, it is possible to monitor dynamic processes which occur in the time scale of several minutes or longer. Two experiments were performed to study mixing processes between CH_4 and N_2. In the first experiment a mixture of 9,8 % CH_4 and 90,2 % N_2 was prepared in the following way: The acoustic cell ($\approx 0,8$ l) and the tube system to the manometer, pump, and gas bottles (all together $\approx 0,2$ l) were filled with the corresponding pressure of CH_4. Then N_2 was added quickly by opening the valve to a high pressure container until the final pressure of 400 Torr was reached in the cell and tube system. Fig. 3 shows the resulting decrease of the resonance frequency caused by the increasing N_2 concentration in the resonator. After about 2 minutes, the measured resonance frequency is nearly constant. Within experimental error, the experimental value of 4245 Hz agrees with the calculated value of 4243 Hz.

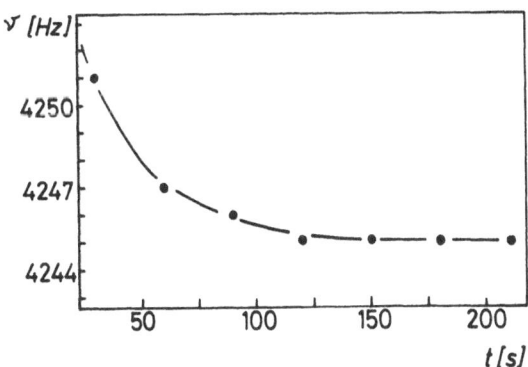

FIGURE 3 : Shift of the resonance frequency as a function of time after mixing 9,8 % CH_4 and 10,2 % N_2.

In the second experiment the same procedure was employed; however, a mixture of 50 % CH_4 and 50 % N_2 with a total pressure of 400 torr was prepared. As can be

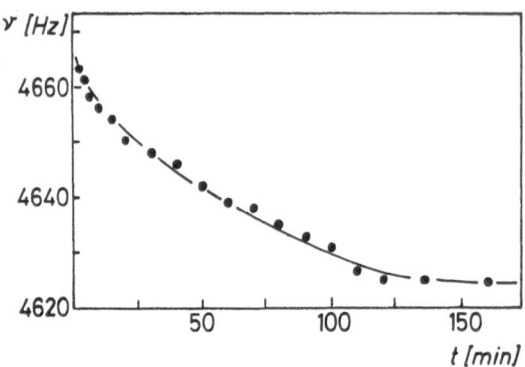

FIGURE 4 : Shift of the resonance frequency as a function of time after mixing 50 % CH_4 and 50 % N_2.

seen in Fig. 4, about 2 hours are needed until a nearly
constant frequency is observed. In contrast to the find-
ing in the first experiment, the resonance frequency of
4624 Hz measured after about 2 hours is still higher
than the calculated value of 4612 Hz indicating incom-
plete mixing. At 296,15 K the resonance frequency of pure
CH_4 is 5321 Hz and of pure N_2 4171 Hz. From these values
one can estimate that a concentration change of 1 % produces a frequency change of about 11 Hz. Thus the difference between experimental and theoretical resonance frequency indicates a deviation of about 1 % from the equilibrium concentration. This suggests that there is still
a somewhat higher N_2 concentration in the tube system
and final equilibration occurs by a very slow diffusion
controlled process.

As shown in ref.[9] the resonance frequency of a resonator
is also changed by relaxation processes. These processes
alter the specific heat ratio and cause the well known
dispersion of the sound velocity measured in ultrasonic
experiments, and the dispersion of the acoustic resonances. This frequency dispersion resulting from vibrational relaxation in CH_4 can be seen in Fig. 5. In this experiment the asymmetric stretching vibration v_3 was excited with a 3,39 µm He-Ne laser. Unfortunately, this

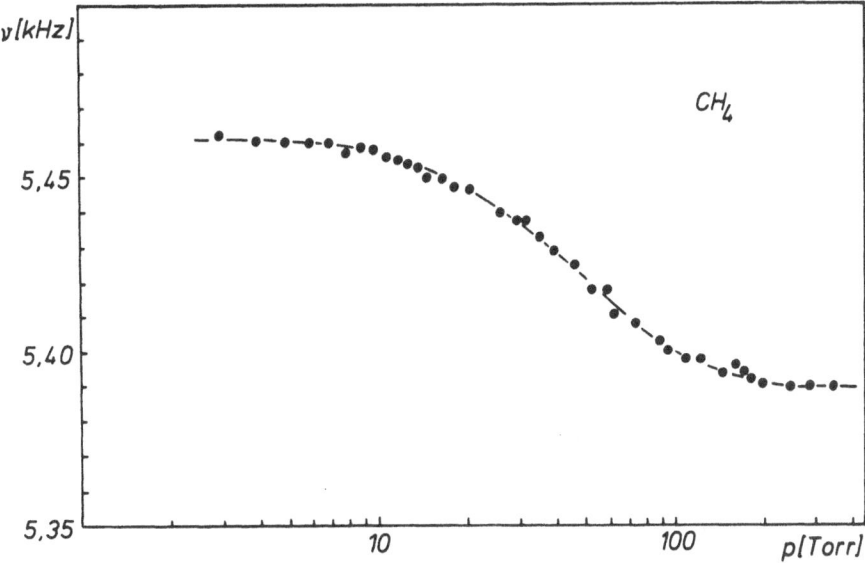

FIGURE 5 : Dispersion of the resonance frequency as a function of pressure for vibrational relaxation in CH_4.

laser line is absorbed only by a small number of molecules. On the other hand, the excitation of a larger number of molecules containing C-H bonds seems to be possible with the He-Ne laser line at 1,15 µm. Photoacoustic experiments are in progress in our laboratory using this wavelength to excite polyatomic molecules[10]. An advantage of the resonant photoacoustic technique is the possibility to determine an accurate value of the heat capacity involved in the relaxation process. The method seems to be promising for vibrational relaxation studies especially at low frequency/pressure ratios. Just recently a resonant tube was developed for the range 0,1 - 2500 Hz/atm to study the role of nitrogen as a sound absorbing constituent of the Earth's atmosphere[11]. Resonant photoacoustic spectroscopy is also a promising technique to study such slow relaxation processes.

ACKNOWLEDGEMENTS

We are grateful to the Deutsche Forschungsgemeinschaft and the Fonds der Chemischen Industrie for research support.

REFERENCES

1   A.G. Bell, Upon the production of sound by radiant energy, Phil.Mag. <u>11</u> 510 (1881)

2   C. F. Dewey Jr., R.D. Kamm, and C.E. Hackett, Acoustic amplifier for detection of atmospheric pollutants, Appl.Phys.Lett. <u>23</u> 633 (1973)

3   R.D. Kamm, Detection of weakly absorbing gases using a resonant optoacoustic method, J.Appl.Phys. <u>47</u> 3550 (1976)

4   E. Nodov, Optimization of resonant cell design for optoacoustic gas spectroscopy (H-type), Appl.Opt. <u>17</u> 1110 (1978)

5   J. Röper and P. Hess, to be published

6   K. Frank and P. Hess, The role of relaxational processes in resonant optoacoustic spectroscopy, Ber.Bunsenges.Phys.Chem. <u>84</u> 724 (1980)

7   L.J. Thomas, M.J. Kelly, and N.M. Amer, The role of buffer gases in optoacoustic spectroscopy, Appl.Phys. Lett. <u>32</u> 736 (1978)

8   M.R. Moldover, M. Waxman, and M. Greenspan, Spherical acoustic resonators for temperature and thermophysical property measurements, High Temperatures - High Pressures <u>11</u> 75 (1979)

9   K. Frank and P. Hess, Accurate measurement of relaxation times with an acoustically resonant optoacoustic cell, Chem.Phys.Lett. <u>68</u> 540 (1979)

10 J. Röper and P. Hess, to be published

11 A.J. Zuckerwar and W.A. Griffin, Resonant tube for measurement of sound absorption in gases at low frequency/pressure ratios, J.Acoust.Soc.Am. <u>68</u> 218 (1980)

PHOTOACOUSTIC
FOURIER TRANSFORM SPECTROSCOPY FOR THE VISIBLE AND THE NEAR INFRARED

D. DEBARRE, A.C. BOCCARA and D. FOURNIER

Laboratoire d'Optique Physique

10, rue Vauquelin, 75231 PARIS CEDEX 05

FRANCE

ABSTRACT

The interest of a multiplex method such as Fourier Transform Spectroscopy over conventional sequential dispersive spectroscopy is discussed for its application to Photoacoustic detection. The expected improvement was calculated to be of about two orders of magnitude, and indeed in some cases a such significative gain has been observed experimentally. A rare earth oxyde powder spectrum will illustrate this point.

Then the use of commercial Fourier Transform Infrared Spectrometer is analysed when using a photoacoustic cell as detector, and compared to step-and-integrated interferometers which are found to be more suitable for this kind of detection.

INTRODUCTION

Photoacoustic spectroscopy has been proved to be a powerful analytical tool for "exotic" samples such as powders, opaque materials, and living tissues. Unfortunately this method, which uses the thermal conversion of the light absorbed by the sample, is much less sensitive than photoelectric detection.

For such a "detector-limited" method it is well-known that a Fourier Transform (FT) spectrometer should be very efficient in improving the signal-to-noise ratio (s/n). The aim of this paper is to confirm this point and to analyse the principal methods of accomplishing Photoacoustic Fourier Transform spectroscopy (PAFT).

In the first section we review briefly the advantages of FT spectroscopy over sequential (including dispersive) methods when coupled with photoacoustic (PA) (and more generaly photothermal) detection. We then analyse, in the second part, the various methods used to construct a PAFT spectrometer.

Our own experimental set-up and some recent results are described
in the third part.

I - <u>COMPARISON BETWEEN SEQUENTIAL AND MULTIPLEX METHODS</u>.

Let us outline the three main advantages of the FT spectrometer
over conventional dispersive spectrometers (1) :
                  - *Fellgett's* advantage is the improvement of the s/n by
a factor about $\sqrt{N}$ when N spectral elements are simultaneously analysed
by the detector during the whole time of the experiment.
                  - *Jacquinot's* advantage deals with the geometrical aper-
ture of the apparatus, which is much larger in the case of the interferometer for an equal spectral resolution.
                  - *Connes'* advantage finally, is associated to the abso-
lute calibration which can be achieved in the case of an interferometer
whose moving mirror is accurately referenced by a laser.

Rather than reviewing the analytical equations, which are developed
in details in many articles (1) and text books (2), let us consider a
specific example which will account for the s/n improvement when going
from a dispersive to a multiplex apparatus. The dispersive set-up is composed of a 450 Watt Xe arc associated with a high luminosity J.Y. monochromator (150 mm diameter holographic grating) and a conventional PA cell.
The home-made interferometer (3) (detailed below) uses the same source
but optical components of 60 mm/diameter. In order to achieve the comparison of these two laboratory apparati, suppose we have to run the PA
spectrum of a "classical" rare earth oxyde powder which exhibits (several
narrow lines spread) from the near UV to the near IR.

The detection is performed at the same frequency with the same PA
cell. In both cases the entire flux was incident on the sample.

If we choose a spectral bandwidth of about 15 cm^{-1}, the average light
power out put is about 75 µW, whereas the total light flux emerging from
our Michelson interferometer is about 1 W for the 1400 spectral elements
(spectral range 4000 to 25 000 cm^{-1}) leading to an average power of
700 µW for the same resolution. This improvement is a consequence of
Jacquinot's advantage, while that of Fellgett leads to an additional
improvement of s/n of $\sqrt{1400/4}$ (1).

The total expected gain is larger than two orders of magnitude and,
in fact, such gains have been observed experimentally (3).

II - VARIOUS METHODS TO CONSTRUCT A PHOTOACOUSTIC FOURIER
TRANSFORM SPECTROMETER

To generate a PA signal one needs an intensity-modulated light beam.
While the conventional dispersive methode generaly uses a chopper or an
intensity modulated source, we have to consider the scanning procedure of
the interferometer in analysing the amplitude and the frequency of the
modulation for each wavelength. One can distinguish two principal scanning procedures for the moving mirror of the Michelson interferometer,
continuous scanning (4) (5) (6) and step and integrate (8) (3).

In the continuous scanning method for a constant velocity v of the
moving mirror, the intensity of the light associated with a wavelength
$\lambda_o = 1/\sigma_o$ is :

$$I = I_o/2 \ (1 + \cos 4\pi\sigma_o vt)$$

Thus, the modulation frequency $2\sigma_o v$ depends on the wavenumber of the
radiation. For instance, for the spectral range considered in the first
paragraph (4000-25 000 cm^{-1}) the frequencies will vary within a ratio
as large as 6. Remembering the importance of the frequency dependence of
the photoacoustic signal (magnitude and saturation effects (7)), one can
conclude that this method is convenient for a crude qualitative approach.
It is clearly possible to improve the quality of the results by analysing
various spectra run at various velocities. Most of the commercial infrared FT spectrometers use this kind of scanning and the PA cell matching
is straightforward. Let us point out, however, that, in principle, the
PA signal phase can be obtained directly from the phase of the FT signal,
but to the best of our knowledge this has not yet been demonstrated.

The step and integrate method seems to be mostly used for the visible
and near infrared for various technical reasons (3). In this method
the optical path difference is no longer a continuous time function but
one of discrete steps. The sampling interval must be chosen, in accordance with the sampling theorem, to match the free spectral range. Often
(8) (3) for He-Ne laser controled interferometers a sampling interval of
6328/4 Å is chosen leading to a free spectral range of 31 605 cm^{-1} convenient for the visible and near infrared.

In this case each sampled value of the optical path difference is
fixed and there is no inherent modulation. Therefore one can create the
same frequency of modulation for each spectral element and use conventional lock-in detection. If an external modulation (chopper or intensity

modulated lamp) is used, the amplitude and frequency are the same for the
whole spectrum, but the recorded interferogram exhibits a significant
DC component and is therefore very sensitive to the intensity fluctuations of the source. In order to avoid this problem, "internal" modulation may be achieved by using an oscillating mirror to modulate the path
difference δ at each step. However, now the modulation amplitude for the
wavenumber σ is proportional to the first order Bessel function
$\delta_1(2\pi\delta_m\sigma)$ where δ_m is the amplitude of the internal modulation. This
envelope function is directly compensated by the normalisation of the
spectrum.

III - EXPERIMENTAL SET-UP AND RESULTS

Fig. 1 is a schematic diagram of our experimental set-up.

The interferometer uses the step-and-integrate method : the moving
mirror displacement is alternately controlled by two feed-back loops, one

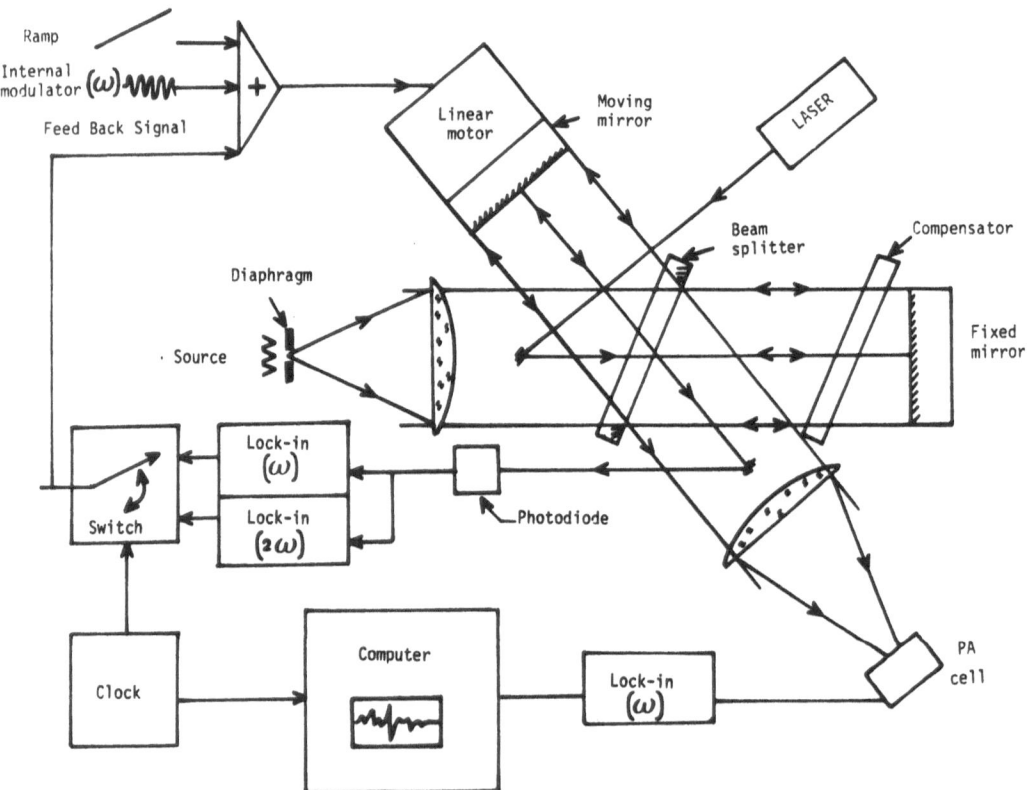

FIGURE 1 - Fourier Transform PA spectrometer : experimental set-up.

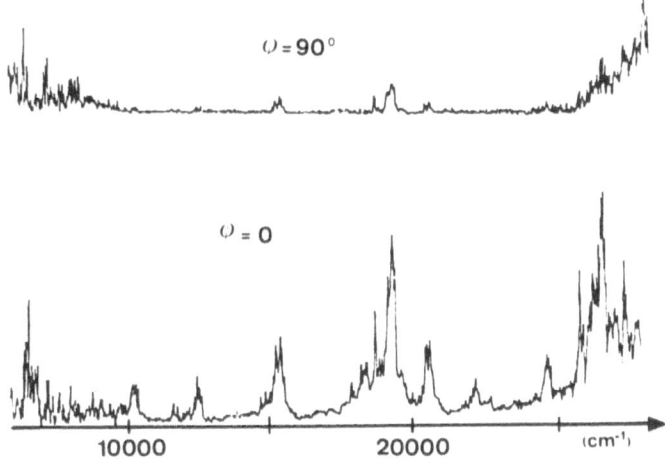

FIGURE 2 - In phase and in quadrature Fourier Transform PA spectra of powdered Er_2O_3.

which holds the system at the top and the bottom of the helium neon laser fringes, the other one at the middle of the fringes (yielding 6328 Å/4 steps in the optical pathway).

The internal modulation can be optimized either to cover the full spectral range (first zero of the Bessel function at about 30 000 cm^{-1}) or to enhance one particular region of the spectrum. A detailed schematic of the experimental set-up can be found in ref. (3).

Fig. 2 shows the room temperature in-phase and quadrature FTPA normalised spectra of erbium oxyde powder at 400 Hz. One can see that, even at this frequency, the strongest lines are still partially saturated.

Fig. 3 shows the good resolution of the apparatus (16 cm^{-1} expected) by detailing a small region of Fig. 2.

Through this example and others data which will be published soon (3) we have demonstrated the superiority of this multiplex method over sequential dispersive methods.

CONCLUSION

Although the FT method, when applied to photoacoustic spectroscopy of solids, seems to be very sophisticated (indirect methods, need for a

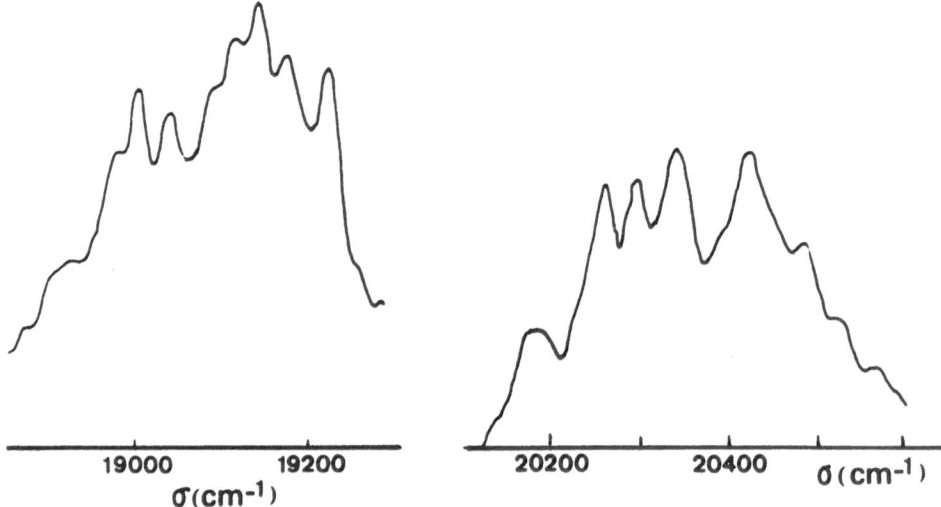

FIGURE 3 - Detail of the in phase Figure 2 spectrum.

computer...) a clear improvement of the data has been proved either in the medium infrared (6) or in the near infrared and visible region (3).

Finally it should be noted that this method can be associated with other kinds of photothermal detections such as photothermal radiometry (9) or "mirage effect" (10), the high sensitivity of the later leading to a large improvement of the s/n ratio. Such studies are underway in our laboratory.

ACKNOWLEDGMENTS

We want to thank D.G.R.S.T. for financial support.

REFERENCES

(1) Aspen International Conference on Fourier Spectroscopy.
    G.A. Vanasse, A.T. Stair, D. Baker Editors. Special reports n° 114
    US Air Force Systems, Command 1971.

(2) J.W.CHAMBERLAIN, The Principles of Interferometric Spectroscopy,
    Wiley, New-York, 1979.

(3) D. DEBARRE, Thèse de 3ème Cycle, Paris 1981.
    and :
    D. DEBARRE, A.C. BOCCARA and D. FOURNIER,
    High Luminosity Visible and Near-Infrared Fourier Transform
    Photoacoustic Spectrometer.
    (To be published).

(4) G. BUSSE and B. BÜLLEMER.
    Use of the Opto Acoustic Effect for Rapid Scan Fourier Spectroscopy
    Infrared Physics, 18, 256, 1978.

(5) B.S.H. ROYCE, Y.G. TENG, J. ENNS.
    Fourier Transform Infrared Photo-Acoustic Spectroscopy of Solids
    1980 IEEE ULTRASONIC SYMPOSIUM.

(6) D.G. MEAD, S.R. LOWRY and R.C. ANDERSON.
    Photoacoustic Infrared Spectroscopy of Some Solids.
    Int. Journal of Infrared and Millimeter Waves (under press).

(7) A. ROSENCWAIG and A. GERSHO
    Theory of the photoacoustic effect with solids.
    J. Appl. Phys. 47, 64 (1976).

(8) L. LLOYD, S. RISEMAN, R. BURNHAM, E. EYRING and M. FARROW
    Fourier transform photoacoustic spectrometer.
    Rev. Sci. Instrum. 51, 1488 (1980).

(9) P.E. NORDAL and S.O. KANSTAD.
    Photothermal Radiometry.
    Physica Scripta 20, 659 (1979).

(10) A.C. BOCCARA, D. FOURNIER and J. BADOZ
     Thermo-optical Spectroscopy : Detection by the "Mirage effect".
     Appl. Phys. Lett. 36, 130, (1980).

PHOTOACOUSTIC SPECTRA OF BILIVERDIN DIMETHYL ESTER

S. E. Braslavsky, R. Ellul, S. Culshaw, I.-M. Tegmo-Larsson, and
K. Schaffner

Institut für Strahlenchemie im Max-Planck-Institut für Kohlenforschung, Stiftstraße 34-36, D-4330 Mülheim a.d. Ruhr, West Germany

Abstract

Biliverdin dimethylester (BVE) in ethanol and BVE incorporated into dipalmitoyl
lecithine (DPL) vesicles were irradiated from 555 to 690 nm with a 15 ns tunable
dye-laser pulse. A spectrum was obtained by plotting the first amplitude of the
photoacoustic (PA) signal, measured with a piezoelectric detector and corrected
for the different laser energies, as a function of the excitation wavelength. The
system was calibrated using $CuSO_4$ either in H_2O-NH_3 or incorporated into DPL-Tris
buffer vesicles. The PA signal of BVE in ethanol, on one hand, showed a "defect"
at ca. 640 nm with respect to the absorption spectrum. This wavelength corresponds to the fluorescence excitation maximum of BVE in ethanol. However, the
magnitude of the difference between the PA and absorption spectra indicates
an additional occurrence of thermally reversible photoisomerizations. On the
other hand, intensity of the PA spectrum for BVE-DPL, measured over a period
of 20 h, decreased with time upon incorporation of the pigment into the vesicles.
The absorption spectrum, showed no such time dependence. The time behaviour
of the PA signal is consistent with our previous observations of an increase
in fluorescence yield with time in BVE-DPL. The PA change is, however, of a
much larger magnitude. These results are interpreted in terms of a chronological
decrease in the radiationless transitions of BVE upon incorporation into the
more rigid lipid bilayer, and an additional increase of a photoisomerization
reaction of BVE which thermally reverts to its original state.

INTRODUCTION

Biliverdin dimethyl ester (BVE), and related bilindione-type
compounds find increasing attention as model chromophores of the
plant photoreceptor phytochrome.[1] This important chromoprotein is
subject to photochromic changes, $P_r \rightleftharpoons P_{fr}$, which have been
interpreted over the years in several ways. They include the
possible involvement of conformational changes of the chromophore,[2] which could be similar to those held responsible for the
solvatochromic, thermochromic, and photochemical behaviour of BVE
dissolved in organic solvents.[3] These properties have been interpreted to arise from a conformational equilibrium between families
of helically "coiled" and "more stretched" forms. This conclusion
was based on absorption, fluorescence and solvent induced circular

dichroism measurements. Moreover, we have proved in a recent study[4] that the facile incorporation of BVE into lipid bilayers is an efficient method to control the conformational equilibrium of this bilidione. We report here the results obtained on photoacoustic measurements of BVE both dissolved in ethanol and incorporated into liposomes. These measurements point to the occurrence of photo-isomerizations of BVE, in addition to the emission and radiationless decay processes. Part of the results were included in reference 4.

BILIVERDIN DIMETHYL ESTER
(BVE)

MATERIALS

BVE was prepared and purified as described previously.[5] Dipalmitoyl lecithine (DPL) (Sigma) and egg yolk lecithine (EYL), prepared as described,[6] were checked for homogeneity by thin-layer chromatography on silica gel ($CHCl_3$-CH_3OH-H_2O 65:25:4). Incorporation of BVE into the lipid bilayer was effected by incubation of BVE into preformed vesicles in potassium phosphate buffer (pH 7.5) (Method 2 in ref. 4).

Cosonication of $CuSO_4$ and DPL was used to produce DPL vesicles with incorporated $CuSO_4$. The preparation was performed in Tris hydrochloride (0.1 M, pH 7.4) and with a concentration of $CuSO_4$ resulting in an A_{660} = 0.2 of the $CuSO_4$-DPL sample. Organic solvents were purified to fluorescence grade.

METHODS

Absorption spectra were recorded on Cary 17 and 219 UV/VIS spectrophotometers using, in the case of BVE incorporated into lipid vesicles, appropriate BVE-free liposome dispersions as a reference.

The PA signals were registered by a piezoelectric (PZT) transducer (Siemens H 42, resonance frequency 300 KHz, Pb-Zr-Ti) placed in a polished stainless steel casing as described by Tam and Patel.[7] The transducer case was attached to the side of a 1 cm quartz-fluorescence cuvette with epoxy resin. The dye-laser beam (15 ns, bandwidth 0.5 cm^{-1}) crossed the cuvette perpendicular to the face to which the PZT transducer was attached. The signal of the transducer was amplified by a Keithley 103 A nanovolt preamplifier, a Tektronix 7A22 plug-in amplifier, and then fed into a Biomation 8100 transient recorder. Pictures were taken of the signal displayed on a screen and the amplitude (H) of the first pressure pulse (9 µs after the laser pulse) was measured (see Fig. 1. The output of the transient recorder was fed into a Digital PDP11/04 computer and then plotted).

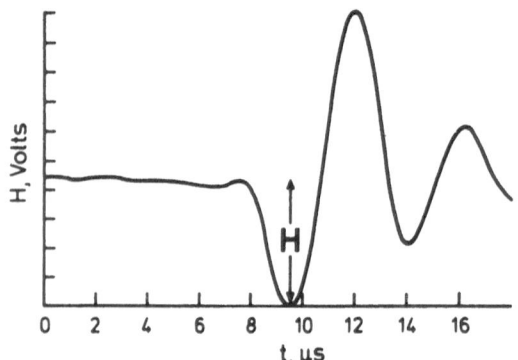

Figure 1.

Computer print-out of the first 18 µs of the ringing PZT signal. BVE in ethanol $(3 \cdot 10^{-5}$ M); λ_{irr} = 640 nm.

The H values were linearly proportional to the dye-laser pulse energy (E) as shown for a typical case in Fig. 2. H/E was proportional to the light intensity absorbed by the BVE solutions up to A = 0.3 (Fig. 3). The H/E values were measured for λ_{irr}=555 to 730 nm. The following dyes dissolved in CH_3OH were used for this purpose: rhodamine 6G (555-570 nm), B (580-595 nm), and 101 (605-615 nm), DCM (620-650 nm), rhodamine 101 + cresyl violet + nile blue (670-680 nm), nile blue (685-690 nm), carbazine 720 (690-730 nm). The dye-laser system (J.K. Lasers) was pumped with the second harmonic (532 nm) of a Nd-Yag laser (J.K. Lasers). All the measurements were performed at room temperature, 19 °C. E was measured with an online photodiode (J.K. Lasers) calibrated against a 14 NO thermopile (Laser Instrumentation).

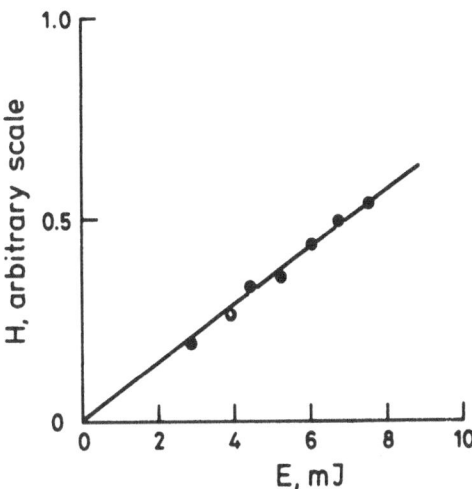

Figure 2. Dependence of the non normalized PA signal (H) on the dye-laser energy for an ethanol solution of BVE ($3 \cdot 10^{-5}$ M); λ_{irr} = 640 nm.

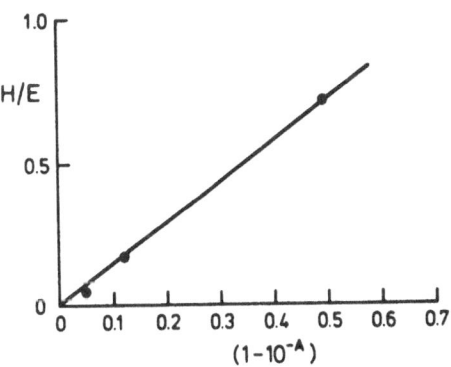

Figure 3. Dependence of the normalized PA signal (H/E) on the light intensity absorbed in ethanol solutions of BVE. λ_{irr} = 640 nm.

RESULTS

The spectrum shown in Fig. 4a was obtained by plotting the normalized PA signal (H/E) obtained with a solution of BVE in ethanol against λ_{irr}. Figures 4b and 4c depict the absorption and corrected fluorescence excitation (λ_{em}=705 nm), respectively, obtained for the same solution and similar to those we reported in previous publications.[3]

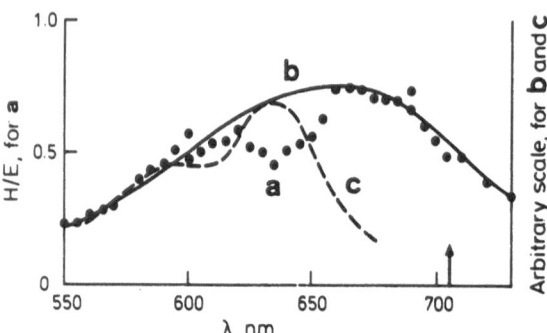

Figure 4. a-Normalized PA signal (H/E) as a function of λ_{irr}; BVE
in ethanol (A_{665} = 0.27). b-Absorption spectrum, and c-corrected
fluorescence excitation of the same solution for λ_{em} = 705 nm
(indicated by the arrow; see ref. 3. Note that absorption and
PA spectra have been normalized to coincide at 665 nm and spectrum
c has been normalized to coincide with spectrum b at 640 nm.

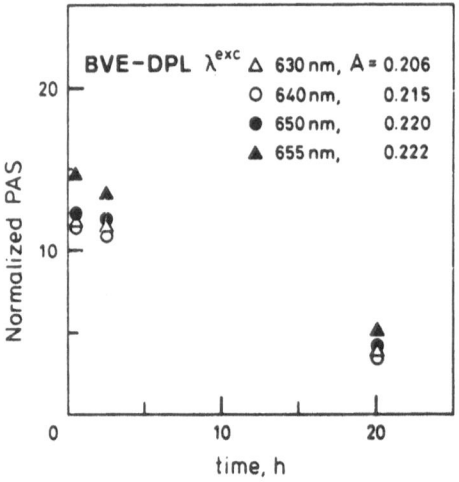

Figure 5. PA signal of BVE-DPL lipsomes prepared by incubation.
Decrease of signal intensity after incorporation (The PA signal of
BVE-EYL liposomes did not show any time dependence).

A spectral analysis of the H/E values obtained from BVE incorporated into DPL and EYL vesicles was more difficult and
produced scattered results. However, for BVE-DPL the corrected
PA signals experienced a time dependence along the whole spectrum.

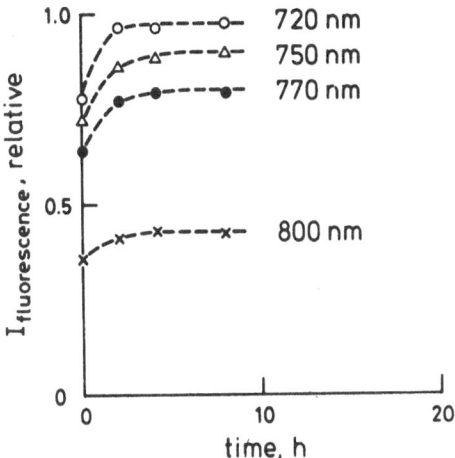

Figure 6. Increase of fluorescence intensity during incorporation of BVE by incubation into preformed DPL liposomes; λ_{exc} 355 nm.

This behaviour is exemplified for four wavelengths in Figure 5. H/E decreased over a period of 20 h after incubation by a factor of about 3, while the fluorescence intensity of the same sample increased only by 5-20% within 2 h.[4] The latter remained constant thereafter (Fig. 6). The H/E values of copper sulfate in DPL liposomes did not change at all with time. This discards progressive liposome aggregation changing the transmission of the acoustic wave, which could have been the origin of the time-dependent PA intensity reduction. Contrary to the observation with DPL, H/E for BVE-EYL remained unchanged from the beginning of incorporation.

DISCUSSION

The comparison of the PA and absorption spectra in Fig. 4 (a and b, respectively) immediately shows that in the 640 nm region, part of the molecules undergo processes other than mere radiationless decay to the ground state. The fluorescence excitation spectrum (Fig. 4c) shows that when excited with radiation of this wavelength some molecules fluoresce. However, a quantitative comparison of the "missing" absorption area in the PA spectrum with the fluorescence quantum yield ($\Phi_F = 5 \cdot 10^{-4}$ for BVE in ethanol and λ_{exc} = 593 nm)[4] indicates that some additional radiationless chemical processes might be occurring. Since no final change could be observed after photolysis of ethanol solutions of BVE at room temperature,[3c] we assign these chemical processes to a thermally reversible photoisomerization process, with a time constant for the reversion longer than ca. 10 μs.[8]

When analysing the results obtained for the BVE-DPL system it is again observed that while the fluorescence intensity increased a mere 5-20% after the onset of incubation (Fig. 6), the final H/E value was only about 30% of the initial value (Fig. 5). Moreover, the time scales of the two observations were different, with only 2 h to reach a constant ϕ_F value. The portion of decrease in photacoustic signal which correlates with the increase in ϕ_F is obviously overridden by a similar but much larger effect. The only plausible origin for this large PA deficit must then again be photochemistry, the nature of which is open to debate.[9]

The fluorescence increase after incubation of BVE with EYL liposomes was of the same magnitude as that observed for DPL (initial ϕ_F = 4·10^{-3}, final ϕ_F = 4.5·10^{-3}).[4] A corresponding change in PA intensity of BVE-EYL would have been too small for our instrumental error, and in fact no significant PA change was observed in this instance. A photochemical and thermally reversible change in conformation of BVE would still be consistent with these observations. At room temperature, the thermal reversion in the "soft" EYL liposomes (transition temperature, T_m = -5 °C), would have to be faster than 10 μs,[8] whereas in the more rigid DPL liposomes (T_m = 42 °C) the reversion would have to be slower. A somewhat similar situation has been encountered previously[3c] in the photochemistry of BVE in ethanol. There, an overall photochemical change could only be observed at 173 K; this thermally reverted on heating to ambient temperature.

The results in ethanol seem to indicate that the photoisomerization occurs preferentially in one wavelength region which happens to be coincidental with the fluorescence excitation band. In DPL, however, the decrease of PAS intensity occurred uniformly over the whole absorption band indicating a nonspecific (with respect to λ) photoisomerization.

The labile photoproduct of BVE in ethanol and in DPL cannot be the so-called stretched conformer(s) observed by us by fluorescence since these are thermally stable.[4] A possibility is, however, that the Z → E isomerized BVE's are produced, which have been described by Falk[10]. These products were photochemically formed when BVE was adsorbed on aluminum oxide, and they slowly re-formed the (all Z)-BVE on standing in solution.

ACKNOWLEDGEMENTS

We are indebted to Drs. C.K.N. Patel and A. Rosencwaig for making available to us information about some of their results prior to publication.

REFERENCES

1. W. Rüdiger, "Phytochrome, a light receptor of plant photomorphogenesis". In "Structure and Bonding". J.B. Goodenough, P. Hemmerich, J. A. Ibers, C.K. Jorgenson, J.B. Neilands, D. Reinen and R.J.P. Willams, Eds; p. 101. Springer Verlag, Berlin, (1980).

2. (a) M.J. Burke, D.C. Pratt, and A. Moscowitz, "Low-temperature absorption and circular dichroism studies of phytochrome". Biochemistry 11, 4025 (1972). (b) R. Pasternak and G. Wagnière, "Possible interpretation of long-wavelength spectral shifts in phytochrome P_r and P_{fr} forms". J. Am. Chem. Soc. 101, 1662 (1979). (c) S.E. Braslavsky, J.I. Matthews, H.J. Herbert, J. de Kok, C.J.P. Spruit, and K. Schaffner, "Characterization of a microsecond intermediate in the laser flash photolysis of small phytochrome from oat".Photochem.Photobiol. 31, 417 (1980).

3. (a) S.E. Braslavsky, A.R. Holzwarth, H. Lehner, and K. Schaffner, "The fluorescence of biliverdin dimethyl ester". Helv. Chim. Acta 61, 2219 (1978). (b) A.R. Holzwarth, H. Lehner, S.E. Braslavsky and K. Schaffner, "The fluorescence of biliverdin dimethyl ester". Liebigs Ann. Chem. 2002 (1978). (c) S.E. Braslavsky, A.R. Holzwarth, E. Langer, H. Lehner, and K. Schaffner, "Conformational heterogeneity and photochemical changes of biliverdin dimethyl esters in solution". Israel J. Chem. 20, 196 (1980).

4.   I.-M. Tegmo-Larsson, S.E. Braslavsky, S. Culshaw, R. Ellul,
      C. Nicolau, and K. Schaffner, "Conformation control by
      membrane of biliverdin dimethyl ester incorporated into lipid
      vesicles". To be published (1981). I.-M. Tegmo-Larsson, S.E.
      Braslavsky, C. Nicolau, and K. Schaffner, "Conformational
      control by membrane of biliverdin dimethyl ester incorporated
      into liposomes, fluorescence and membrane-induced circular
      dichroism". Biophys. Struct. Mechanism $\underline{6}$ (Suppl.), 112 (1980).

5.   H. Lehner, S.E. Braslavsky, and K. Schaffner, "Isolation,
      characterization and solution conformation of biliverdin
      dimethyl ester and its XIIIα isomer". Liebigs Ann. Chem. 1990
      (1978).

6.   N.S. Singleton, M.S. Gray, M.L. Brown, ana J.L. White,
      "Chromatographically homogeneous lecithine from egg phospho-
      lipids" J. Am. Oil Chem. Soc. $\underline{42}$, 53 (1965).

7.   (a) A.C. Tam and C.K.N. Patel, "Ultimate corrosion-resistant
      optoacoustic cell for spectroscopy of liquids". Optics Lett.
      $\underline{5}$, 27 (1980), and references therein. (b) C.K.N. Patel,
      private communication.

8.   The limit calculated for a discrimination between the de-
      activation of the initially formed singlet state and a sub-
      sequent thermal deactivation of a photoisomer would be
      essentially given by a combination of the detector rise time
      and the transient time of the acoustic wave across the
      exciting beam, i.e., ca. 10 μs.

9.   Photochemical experiments with BVE in liposomes are in
      progress.

10.   H. Falk, G. Grubmayr, E. Haslinger, T. Schlederer, and
      K. Thirring, "The diasteromeric (geometrically isomeric)
      biliverdin dimethyl esters - Structure, configuration and
      conformation". Monatsh. Chem. $\underline{109}$, 1451 (1978).

QUANTITATIVE PHOTOACOUSTIC SPECTROSCOPY
OF PHENOLRED POTASSIUM SALT IN POLYVINALALCOHOL

W. Görtz and H.-H. Perkampus
Institut für Physikalische Chemie, Lehrstuhl I
der Universität Düsseldorf

ABSTRACT

Some working parameters of a gas-coupled doublebeam photoacoustic spectrometer are described:
- efficiency of the acoustic isolation
- calibration of the photoacoustic cells
- sensitivity of the system

Phenolred potassium salt solutions in polyvinylalcohol with
sample thickness between 10 μm and 200 μm and an optical density between 10 cm-1 and 2000 cm-1 are investigated with transmission and photoacoustic spectroscopy. Results are reported
and compared with the ROSENCWAIG-GERSHO-theory.

INTRODUCTION

The relation between the photoacoustic (PA) signal and the
optical properties of the sample was investigated by several
authors using thermally thick dye solutions (1-5). In quantitative studies such liquids show some disadvantages, which can
be overcome, if transparent polymers are applied as a dye matrix:

- Transmission (TM) measurements of solutions with optical
  densities greater than 50 cm^{-1} require special small cuvet-
  tes; thus in most investigations the optical density is
  calculated from the dye concentration. This is a source of
  uncertainty, because LAMBERT-BEER's law might not be valid
  in concentrated solutions.

- Deviations from the ROSENCWAIG-GERSHO (RG) theory due to
  acoustic coupling are smaller if a glass like matrix is
  used (4).

- In a gas-coupled system vaporization of the solvent may
  cause changes in the thermal properties of the gas;
  the microphone sensitivity also might be affected.

- Polyvinylalcohol (PVA) is a well known material in optical spectroscopy (6). It has a good solubility for most polar dyes and preparation of samples with variing thickness and optical density is easy. This report is concerned with a first attempt to apply this system for quantitative PA spectroscopy.

INSTRUMENTATION

A conventional double-beam PA-spectrometer was used. The light source is a 450 W Xe-HD arc in a PRA ALH 220 housing, equipped with a PRA 301 system as power supply. This allows electronic modulation from 10 s^{-1} to 10.000 s^{-1}. A feedback-system keeps the intensity of the emitted light constant within ±1%. The Zeiss MB3-gratingmonochromator has a workingrange from 300 nm to 900 nm. For mechanical chopping from 10 s^{-1} to 150 s^{-1} a rotating mirror-blade is mounted on a Ithaco 383 chopper; this allows a 1:1 beam-splitting without intensity losses on the sample. A reference cell filled with a carbon-black sample is calibrated to determine the intensity of the incident light. The sample cell employed was made from steel and is furnished with an exit window to allow quantitative measurements of optically transparent samples; its volume is 1.6 cm^3. Signals are detected by Bruel & Kjaer 4166 microphones and analysed by Ithaco lock-in amplifiers.

The sample cell is positioned inside a glove-box. This arrangement allows investigations of air- and humidity-sensitive samples and is also efficient as isolation against interfering air-sound. A cast-iron plate in sand is used to isolate the sample cell against building vibrations. The efficiency of the acoustic isolation is shown in fig.1. The standard deviation of the microphone signal in the sample cell was determined inside and outside the glove-box at different modulation frequencies. As transient noise was not taken into account, the data represent only the lower limit for the noise level disturbing PA-measurements. The acoustic isolation damps the noise level for most frequencies investigated. No effect can be seen at 50 s^{-1}, where electrical noise is dominant. At 39 s^{-1}, the resonance frequency of the glove-box, the noise inside the box

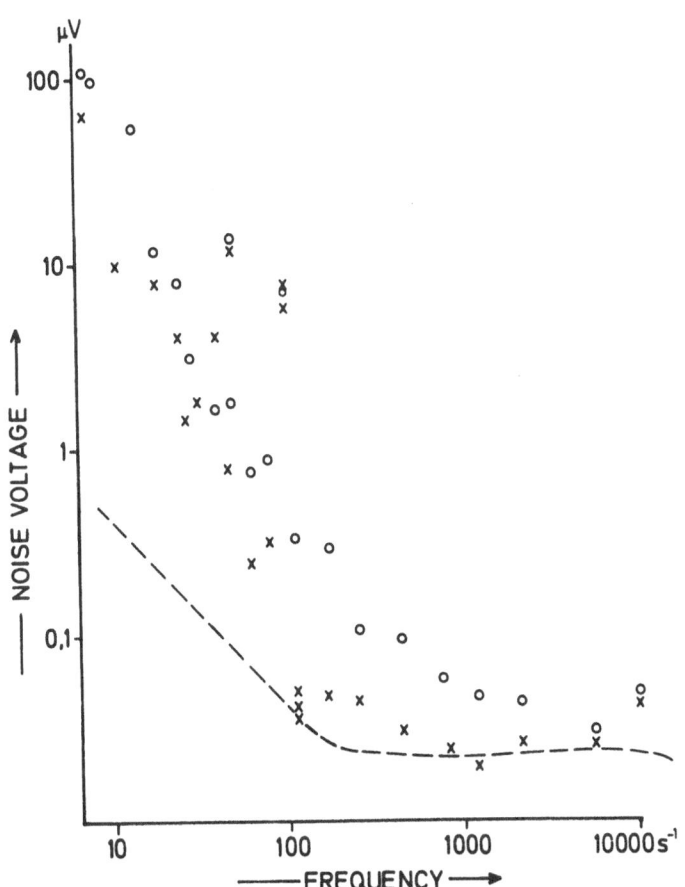

FIGURE 1

Microphone noise voltage in the sample cell vs frequency;
measured inside (x) and outside (o) the glove box

is increased. The dashed line represents the microphone/pre-
amplifier noise in accordance with the data supplied by the
manufacturer (7). These data are corrected for the utilized
bandwidth of 0.1 s^{-1}. Fig.1 indicates that additional efforts
in acoustic isolation will bring no significant improvement
for PA spectroscopy at frequencies greater than 100 s^{-1}. The
best signal to noise ratio is found at 105 s^{-1}, with an absolute noise level of $2.5 \cdot 10^{-6}$ Pa at 1 s^{-1} bandwidth. The detection limit for measurements on PVA corresponds to an optical density of 2 cm^{-1}.

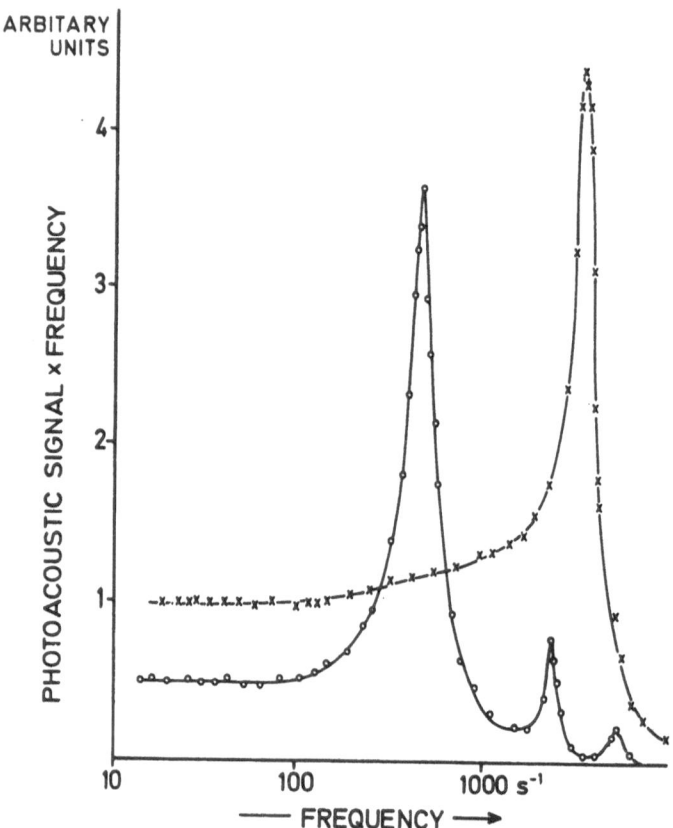

<u>FIGURE 2</u>

(PA signal · frequency) vs frequency; for sample (x)
and reference cell (o)

Acoustic resonance effects in PA cells contribute to a nonlinear frequency dependence of the PA signal; therefore it is
necessary to calibrate the cells:
For a thermally thick carbon-black sample the amplitude of the
PA signal is measured in dependence of the modulation frequency; deviations from the non-resonant behaviour are shown in
fig.2. In the sample cell an acoustic enhancement of the PA
signal is observed at frequencies greater than 150 s^{-1}. Correction factors that consider the influence of frequency on the
cell sensitivity can be estimated from the graph shown. Errors
may arise in the frequency range around 3500 s^{-1}, because the
resonance frequency may drift when a sample change alters the
cell volume.

EXPERIMENTAL

Aqueous solutions of PVA (WACKER 5/20) were mixed with different amounts of phenol red potassium salt. Well defined volumes of these mixtures were deposited on cover glasses and
dried on a balanced table. As PVA films are hygroscopic their
thermal properties change in relation to air humidity. Therefore the samples were stored in an exsiccator until their
weight was constant. The sample thickness was determined by
gravimetry assuming a density of 1.27 g $\cdot$ cm^{-3} (8).

Different methods were used to determine the absolute intensity of the incident light at the emission maximum of the lamp
monochromator combination (475 nm, bandwidth 10 nm). A calibrated pyroelectric detector measured 2.2 ($\pm$0.2) mW. With a
ferrioxalat actinometer (9) 2.0 ($\pm$0.3) mW were found. Compared
with the PA signal of a 2 mW HeNe laser the estimated intensity was 1.9 mW.

All spectroscopic measurements were carried out at the absorption maximum of the phenol red potassium salt (571 nm), using
a spectral bandwidth of 10 nm. A Zeiss PMQ 4 spectral photome-
ter was used for TM measurements. The extinction was determined against a dye-free PVA film of the same thickness as the
sample under study.

RESULTS AND DISCUSSION

The PA signals observed varied between 1 $\cdot$ 10^{-5} and 4 $\cdot$ 10^{-4}
Pa. For a quantitative evaluation, especially at lower signal
level, a correction for coherent acoustic noise is necessary
(10). This was achieved by subtracting the lock-in signal at
the same phase angle but with no light incident on the sample.
A typical acoustic background signal was 5 $\cdot$ 10^{-6} Pa. Another
source of error, that becomes important at low signal levels
arises from unspecific absorption at the cell walls and windows (10). The magnitude of this signal depends on cell design,
as well as on the amount of light scattered by the sample. To
take this problem into account the PA signal of the dye-free
PVA-film, typically 8 $\cdot$ 10^{-6} Pa, was subtracted from the observed signal with respect to variing phase angles.

Fig. 2 shows the PA signal at different sample thickness for
an optical density of 55 cm^{-1}. At 105 s^{-1} modulation frequency only a slight reduction of the PA signal is observed variing
the sample thickness from 200 μm to 10 μm. In parallel no change of phase angle took place. Similar results were obtained at
optical densities of 15 cm^{-1}, 150 cm^{-1} and 500 cm^{-1}. At
12.5 s^{-1} the PA signal shows a more distinct dependence on the
sample thickness. The slope of the upper curve in fig.3, however shows that even for the 13 μm thick sample the thermally
thin case is not realized. Thus the PA signal depends on the
thermal properties of sample and backing material as well.

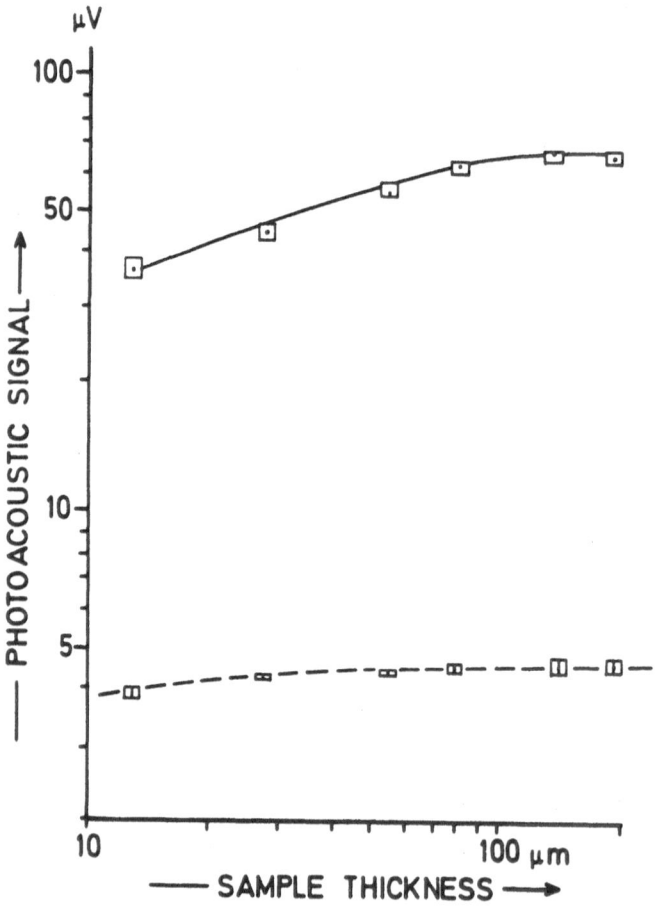

FIGURE 3

PA signal vs sample thickness; optical density: 55 cm^{-1},
frequency: 105 s^{-1} (---) and 12.5 s^{-1} (———)

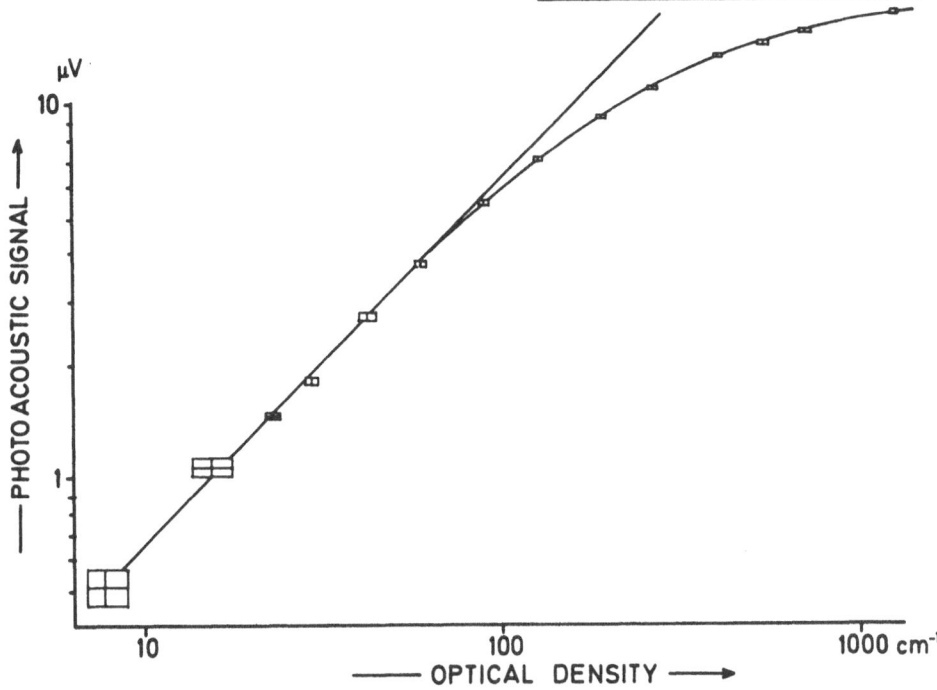

FIGURE 4

PA signal vs optical density; frequency: 105 s^{-1};
optical densities above 1000 cm^{-1} are estimated from
the dye concentration.

As it turned out that the optical homogeneity of the PVA films
is best at a sample thickness of about 30 μm, this thickness
and 105 s^{-1} modulation frequency were selected to check the
relation between PA signal and optical density. 3 to 5 samples
of similar optical properties were investigated. The results
are shown in fig.4. All data obtained are within the rectang-
les in the diagramm. Deviations in TM measurements are typically within ±5%, whereas for PA spectroscopy a reproducibility
up to ±1% was found. This advantage of PA over TM spectroscopy
could be explained by a roughness of the PVA surface, that
causes scattering as well as deflections of the transmitted
light.

The curve in fig.4 shows a behaviour as it is predicted by the
RG-theory. For the optically transparent samples the PA signal
is proportional to the optical density; no nonlinearities due
to thermoelastic effects can be seen. Going to optical densi-

ties greater than 200 cm^{-1} the PA signal starts to saturate; this goes parallel with a change in the phase angle. For the saturated signal an intensity of 19.0 ($\pm$0.2) μV was determined. As fig.3 indicates that the films are thermally thick, the thermal properties of PVA can be calculated by application of the cases 1c and 2b of the RG-theory (2). For the opaque case the following expression is obtained:

$$S^{sat} = \frac{I}{f} \cdot \frac{\sqrt{a_s}}{K_s} \cdot c \tag{1}$$

S^{sat}, saturated PA signal; I, light intensity; f, modulation frequency; a_s, thermal diffusivity, K_s, thermal conductivity; c, constant for standard conditions and given instrumentation.

The optically transparent case is described by:

$$S^{trans} = \frac{I \cdot \beta}{f} \cdot \sqrt{\frac{2}{f \cdot \pi}} \cdot \frac{a_s}{K_s} \cdot c \quad \text{with} \quad a_s = \frac{K_s}{\rho \cdot C_s} \tag{2}$$

S^{trans}, PA signal for the transparent sample.
C_s, heat capacity. ρ, density. ß, optical density.

From the graph in fig.4 and equation 2 the heat capacity of the PVA is calculated as C_s = 2.6 ($\pm$0.3) J/g·K. The accuracy of this determination is mainly restricted by the uncertainty in measuring the incident light intensity. Though no data could be found in literature the result seems quite reasonable when it is compared to the heat capacities of liquid ethanol 2.5 J/g·K and methanol 2.6 J/g·K (8).
From equations 1 and 2 the thermal properties of the PVA are calculated as follows:

$$K_s = 6.8 \cdot 10^{-3} \ (\pm \ 0.8 \cdot 10^{-3}) \quad J \cdot cm^{-1} \cdot s^{-1} \cdot K^{-1}$$

$$a_s = 2.0 \cdot 10^{-3} \ (\pm \ 0.15 \cdot 10^{-3}) \quad cm^2 \cdot s^{-1}$$

The determination of the thermal diffusivity is more accurate, because it does not depend on light intensity; this becomes obvious when equation 1 is divided by equation 2.

The resulting thermal diffusion length for 105 s^{-1} modulation frequency is μ_s = 25 μm. This is in some contradiction to the statement that PVA films of 30 μm are thermally thick. More detailed investigations, especially on the relation between PA signal and the thermal properties of the backing material, are necessary to establish the results described. PVA films as well as dye coated glass samples (11) will be a useful tool in such studies.

ACKNOWLEDGEMENTS

This work was supported by Ministerium für Wissenschaft und Forschung des Landes Nordrhein-Westfalen and by Fonds der Chemischen Industrie.

REFERENCES

1   McClelland, J.F.; Kniseley, R.N., Signal saturation effects in photoacoustic spectroscopy with applicability to solid and liquid samples, Appl.Phys.Lett.,28,467(1976)

2   Rosencwaig, A.; Photoacoustics and photoacoustic spectroscopy. John Wiley & Sons, New York, Chichester, Brisbane, Toronto, 1980

3   Wetsel, G.C.; McDonald, A.F., Photoacoustic determination of absolute optical absorption coefficient, Appl.Phys.Lett. 30, 252(1977)

4   McDonald, A.F.; Wetsel, G.C., Generalized theory of the photoacoustic effect, J.Appl.Phys. 49, 2323(1978)

5   Poulet, P.; Chamborn, J., Quantitative photoacoustic spectroscopy applied to thermally thick samples, J.Appl. Phys. 51, 1738(1980)

6   Hanle, W.; Kleinpoppen, H.; Scharmann, A., Dichroismus und Lumineszenzpolarisation an verstreckten Polyvinylalkohol-Folien, Z.Naturforschung 13a, 64(1958)

7   Tarnow, V.; Thermisches Rauschen in Mikrofonen und Vorverstärkern, Bruel & Kjaer Technical Review, 3, 3(1972)

8   Weast, R.C. Handbook of Chemistry an Physics, Cleveland, 1971

9   Parker, C.A., A new sensitive chemical actinometer, Proc. Roy.Soc. A, 220, 104(1953)

10 Munroe, D.M.; Reichard, H.S., Practical PAS of solids, PAR application note 147, 1976

11 . Hursh, D.; Kuwana, T.,Photoacoustic and spectrophotometric quantitation of copper phtalocyanine films, Anal. Chem. 52, 646(1980)

IN SITU PHOTOACOUSTIC SPECTROSCOPY OF
COPPER ELECTRODES IN SOLUTION

Ulrich Sander and Jürgen K. Dohrmann

Institut für Physikalische Chemie der Freien Universität Berlin
Takustraße 3, D-1000 Berlin 33, Germany

Hans-Henning Strehblow

Institut für Physikalische Chemie II, Universität Düsseldorf
Universitätsstraße, D-4000 Düsseldorf, Fed. Rep. of Germany

ABSTRACT

A double-beam photoacoustic spectrometer and a photoacoustic
cell for in situ studies of electrode processes are described.
The photoacoustic spectra of pure bulk copper and of an oxide-
covered electrode in contact with potassium hydroxide solution at
various potentials are presented. A relation between the photoacoustic spectrum and the reflection spectrum is given. The measured photoacoustic spectrum of the pure copper electrode is in
agreement with the reflection spectrum calculated from known values of the optical constants (n and k) of copper and the electrolyte.

INTRODUCTION

Since photoacoustic spectra can be measured easily, we decided to
examine if PAS can be applied to the in situ study of the electrode/elec-
trolyte interface simultaneously with electrochemical measurements. Our
ultimate goal is to obtain information on the optical properties of electrodes and of surface layers formed under various electrochemical conditions. The optical constants of such systems are usually measured by means
of reflection spectroscopy[1] or ellipsometry[2].

In the following we report the result of an in situ photoacoustic
study on copper in contact with aqueous potassium hydroxide solution and
of thin oxide layers on copper.

EXPERIMENTAL

We use a cell assembly with the microphone located at the rear side
of a thin copper foil immersed in the electrolyte (Fig. 1). An electret
microphone, M, is fitted into a metal tube which is surrounded by a tef-

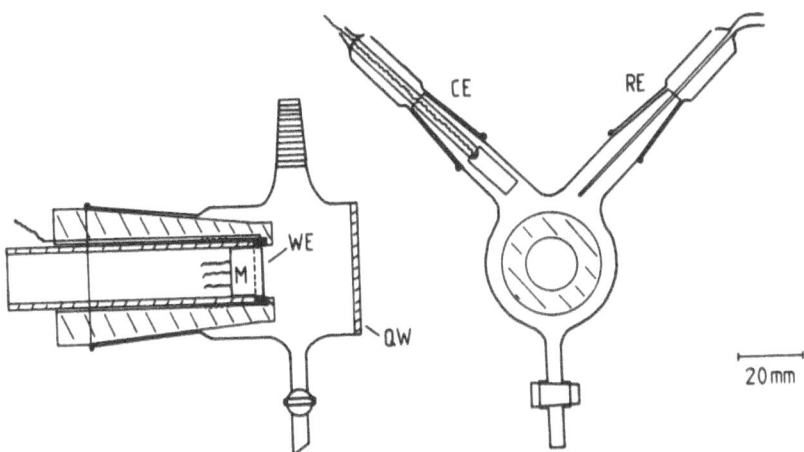

FIGURE 1   Microphone-working electrode assembly and electrolyte vessel
for in situ PAS studies. (M) microphone; (QW) quartz window;
(WE) working -; (CE) counter -; (RE) reference electrode.

lon joint. The circular 0.1 mm thick copper foil is mounted between two
silicon rubber rings just in front of the microphone. Electrical contact between the electrode and the current source (a potentiostat) is
made by a platinum ring from the rear side of the electrode. The electrochemical cell is a glass vessel with a volume of ca. 50 ml containing
the working electrode-microphone assembly, WE, a platinum counter electrode, CE, and a reference electrode, RE. The light enters the cell
through a quartz window, QW, opposite the working electrode. Further details have been described elsewhere[3].

A PAS technique similar to ours has been used independently by
Honda and coworkers for the investigation of the electrode/electrolyte
interface[4,5] and by Kanstadt and Nordal for the study of the solid gas
interface[6].

Fig. 2 shows the block diagram of our double-beam photoacoustic
spectrometer. The light source is a 450 Watt high-pressure xenon arc
lamp. The light is dispersed by a high-throughput monochromator and is
focused onto the electrode which forms part of the photoacoustic cell, A.
Ca. 7 % is reflected by a quartz plate and is focused onto the photoacoustic reference cell, B, containing a carbon powder sample. The microphone signals from both cells are detected by two lock-in amplifiers and
are fed into a ratiometer which provides an output proportional to the
photoacoustic signal from the sample and normalized to the intensity of
the incident light. The signal is recorded either digitally or with an
X-Y recorder. For further details see Ref. 3.

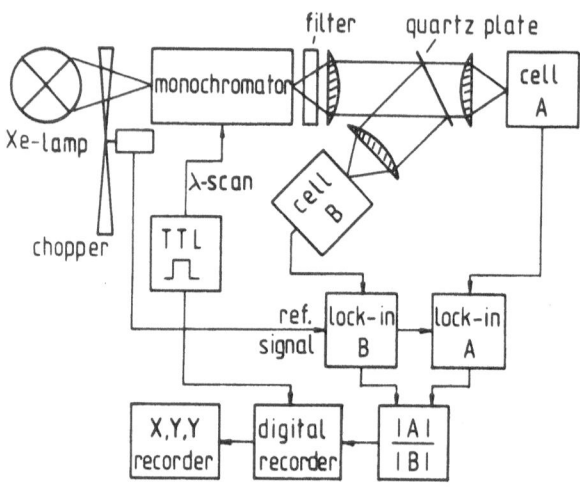

FIGURE 2   Block diagram of the double-beam photoacoustic spectrometer.

RESULTS AND DISCUSSION

The PAS spectra from the copper electrode surface taken at various potentials during potentiostatic polarization in 0.1 M potassium hydroxide solution are presented in Fig. 3.

Spectrum a refers to an electropolished copper electrode freed of surface oxides by extensive prepolarization at -0.9 V_H against the standard-hydrogen-electrode. The spectrum is closely related to the spectral dependence of the reflectance of bulk copper. This will be discussed later.

Spectra b and c have been measured after the formation of oxide or hydroxide layers at appropriate electrode potentials. Spectrum b was taken at a potential of +0.6 V_H where the electrode is covered by a transparent layer of copper(I) oxide and copper(II) oxide or hydroxide[7]. The thickness of the layer as determined by chronocoulometry was ca. 4 nm. Comparison of spectra a and b shows an onset of the PAS signal at longer wavelength for the oxide-covered electrode and a generally larger absorption. Spectrum c was measured at -0.35 V_H after the reduction of copper(II) oxide to copper(I) oxide. Here the electrode is covered by Cu_2O with a thickness of 2.6 nm. The PAS signal between 300 and 500 nm is much stronger than that of spectra a and b. This is due to a larger absorption of the Cu_2O layer.

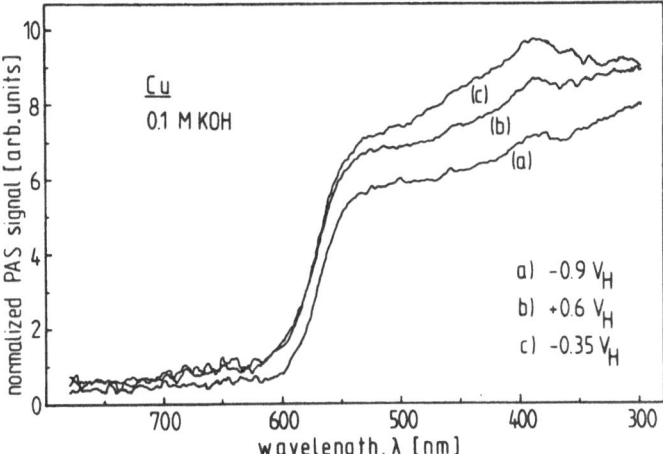

FIGURE 3   In situ PAS spectra from the copper electrode in 0.1 M KOH at
            various potentials:
            (a) oxide-free copper surface at −0.9 V_H;
            (b) 4 nm $Cu_2O/CuO,Cu(OH)_2$ layer on copper at 0.6 V_H;
            (c) 2.6 nm Cu_2O layer on copper at −0.35 V_H.
            Spectral bandwidth 10 nm, modulation frequency 20 Hz.

The PAS powder spectrum of Cu_2O shows an onset of absorption at
640 nm. This is in accord with our PAS data of the thin surface film.
However, a maximum at 400 nm as seen in spectrum $\underline{c}$ does not occur in the
photoacoustic, reflectance or transmittance spectra[8-10] of bulk Cu_2O.

Since the oxide layers are transparent, spectra $\underline{b}$ and $\underline{c}$ represent
the PAS response to both the oxide layers and the copper surface. The
analysis of the photoacoustic spectra of a system made up of three, or
even four, phases (copper/surface oxide(s)/electrolyte) is a difficult problem. Here, we shall not dwell on this matter further. At least, the spectra indicate that a surface phase consisting of ca. 10 monolayers of copper oxide can be detected by in situ measurements.

We have tried to quantitatively compare the photoacoustic spectra of
two-phase systems with their reflection spectra. This can be done by the
following considerations:

From conservation of energy the sum of the fractions of light which
are reflected (R), absorbed (A) or transmitted (T) must be unity

$$1 = R + A + T. \tag{1}$$

For an optically opaque sample – as in the present case of the copper
electrode – T equals zero.

The fraction of light reflected from the surface can be calculated by use of Fresnel's Equation. For normal-incidence, one obtains[11]

$$R = \frac{(n_2 - n_1)^2 + k_2^2}{(n_2 + n_1)^2 + k_2^2} \quad , \tag{2}$$

where n_2 and k_2 are the index of refraction and the extinction coefficient, respectively, of the electrode and n_1 is the index of refraction of the (non-absorbing) electrolyte. The absorbed part of the light, A, is converted into heat, H,

$$H = \eta A \quad ; \quad \eta \leq 1. \tag{3}$$

The conversion efficiency, η, equals one if other dissipation processes, e.g. luminescence, conversion into electrical or chemical energy, can be excluded.

Since the photoacoustic signal, P, is proportional to the heat, the quantity ηA can be measured if it is possible to calibrate the photoacoustic cell. This can be done by measuring the PAS signal, P_o, from a sample which converts all the incident light into heat and is thermally identical to the sample to be investigated. In this case,

$$\eta A = P/P_o \quad . \tag{4}$$

From Eq. (1) and T = 0 one obtains

$$P/P_o = \eta(1 - R). \tag{5}$$

We have measured P_o in the following manner: A black surface layer of copper oxide ($CuO/Cu(OH)_2$) was deposited on the electrode by anodic oxidation immediately after recording the spectrum, P, of the oxide-free electrode. Then, the spectrum of the oxide-covered electrode was measured. It was taken to correspond to P_o. Since the thermal diffusion length at the modulation frequency of 20 Hz used in the experiments was much larger (~1.4 mm) than the thickness of the electrode (0.1 mm) and the surface layer was more than two orders of magnitude thinner than the electrode, it seems reasonable to assume that the surface layer does not influence the thermal properties of the sample. The required ratio P/P_o was obtained by dividing the spectra of the oxide-free and oxide-covered electrodes. This ratio should correspond to $\eta(1 - R)$ in Eq. (5).

Fig. 4 shows the photoacoustic spectrum of the pure copper electrode in 1 M potassium hydroxide solution normalized to P_o measured in the same solution. The circles represent the quantity $(1 - R)$ calculated by use of Eq. (2). The optical constants of copper were taken from the work by Johnson and Christy[12], who determined n_2 and k_2 for vacuum evaporated films by transmission and reflection spectroscopy. The agreement between

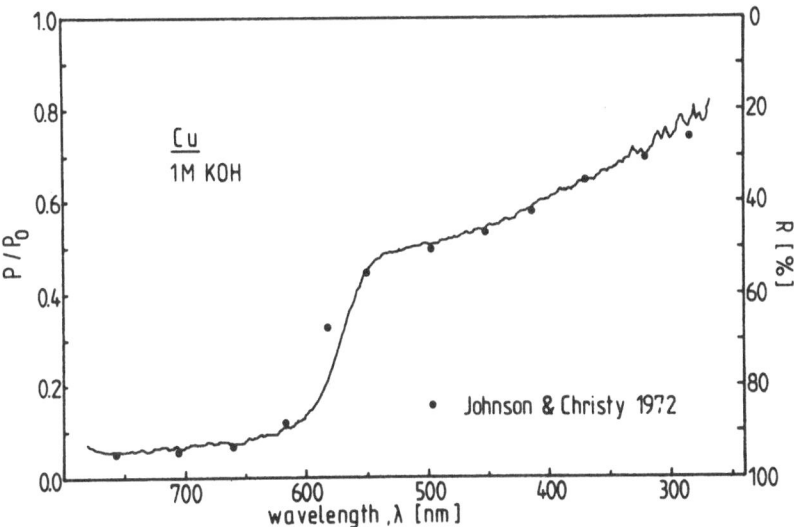

FIGURE 4   In situ PAS spectrum of copper in 1 M KOH at −0.5 V_H normalized
            to the spectrum of CuO and reflection spectrum of copper (●)
            calculated by Eq. (2) with n_2, k_2 from Ref. 12 and n_1 = 1.333.

P/P_0 and (1 − R) is excellent except at the wavelength of 580 nm. Although the reasons for the deviation at 580 nm are not fully understood,
we conclude that the efficiency of conversion of the absorbed light into
heat is 100 % for the copper electrode. It is interesting to note that
the PAS spectrum from copper powder exposed to air – which has been
published by Rosencwaig[13] – shows a hump at wavelengths near 400 nm. This
feature is missing in our spectrum. This is clearly due to the fact
that the electrode was oxide-free in our experiment, while the presence
of oxide films cannot be excluded for copper powder exposed to air.

Thin transparent surface layers could be investigated in the same
manner. Instead of Eq. (2), the corresponding relation for the three-
phase system electrode/surface layer/electrolyte would be applied. In
principle, it should even be possible to determine the optical constants
of electrodes and surface layers by in situ PAS. Similar to the procedures used in reflection spectroscopy, the photoacoustic spectra would
have to be measured at different angles of incidence or would have to be
analyzed by use of the Kramers-Kronig-relation[1]. We do not think, however,
that PAS has any advantages over the well-established reflection techniques, except perhaps with strongly light scattering samples.

There is a more interesting point, however. For a luminescent sample
or for a photovoltaic device, the conversion efficiency η in Eq. (5) must
be less than unity. A simultaneous measurement of photoacoustic and reflection data would thus result in the luminescence quantum yield or in
the energy conversion efficiency of a photovoltaic or photoelectrochemical
device. We are presently working in this direction.

REFERENCES

(1) J.D.E. McIntyre, Specular Reflection Spectroscopy of the Electrode-
    Solution Interphase, in "Advances in Electrochemistry and Electroche-
    mical Engineering", Vol.9, R.H. Muller,Ed., p.61, John Wiley and Sons,
    New York, 1973.

(2) J. Kruger, Application of Ellipsometry to Electrochemistry, in "Ad-
    vances in Electrochemistry and Electrochemical Engineering", Vol.9,
    R.H. Muller, Ed., p.227, John Wiley and Sons, New York, 1973.

(3) U. Sander, H.-H. Strehblow, and J.K. Dohrmann, In Situ Photoacoustic
    Spectroscopy of Thin Oxide Layers on Metal Electrodes. Copper in Al-
    kaline Solution, J. Phys. Chem., 85, (1981) in press.

(4) A. Fujishima, H. Masuda, and K. Honda, Studies of Electrode Surface
    Changes in situ by Photoacoustic Spectroscopy, Chem. Lett., 1063, 1979.

(5) H. Masuda, A. Fujishima, and K. Honda, In Situ Measurement of Elec-
    trode Surface Change by Photoacoustic Spectroscopy, Bull. Chem. Soc.
    Jpn., 53, 1542 (1980).

(6) S.O. Kanstadt and P.E. Nordal, Infrared Photoacoustic Spectroscopy of
    Solids and Liquids, Infrared Phys., 19, 413 (1979).

(7) H.-H. Strehblow and B. Titze, The Investigation of Passive Behaviour
    of Copper in Weakly Acid and Alkaline Solution and the Examination
    of the Passive Film by ESCA and ISS, Electrochim.Acta, 25, 839 (1980).

(8) P.W. Baumeister, Optical Absorption of Cuprous Oxide, Phys. Rev., 121,
    359 (1960).

(9) S.P. Tandon and J.P. Gupta, Diffuse Reflectance Spectra of Cuprous
    Oxide, Phys. Status Solidi B, 37 43 (1970).

(10) S. Brahms and S. Nikitine, Intrinsic Absorption and Reflection of Cu-
     prous Oxide in the 2.5 to 6.5 eV Region, Solid State Comm., 3, 209 (1965).

(11) Bergmann-Schäfer, Lehrbuch der Experimentalphysik, Vol.3 Optik, H.
     Gobrecht, Ed., 7th ed., p.480, de Gruyter, Berlin, New York, 1978.

(12) P.B. Johnson and R.W. Christy, Optical Constants of the Noble Metals,
     Phys. Rev. B, 6, 4370 (1972).

(13) A. Rosencwaig, Solid State Photoacoustic Spectroscopy, in "Optoacou-
     stic Spectroscopy and Detection", Yoh-Han Pao, Ed., p.194, Academic
     Press, New York, 1977.

3. Photoacoustic Detection and Monitoring

OPTOACOUSTIC RELAXATION OF PERIODICALLY IRRADIATED
SOLUTIONS

Egge Hey and Klaus Gollnick[+)]

Max-Planck-Institut für Kohlenforschung,

Abt. Strahlenchemie, Mülheim-Ruhr, Germany

ABSTRACT [++)]

Some time ago, 1964 - 1967, we developed a relaxation method
which permits to determine radiationless transition rates in a direct
way [1]. Intensity - modulated light is employed to produce excited singlet
and triplet states of the absorbing molecules; radiationless transitions
between excited states as well as between excited states and the ground -
state produce pressure waves which are measured by means of a micro -
phone. With eosin in methanol at room temperature, the resultant energy dispersion curve as a function of the modulation frequency allowed to
obtain a rate - constant of $1.25 \cdot 10^8$ sec^{-1} for the radiationless inter -
system crossing from the lowest excited singlet state to the triplet state
and a rate - constant of 900 sec^{-1} for the radiationless transition from
the triplet state to the ground - state. Internal conversion, i.e. radiationless transition from the lowest excited singlet state to the ground-state,
appears to be negligible in comparison with intersystem crossing and
fluorescence, the competing radiative process that extinguishes the low -
est excited singlet state.

+) Author to whom correspondence should be directed at Institut für Or-
ganische Chemie der Universität, Karlstr. 23, D-8000 München 2.

++) For the full version see: E. Hey and K. Gollnick, J. Photoacoustics,
in press.

1. a. E. Hey and K. Gollnick, Int. Conf. Photochemistry, Munich 1967,
Sept. 6 - 9; Preprints Part II, p. 465 - 481; abstracted in Ber.
Bunsengesellschaft 72, 263 (1968).
   b. E. Hey, Dissertation, University of Heidelberg, 1967.

PHOTOACOUSTIC ANALYSIS OF CHEMICAL REACTIONS:
H-D EXCHANGE IN THE SYSTEM H_2S-D_2S

R. Kadibelban and P. Hess

Institut für Physikalische Chemie
der Universität Heidelberg
6900 Heidelberg 1, F.R.G.

ABSTRACT

A photoacoustic apparatus is described consisting of a
TEA CO_2 laser as excitation source, two photoacoustic
cells, and for signal processing a transient recorder
and a digital signal analyser. Single pulse analysis as
well as averaging of the acoustic signals of several
laser pulses could be performed. First results are reported for the system H_2S-D_2S. Photoacoustic spectra
recorded for D_2S and H_2S-D_2S mixtures indicate that an
interference free excitation of D_2S is possible at 10,7
/um. Excitation at this wavelength was employed to monitor the D_2S concentration in H_2S-D_2S mixtures and to
determine the equilibrium constant for the hydrogen-
deuterium self-exchange reaction. The value obtained by
this photoacoustic method is compared with theoretical
calculations and the result of a mass spectroscopic
analysis.

INTRODUCTION

The photoacoustic effect was already applied to gas analysis and vapor detection in 1938 by Veingerov[1]. The
wide-spread use of this effect to detect pollutants and
to measure concentrations of infrared-absorbing species
started in 1968 with the first application of a laser as
radiation source[2]. The progress achieved in infrared
photoacoustic vapor detection with laser sources is discussed in a recent review article by Claspy[3].

In most photoacoustic measurements the cw radiation of a
He-Ne, CO or CO_2 laser is modulated with a chopper. In
this work a pulsed TEA CO_2 laser is used as excitation

source. This allows not only averaging of several pulses
to eliminate fluctuations in the acoustic signal, but
also a pulse to pulse analysis to detect kinetic processes.

The experimental setup and performance of the experiments
are described in detail in the next section. As a first
application the hydrogen-deuterium self-exchange reaction
in the system H_2S-D_2S is studied:

$$H_2S + D_2S = 2 \text{ HDS} \qquad\qquad (1)$$

In this work equilibrated mixtures were analysed and a
photoacoustic determination of the equilibrium constant
of this exchange reaction is reported. The result is compared with theoretical calculations and a mass spectroscopic analysis of the equilibrium isotope effect. The
accuracy and sources of error of the photoacoustic
method are discussed.

METHOD

Fig. 1 shows a schematic diagram of the experimental arrangement. A Lumonics 102 TEA CO_2 laser was used as excitation source. A typical pulse had a width of about
0,2 µs followed by a tail of 1,5 µs and an energy of
about 50 mJ/pulse. The apparatus was equipped with two
nearly identical photoacoustic cells in tandem. The first,
equipped with NaCl windows and filled with a constant
pressure of about 16 Torr D_2S, was used as a reference to
monitor the laser power. The second, equipped with CaF_2
windows, was the reaction and analysis chamber. The externally biased condenser microphones were mounted perpendicular to the laser beam. These home-made microphones
were of the type described in ref.[4] and had a diameter of
2,5 cm. The microphone signals were preamplified and fed
into a Biomation 8100 transient recorder. A Tracor
TN-1710 digital signal analyser together with a floppy

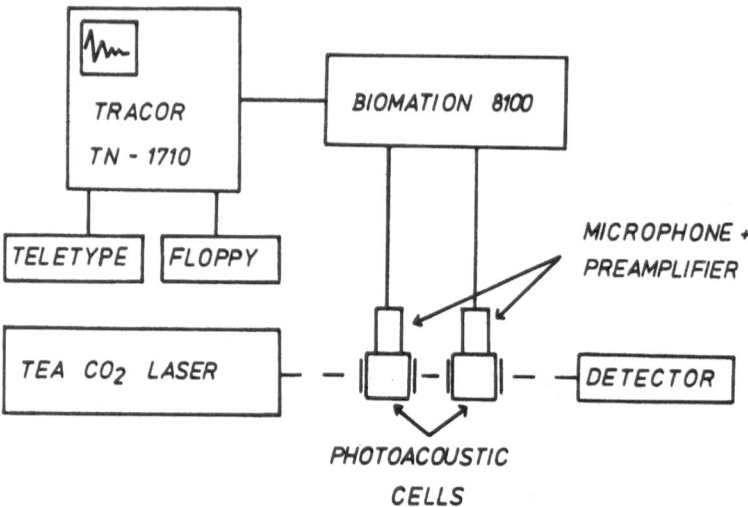

FIGURE 1 : Schematic diagram of the experimental setup.

disk system and a teletype were used for further signal processing.

The photoacoustic signal from the second cell filled with the gas under investigation was divided by the signal obtained from the reference cell. In a typical experiment the signals of about ten laser pulses were averaged and normalized to reduce the effect of fluctuations in the acoustic signal.

All experiments were performed at room temperature 295 $\pm$ 2 K . The pressure was measured with a MKS Baratron pressure transducer type 220.

The H_2S gas had a purity of 99,5 % and was supplied by Baker. The D_2S sample was delivered by Merck, Sharp and Dohme with 97 atom % D. Before use, the gas was further purified. The sample was cooled to 77 K to remove non-condensable impurities as H_2 and D_2 by pumping.

RESULTS

A level scheme of the vibrational modes of H_2S, HDS, and D_2S is shown in Fig. 2 . From the infrared spectra one ex-

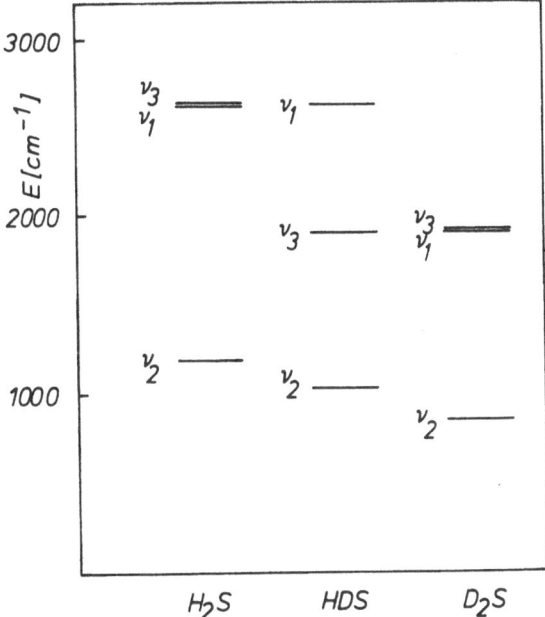

FIGURE 2 : Energy level diagram for H₂S, HDS, and D₂S. The vibrational frequencies were taken from ref.7.

pects that the bending mode ν_2 in D_2S can be excited with a CO_2 laser in the region 10 /um - 11 /um [5]. Photoacoustic spectra recorded for D_2S and a 1:1 mixture of H_2S and D_2S after equilibration indicate, that an interference free excitation of D_2S is possible at 10,7 /um. At smaller laser wavelengths vibrational excitation of HDS occurs and may interfere with D_2S absorption [6]. H_2S does not absorb in this spectral region and, therefore, excitation at 10,7 /um was selected to monitor the D_2S concentration in H_2S-D_2S mixtures by comparing the corresponding photoacoustic signal of the D_2S sample and the mixture under investigation.

A simple analysis of the acoustic signal is to determine the amplitude of the first peak. More reliable results are obtained, however, by integrating a certain number of peaks of the microphone signal. In the present work the peaks corresponding to the first two pressure pulses were integrated. In this way pressure effects which may arise

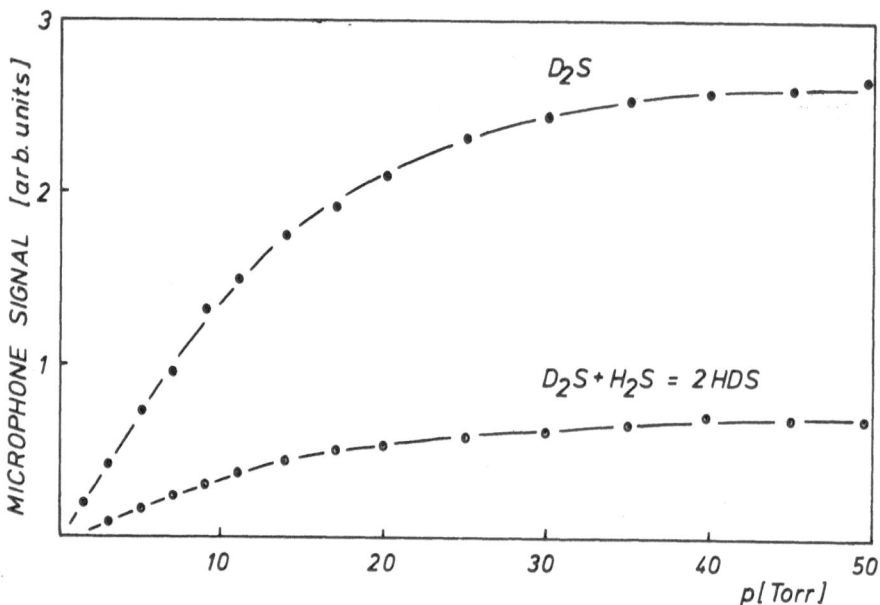

FIGURE 3 : Normalized, integrated microphone signals
for D_2S (97 atom % D) ● and an equilibrated 1:1 mixture of H_2S and D_2S ⊙ as a function of pressure. In
all experiments D_2S was excited at $10,7 \mu m$.

from window and wall absorption as well as reflections
can be minimized. The integration procedure leads to a
smaller scattering of the data than the amplitude analysis and to a somewhat higher value of the equilibrium
constant for the system H_2S-D_2S.

In Fig. 3 the result of a series of photoacoustic measurements is presented for D_2S (97 atom % D) and a 1:1 mixture of H_2S and D_2S after equilibration exciting at 10,7
μm. Each point was obtained by averaging and normalizing
the integrated signal of about ten laser shots. Despite
the fact that we find no linear variation of the microphone signal with pressure, the photoacoustic signal observed for the equilibrated 1:1 mixture is always about one
fourth of the D_2S signal in the pressure range studied.

If it is true that an interference free excitation of D_2S
is possible at $10,7 \mu m$, an accurate value of the equilibrium constant can be determined from these measurements.

The analysis of the data shown in Fig. 3 yields a value
of 3,8 $\pm$ 0,4 for the H-D exchange reaction (1).

DISCUSSION

Hydrogen-deuterium exchange has been extensively studied
in systems as H_2-D_2 and H_2O-D_2O, experimentally and theoretically. For the system H_2S-D_2S only one direct measurement of the equilibrium constant is known to us, using
mass spectroscopy[8]. The value obtained by this method is
3,88 $\pm$ 0,03 at 297 K . According to the rule of the geometric mean which assumes a random distribution of the
hydrogenic species between the molecules, a value of four
is expected. A calculation of the equilibrium constant
from spectroscopic data reported in ref.[8] yielded the
somewhat smaller theoretical value of 3,92 for the system
H_2S-D_2S. The result of 3,8 $\pm$ 0,4 found by photoacoustic
detection of the D_2S concentration is in good agreement
with theory and mass spectroscopic analysis although the
deviations from the mean value are much larger than for
the latter method.

The main sources of error employing the photoacoustic
method are 1. Window and wall effects: Heating of the
cell windows and walls and desorption of adsorbed species from these surfaces may influence the acoustic signal. These effects are at least partly eliminated in the
present experiments by comparing signals obtained under
similar conditions and analysing only the first two pressure pulses. 2. Kinetic effects: In a H_2S-D_2S mixture
the D_2S (H_2S) concentration decreases due to H-D exchange
until equilibrium is reached. This kinetic process can be
monitored applying a pulse to pulse detection. In preliminary kinetic experiments it was found that equilibrium
was established after about ten minutes in the 1:1 mixture. Therefore, the measurement of equilibrium concentrations was started after a reaction time of about 20
minutes. Details of the kinetic experiments will be re-

ported elsewhere[9]. 4. Surface reactions: During mixing
and reaction time, H-D exchange also occurs with adsorbed
molecules (e. g. H_2O) on the surfaces of the photoacoustic cell and the gas handling system. This heterogeneous
exchange process may be even faster than the gas phase
reaction and can cause systematic errors.

The list of possible errors involved in photoacoustic detection explains the large error limits. Nevertheless, an
accurate determination of equilibrium constants is possible with the photoacoustic method described here, if the
signal of several laser pulses is averaged and the measurements are extended over a large pressure range. An advantage of photoacoustic detection is the possibility to
also study the kinetics of the isotope exchange reaction
in the system H_2S-D_2S for the first time.

ACKNOWLEDGEMENTS

We are grateful to the Deutsche Forschungsgemeinschaft
and the Fonds der Chemischen Industrie for research
support.

REFERENCES

1   M. Veingerov, Gas analysis, Dokl. Akad. Nauk. SSSR 19
    687 (1938)

2   E.L. Kerr and J.G. Atwood, Laser illuminated absorp-
    tivity spectrophone, Appl.Opt. 7 915 (1968)

3   P.C. Claspy in Optoacoustic spectroscopy and detection
    ed. Yoh-Han Pao, p. 133, Academic Press, New York (1977)

4   R. Kadibelban, R. Ahrens-Botzong, and P. Hess, Vibra-
    tional relaxation of CF_4 at low temperatures, Chem.Phys.
    Lett. 46 563 (1977)

5   R.E. Miller and D.F. Eggers Jr., Analysis of the ν_1
    and ν_2 fundamentals in deuterium sulfide, J. Chem.
    Phys. 45 3028 (1966)

6   A.H. Nielsen and H.H. Nielsen, The infrared absorp-
    tion spectrum of the deuterium sulfides, J. Chem.
    Phys. $\underline{5}$ 277 (1937)

7   J.W. Nibler and G.C. Pimentel, Force constant display
    of unsymmetric molecular isotopes of H_2O, H_2S, H_2Se,
    and HCCH, J.Mol.Spectrosc. $\underline{26}$ 294 (1968)

8   J.W. Pyper and R.S. Newbury, Hydrogen-deuterium self-
    exchange in hydrogen sulfide and hydrogen selenide as
    studied with a pulsed-molecular-beam quadrupole mass
    filter, J.Chem.Phys. $\underline{52}$ 1966 (1970)

9   R. Kadibelban and P. Hess, to be published.

PA-METHOD MONITORS EXHAUST GAS CONCENTRATIONS

K.Stephan, W. Hurdelbrink
Institut für Technische Thermodynamik
und Thermische Verfahrenstechnik
Universität Stuttgart
Pfaffenwaldring 9
D 7000 Stuttgart 80

ABSTRACT

For concentration measurements of pollutants in hot flames
and in the exhaust gas of a coal/oil power plant a photoacoustic gas analyzer was developed. Slowly continuous
tuning of the IR-diode laser line over the absorption lines
of the contaminants in the air-tight, nonresonant (100 mm)
PA cell and chopping the laser CW radiation with constant
frequency produces a wavenumber- and concentration dependent signal. By correlating these data with the calibration
data of the pure components it is possible to detect and
identify the different molecules of components such as
CO_2, CO, NO_x, SO_x, H_2S, HCN, HCl and Hydrocarbons (e.g. 3.4
Benzpyrene) and to measure their concentrations in the ppm-
ppb-range.

INTRODUCTION

For pollutant analysis in exhaust gas a convenient measuring
technique is the photoacoustic effect discovered by Alexander Graham Bell (1) 1880. The gas sample is induced into an
air-tight cell and is irradiated with infrared light which
is periodically intensity modulated. As energy is absorbed
by the gas, the temperature changes periodically and inso-
far also the pressure in the cell. The pressure fluctuation
is detected by a microphone in the wall and converted to an
electric signal.
    As mentioned above this technique is not a new one. As
during Bell's time sensitive microphones did not exist, the
effect was soon forgotten. It was not until 1938 that the
effect was rediscovered by K.F. Luft (2) who developed then
his ultrared-absorption system (URAS), a nondispersive instrument, to analyze gases in the near infrared. Nowadays
not only sensitive condensor microphones are available, but
also phase sensitive amplifiers which allow measuring
strongly noise-covered signals. The situation for IR spectroscopy of molecules changed completely when infrared
wavelength tunable laser sources were available. One of the
first important works in this field is that of L.B. Kreuzer (3), in 1971. In the same year the first work on photoacoustic exhaust gas analysis was published: L.B. Kreuzer

and C.K.N. Patel (4) developed a photoacoustic system to measure the air pollution caused by nitric oxide.

The object of our research is to develop a concentration measuring system which allows to obtain concentration data for the layout of combustion chambers. Another research is to find methods and techniques to minimize the emission of pollutants as carbon-, nitric- and sulfur oxid as well as for organic gases. In addition because of the toxicity (e.g. carcinogenesis) these air pollutants have specified maximum safe levels and must be monitored.

METHODS

I. The photoacoustic effect

Irradiating solid, liquid or gaseous substances with coherent, periodically intensity modulated light of a particular wavelength causes pressure fluctuations when energy from the light is absorbed. The fluctuation amplitude depends on the absorption coefficient of the sample, and the frequency is the same as the modulation frequency of the light. As a detector for the pressure fluctuations a very sensitive condensor microphone is used. The conversion of the irradiation energy into an acoustic signal is called photoacoustic effect.

The exact process for a gas sample is as follows: Absorbing power from the radiation converts the molecules from the ground state into an excited state. In order to reach the stable ground state, an excited molecule has three channels open (5):

1.) It may fluoresce, removing the absorbed energy from the cell.
2.) It may activate a chemical reaction.
3.) or it may lose the absorbed energy by a radiationless process. Collisional energy transfer increases the translation-, rotation- and vibration energy of the surrounding molecules. When thermal equilibrium is reached the energy is split equally among the degrees of freedom. The increasing of the translation energy means higher temperature of the gas, or as the gas density in the cell is constant, an increasing of the gas pressure. Equilibrium is reached after about 10^{-5}s. When the irradiation is periodically intensity modulated, pressure fluctuations occur and therefore at the output of the microphone an AC signal can be detected.

The photoacoustic method has some advantages compared to other methods (e.g. spectrophotometry):

1.) The signal is proportional to the irradiation energy. The photoacoustic method makes it possible to measure directly the amount of energy absorbed by the molecules of the sample, because in the infrared almost all absorbed energy is converted into heat by the third process of collisional energy transfer. Therefore the photoacoustic measurement is far more exact and sensitive than the results of a spectrophotometer which is a differential instrument.

2.) The microphone response is totally independent of
    the radiation wavelength, whereas a photodetector has
    a different spectral response.
3.) Substances of any state of matter can be analyzed
    (even powder- or pastelike substances) without destroy-
    ing their structure. This opens up a large field of
    different applications for the photoacoustic method.

II. Pollutant analysis of exhaust gas

Irradiating a sample with light of a particular wavelength
and measuring the absorbed energy as the wavelength is tuned
continuously produces the characteristic spectral absorption
profile of the sample. Doing this one finds that most of the
gases have absorption lines in the infrared range from 2 to
15 µm. Here every gas has its so-called fingerprint. The only
exceptions are the diatomic homopolar molecules such as O_2
or N_2 and those cases when no changing of the dipole moment
of the molecule occurs (e.g. CO_2 ν_1).
    The infrared spectral analysis of exhaust gas causes
problems when CO_2 and H_2O are present in the mixture. At
atmospheric pressure the absorption lines overlap due to
pressure broadening, wheras at low pressure (< 5 Torr) the
line width at half-value is Doppler limited and 10^{-3} cm^{-1}
narrow. The different lines are no longer overlapped but
discret, and hence accurate concentration measurements of
multicomponent mixtures are possible. But the instrument for
this purpose must have a spectral resolution better than
10^{-3} cm^{-1}. After increasing the sensitivity by applying the
photoacoustic method (6÷11) the adequate radiation source
for the IR spectral analysis is a laser, increasing again
both sensitivity and spectral resolution.

A. The IR laser radiation source

A suitable radiation source for IR spectroscopy of molecules
1) covers the spectral range 2 ÷ 15 µm
2) has a line width < 0.001 cm^{-1}
3) has a high spectral irradiance
4) is continuously tunable over several wavenumbers
5) is easy to handle

The radiation source we chose for our experiments and
measurements is an IR diode laser (12,13). This laser
meets the conditions stated above in an almost perfect
way: Diode lasers cover the spectral range 2.8 ÷ 30.3 cm^{-1}.
The linewidth is 10^{-4} cm^{-1} and the multimode power is 4 mW.
That is equal to a spectral irradiance of $4 \cdot 10^4$ W/cm^{2}nm.
By means of the diode current one mode is continuously
tunable over some wavenumbers before mode hop occurs, as
the gain profile and the resonator modes are not synchronously shifted. The total tuning range of the laser by
means of the temperature is about 100 cm^{-1}. Diode lasers
nowadays are easy to handle. They are mounted onto the cold-
head of a cryo-cooler and operated at temperatures 12÷92 K.
Temperature is controlled better than $5 \cdot 10^{-4}$ K with a silicon
diode sensor. Controlling the laser temperatur is essential.

Operating the laser the output power of the intense IR radiation is split among three or four modes. As we need only radiation of one particular wavelength the modes have to be selected. This is achieved by means of a Czerny-Turner 0.5 m monochromator. Because the photoacoustic signal has to be normalized due to intensity variations of the laser the intensity is measured with a liquid nitrogen cooled IR-detector. An important advantage of the diode lasers is their high tuning rates over a small spectral range of $1 \div 2$ cm-1. Up to 10^4 repetitions /s can be obtained. As the linewidth of a mode is so narrow PA derivative spectroscopy can be done, which on its part has some very useful advantages of PA intensity modulated spectroscopy (10,14).

B. The nonresonant PA-cell

The most important part of the PA spectrometer is the photoacoustic detector, the cell. About optimal photoacoustic detector design (resonant or nonresonant type cells) the results of several authors were published ($15 \div 19$).
     The design of our detector was based on the work of L.B. Kreuzer (16). The cell is 102.5 mm long and 1.6 mm in diameter. It is a nonresonant type cell. The length of the cell is chosen in that way that it equals the confocal parameter of the laser beam.

$$L = b = 2\pi w_0^2 / \lambda \qquad (1)$$

with L cell length, b confocal parameter, w_0 beam radius in the middle of the cell length and λ irradiation wavelength. Choosing L in this way the laser beam is well focussed (20). Kreuzer (16) shows that the signal/noise-ratio increases with decreasing cell diameter. On the other hand the adiabatic condition for sound generation must be met:

$$\lambda_s \gg 2 \cdot 10^{-3} \, \kappa / (c_s \, \rho \, C_v) \qquad (2)$$

with c_s speed of sound, λ_s wavelength of sound and the gas properties κ thermal conductivity, ρ gas density and C_v specific heat at constant volume.

In addition the cell diameter must be greater than the thermal diffusion length μ of the sample gas:

$$d > \mu = \sqrt{0.001 \, \kappa / (\pi \rho C_v \omega)} \qquad (3)$$

with $\omega = 2\pi f$ s^{-1} chopper frequency.
The longest drill with a small diameter that was available was an "aircraft extension drill", 6" long and 1/16" ϕ. In this way both conditions are met.
The photoacoustic cell is obtained by drilling a hole into a massiv copper block. At the ends of the cell inlet and outlet valves are located. In the middle a 1" microphone with pressure equalization to the backside of the diaphragm is coupled to the cell cavity. By means of a static pressure sensor and a resistance temperature meter filling pressure and temperature are controlled and monitored. Each connexion to the cell is performed in such a way that the resulting

dead volume is a minimum. In order to achieve minimum intensity losses for the laser beam when it is focussed into the
cell the ends of the detector are closed by KRS-5 brewster
windows. Using Kreuzer's (16) method of detector optimization the minimum detectable power of our cell is:

$$NEP = 3.1 \cdot 10^{-13} \ W/\sqrt{Hz} \qquad (4)$$

C. Measuring system and concentration measurements

The laser modes are selected in the monochromator. Hence only
radiation of one mode appears at the outlet slits. The radiation is chopped by a perpendicular coated mirror and focussed
into the cell. After exciting the test sample the beam leaves
the cell through the exit brewster window. For a concentration measurement the laser line is slowly continuously tuned
over the absorption line of interest. The pressure signal is
detected by the microphone and measured by a lock-in analyzer.
Synchronously the laser intensity that reaches the IR photo-
detector via the mirror chopper is measured by a second lock-
in voltmeter. Both signals are digitized and stored. Pressure
and temperature are controlled all the time.
     For the calibration of the system calibration gases of
high purity are available. The photoacoustic detector is connected to a high vacuum (10^{-10}mbar) system as well as to a
gas filling station. Calibration data for the pure components
are obtained in the same way as mentioned above. These data
are stored on a hard disk containing the spectra library.
By correlating the analysis data with the corresponding data
of the pure components identification of the different molecules of a gas mixture and concentration measurements are done
very fast.

CONCLUSION

The great advantage of the photoacoustic IR spectroscopy is
due to two facts:

  1.) The main process for an excited molecule to reach
      its ground state is collisional energy transfer,
      rather than fluorescence or activation of chemical
      reactions.
  2.) The amount of absorbed energy can be measured di-
      rectly. At linear absorption conditions this amount
      is proportional to the density of the molecules and
      to the incident spectral irradiance.

By combining continuously tunable IR lasers such as diode
lasers with the photoacoustic measuring technique sensitivity and spectral resolution are increased several orders of
magnitude compared to a spectrophotometer with a continuum
IR source and monochromator. Mimimum detectable concentrations easily reach the ppb-range provided that the PA detector is well designed and signal/noise-ratio is optimized.
For our PA spectrometer we found the diode laser to meet each
of the particular conditions for the amplitude modulation
method as well as for the wavelength modulation method (deri-

vative spectroscopy). With a computer controlled data acqui-
sition system the pressure signal is obtained, and a short
time after that the correlation with the library data of the
pure components gives the concentration values for each component.
   At present the PA detector has to be mounted onto an
optical bench and the laser beam has to be adjusted. So we
hope to obtain the first results with our PA system soon.

REFERENCES

(1)   A.G. Bell, Phil. Mag., 11, 510 (1881)

(2)   K.F. Luft, Zeitschr. Tech. Phys., 24, 97 (1943)

(3)   L.B. Kreuzer, Ultralow Gas Concentration Infrared
      Absorption Spectroscopy, J. Appl. Phys., 42, 2934 (1971)

(4)   L.B. Kreuzer, C.K. Patel, Nitric Oxide Air Pollution:
      Detection by Optoacoustic Spectroscopy, Science, 173,
      45 (1971)

(5)   W.R. Harshbarger, M.B. Robin, The opto-acoustic effect:
      Revival of an old Technique for Molecular Spectroscopy,
      Acc. Chem. Res., 6, 329 (1973)

(6)   C.F. Dewey, R.D. Kamm, C.E. Hackett, Acoustic amplifier
      for detection of atmospheric pollutants, Appl. Phys.
      Lett., 23, 633 (1973)

(7)   L.B. Kreuzer, Laser Optoacoustic Spectroscopy-A New
      Technique of Gas Analysis, Anal. Chem., 46, 241 (1974)

(8)   J. Gelbwachs, Limitation of Optoacoustic Detection
      of Atmospheric Gases by Water Vapor Absorption,
      Appl. Opt., 13, 1005 (1974)

(9)   P.D. Goldan, K. Goto, An acoustically resonant system
      for detection of low level infrared absorption in
      atmospheric pollutants, J. Appl. Phys., 45, 4350 (1974)

(10) C.K. Patel, Optoacoustic spectroscopy applied to the
      detection of gaseous pollutants, ACS Symp. Ser., 94
      (Monit. Tox. Subst.), 177 (1979)

(11) P.C. Claspy, Optoacoustic Spectroscopy and Detection,
      Y.H. Pao, 134, AP New York (1977)

(12) E.D. Hinkley, Tunable infrared lasers and their appli-
      cations to air pollution measurements, Opto-Electr.,
      4, 69 (1972)

(13) K.W. Nill, Tunable Infrared Lasers, Opt. Eng., 13,
      516 (1974)

(14) C.F. Dewey, Optoacoustic Spectroscopy and Detection,
      Y.H. Pao, 65, AP New York (1977)

(15) C.F. Dewey, Optoacoustic Spectroscopy, Opt. Eng., 13, 483 (1974)

(16) L.B. Kreuzer, Optoacoustic Spectroscopy and Detection, Y.H. Pao, 1, AP New York (1977)

(17) L.G. Rosengren, Optimal optoacoustic detector design, Appl. Opt., 14, 1960 (1975)

(18) C.K. Patel, R.J. Kerl, A new optoacoustic cell with improved performance, Appl. Phys. Lett., 30, 578 (1977)

(19) E. Nodov, Optimization of resonant cell design for optoacoustic gas spectroscopy (H-type), Appl. Opt. 17, 1110 (1978)

(20) H. Kogelnik, T. Li, Laser Beams and Resonators, Appl. Opt., 5, 1550 (1966)

(21) N.C. Fernelius, Helmholtz Resonance Effects in Photoacoustic Cells, Appl. Opt., 18, 1784 (1979)

APPLICATION OF PAS ON THE LOCAL AND SPECTRAL

IDENTIFICATION OF ADSORBED MOLECULES

S. Schneider and U. Möller

Institut für Physikalische und Theoretische Chemie
Techn. Universität München, D 8046 Garching, FRG

H. Coufal

Physik Department E 13, Technische Universität München
D 8046 Garching, FRG

A B S T R A C T

It is demonstrated that photoacoustic spectroscopy (PAS)
is a valuable tool in analysing the local distribution
of dyes adsorbed on thin layer chromatography (TLC) plates
or proteins bound to ultrathin layers of polyacrylamide
gels used for isoelectric focusing (PAGE-IEF substrates).
PA-spectra taken at different locations can help to identify the adsorbate by comparison with PA-spectra of
possible candidates adsorbed on the same substrate.

INTRODUCTION

Among other potential applications of photoacoustic spectroscopy proposed in the past, spectroscopy of adsorbed
species has been tested in a crude, purely qualitative
manner [1]. The problems which arise, when quantitative
measurements are required, are twofold:

(i)   the appearance of the spectra is to a large extent
      determined by the particle size as is well known
      from reflexion spectroscopy [2]. Differences in the
      procedure of preparation may therefore cause pro-
      nounced changes in the spectra.

(ii) due to specific interactions between surface and
      adsorbate the rates for the deactivation of the ex-
      cited state can be altered drastically (e.g. the
      fluorescence quantum yield) which in turns causes
      also changes in the photoacoustic spectra as has been
      demonstrated in an earlier paper [3].

In this report, we therefore will describe an application which in part avoids the above mentioned problems in as far as the goal of the study is a mapping of the local distribution of different dyes adsorbed on thin layer chromatography plates (TLC-plates) or ultrathin layers of polyacrylamide gels used for isoelectric focusing [4] (PAGE-IEF substrates). Because of the high local resolution, PA-spectra taken at various locations can help to identify the species by comparison with PA-spectra of possible candidates adsorbed on the same substrate. An easy and rapid way of sanning both types of substrates without further preparative work seems interesting to us because it allows to find faster the proper solvent, which separates the components (TLC) or to map the distribution without staining the proteins when uv-light is applied (PAGE-IEF).

EXPERIMENTAL

Figure 1 displays a schematic of the experimental set up used for this study. Its components are described in more detail in an other contribution in this volume (reference 3). The major change performed is the introduction of a beam steering optics which allows to scan the laser beam across the sample. It consists of a prism mounted on a translation stage which in turn is driven by a computer

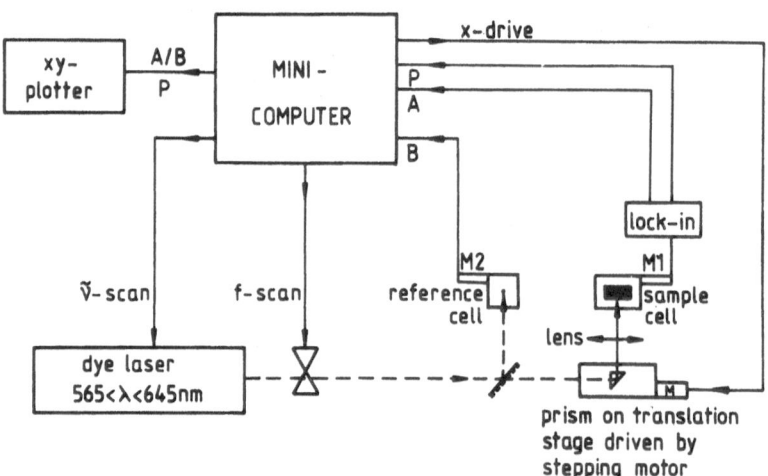

FIGURE 1   Schematic of experimental set up.

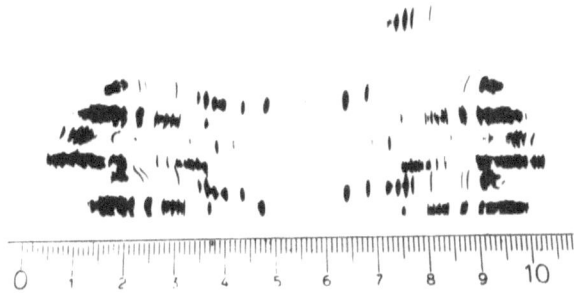

FIGURE 2

Photograph of
PAGE-IEF substrate
with separated
proteins stained by
"Serva Blue"

controlled stepping motor. The local resolution can be
increased by focusing the laser beam onto the sample, but
usually one has to find a compromise because of the finite
photochemical stability of most adsorbed dyes.

The protein samples bound to PAGE-IEF substrates
were provided by courtesy of Dr. Köst (Biol. Institut,
Universität München). Figure 2 presents a photograph of a
whole substrate with a great number of resolved proteins
(true seize is ca. 10 x 3.5 cm). The smaller fraction is
the actual sample used for the scanning experiment. At
present the cell described in reference 5 is used, but
work is in progress to build a new cell capable of housing
substrates of larger size For the same reason, the dyes
under study were adsorbed from ethanolic solution onto
fragments of TLC-plates supplied by Merck & Co.

RESULTS

The PA-spectra of three different dyes adsorbed on TLC-
plates are displayed in figure 3. Curve a originates from
the widely used mode-locking dye DODCI (3,3'-diethyldicar-
bocyanine iodide). In an ethylene-glycol solution, the
maximum of the PA-spectrum is around 595 nm in fair agreement with the optical absorption maximum[3]. The diffuse
reflection spectrum of DODCI adsorbed on silica, on the
other hand, has a minimum around 560 nm[6]; adsorbed on SiO_2,
the minimum is shifted to 530 nm. Since adsorbed DODCI is
photochemically very unstable, additional measurements
were performed with the reasonable stable triphenylme-

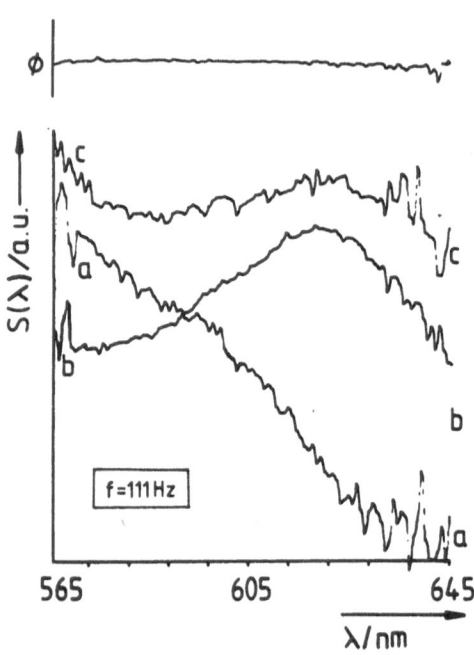

FIGURE 3
PA-spectra of DODCI (a),
malachite green (b)
and crystal violet (c)
adsorbed on TLC-plates
(f = 111 Hz).

thane dyes malachite green (curve b) and crystal violet
(curve c). In ethanol solution these dyes show maxima
of the longwavelength absorption band around λ = 620 nm
(malachite green) and λ = 590 nm (crystal violet), respectively. Although adsorbed on the same substrate, this
difference in the absorption maxima disappears in the
PA-spectra. The fact that the absorption band of crystal
violet has a larger fwhm is, on the other hand, well docu-
mented in the PA-spectra.

Figure 4 shows the modulation frequency dependence of
both amplitude and phase of the PA-signal (a: malachite
green, b: crystal violet). Since for very low frequencies
(f $\leq$ 20 Hz) saturation effects must be expected, mapping
experiments should be performed with frequencies above
100 Hz.

The result of a mapping experiment using light of
different wavelength for excitation is seen in figure 5.
The sample is a piece of a TLC-plate onto which two drops
of a solution of DODCI and malachite green, respectively,
were pipetted. The verticale scale is different for each

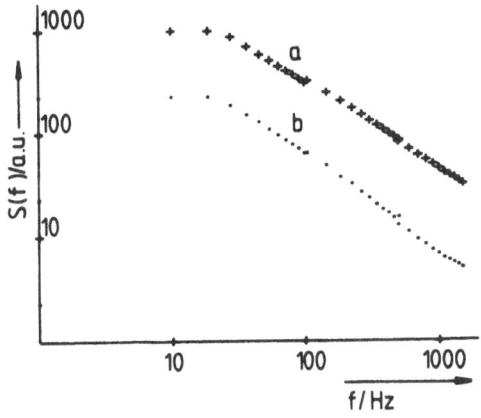

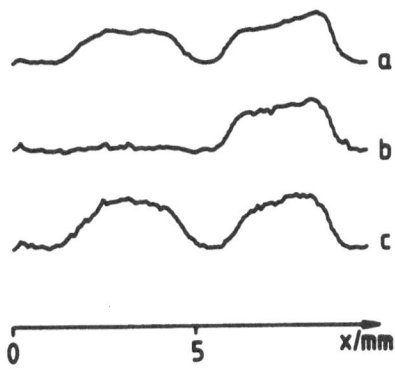

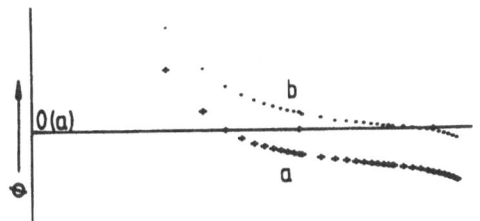

FIGURE 5   Mapping of dye
absorbance at various probing
wavelengths: λ = 605 nm (a),
λ = 640 nm (b) and
λ = 570 nm   (c).   Adsorbed
dyes are DODCI and malachite
green (f = 110 Hz)

FIGURE 4   Frequency dependence
of PA-signal of malachite green (a)
and crystal violet (b) adsorbed on
TLC-plates. (λ = 620 nm)

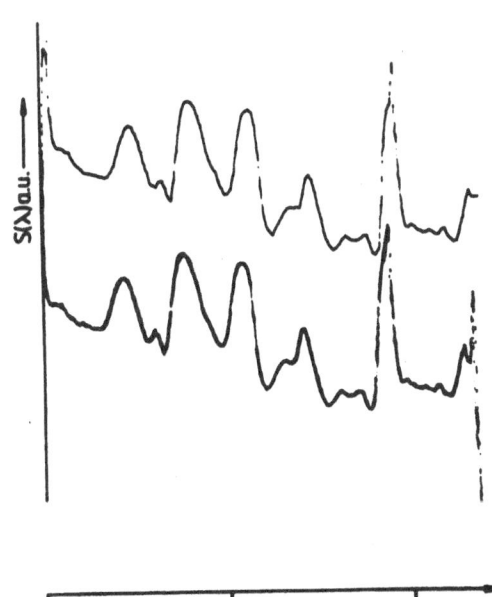

FIGURE 6
Mapping of the blue-stained
PAGE-IEF substrate which is
shown in fig. 2.
(f = 225 Hz; λ = 605nm)

trace since the output of the laser drops of course
drastically towards the ends of the tuning range.

Analogous mapping experiments have been performed
with two colored biliproteins i.e. Phycoerythrin and
Phycocyanine adsorbed on PAGE-substrates. It was found that
the chromophores bound to proteins can take a rather high
illumination level without deterioration.

In figure 6 finally the mapping of the blue-stained
sample described above (figure 2) is displayed. The bottom
trace is a superposition of two consecutive runs aimed to
demonstrate the accuracy achieved with respect to both
local resolution as well as signal amplitude.

DISCUSSION

The PA-spectra displayed for the three different dyes
demonstrate clearly that an identification of adsorbed
molecules by their PA-spectrum alone is nearly impossible.
Spectra taken from different spots on the same substrate,
however are very much alike. An identification by comparison with test spectra of compounds with the assumed chemical structure is therefore feasible.

As mentioned above, an important problem in the application of thin layer chromatography is to find the right
solvent, or with other words, to decide, wether a colored
zone stems from one species only or from several unresolved
compounds. Scanning this zone repeatedly with light of
different color reveals without fail the existence of more
than one species. In case of figure 5, it is easy to derive
that the drop of solution on the left contained DODCI
whilst the signal on the right hand side originates from
malachite green.

Another important advantage of an evaluation of TLC-
plates by using   photoacoustic spectroscopy is the fact
that the separated fractions must not be dissolved again.
Very often, there is an equilibrium established between
different isomeres in solution, i.e. different zones on

the TLC-plates can lead to the same absorption spectra in solution.

The evaluation of PAGE-IEF-substrates by means of photoacoustic spectroscopy has similar advantages. A rough classification of the adsorbed proteins can be achieved easily by scanning the unstained substrates with light of different wavelength (in this case a conventional setup with Xe-lamp and monochromator should be used). In contrast to the use of a densitometer, no restrictions are placed upon the choice of the applied light wavelength by the transmission and / or scattering properties of the substrate. Furthermore, the sensitivity of the method will become independent of the staining efficiency. Staining is necessary mainly to increase the contribution of the protein to the optical density at the probing wavelength above the one of the substrate. As a "surface" method, photoacoustic spectroscopy will see the top layer only and therefore recommend itself for samples with low optical density. Once the location of the protein to be identified is found by a scan at the appropriate wavelength, the PA-spectrum can be recorded for comparison.

As has been mentioned above, work is in progress to build a new photoacoustic cell which allows the scanning of somewhat larger substrates by means of a "broad-band" PA-spectrometer using a Xe-lamp and a monochromator as source for the exciting light. At the same time we pursue a proposal by Radola [4] to reduce the seize of the substrate needed for isoelectric focusing in order to develop a true micromethod. The results will be published elsewhere.

REFERENCES

1) For a review see: A. Rosencwaig
   Photoacoustic spectroscopy
   Adv. Electronics and Electron Physics <u>46</u>, 208 (1979)

2) For a review see: G. Kortüm
   Reflexionsspektroskopie, Grundlagen, Methodik
   und Anwendungen
   Springer, Berlin (1969)

3) S. Schneider, U. Möller, H. Coufal
   Photoisomerisation of DODCI studied by
   Photoacoustic spectroscopy. This volume

4) B.J. Radola
   Ultrathin-layer isoelectric focusing in 50-100 μm
   polyacrylamide gels on silanized glass plates or
   polyester films
   Electrophoresis 43 (1980)

5) H. Coufal, U. Möller, S. Schneider
   Photoacoustic cells for measurements at various
   temperatures and pressures - design and characterisation - . This volume

6) The diffuse reflection spectra were recorded for us
   by Prof. Knözinger, Universität München.

ACKNOWLEDGMENT

Financial support by "Deutsche Forschungsgemeinschaft"
and "Fonds der Chemischen Industrie" is   gratefully
acknowledged.

The authors thank Dr. H.P. Köst for providing the protein
samples on PAGE-IEF substrates.

4. Thermal Applications

OPTICAL ABSORPTION COEFFICIENT AND THERMAL

PROPERTIES OF LIQUIDS,MEASURED BY PHOTOACOUSTIC

SPECTROSCOPY

P.POULET, J. CHAMBRON

  Institut de Physique Biologique - FACULTE DE MEDECINE -

  4, rue Kirschleger - 67085 STRASBOURG CEDEX - FRANCE

R. UNTERREINER

  Département Génie Electrique - Institut National des

  Sciences Appliquées - 20 Avenue Albert Einstein -

  69621 VILLEURBANNE CEDEX - FRANCE

ABSTRACT

The application of Rosencwaig and Gersho's theory (RG) to thermally thick samples allows the measurement of two thermal properties - diffusivity and effusivity -, and of the optical absorption coefficient of liquids.
At a wavelength at which optical absorption can be measured by
conventional methods, the study of the amplitude of the photoacoustic signal as a dependent variable of the modulation frequency, or simultaneous measurement of phase and amplitude at
a given frequency is sufficient to obtain the thermal properties and to calculate the optical absorption spectrum from the
photoacoustic spectrum.
We shall illustrate the application of the RG theory to the
measuring of the optical absorption spectra of various dyes in
different solvents, and of the thermal properties of these
solvents.
We shall also demonstrate the concordance of the experimental
results with that part of the theoretical expression of the
photoacoustic signal which is a dependent variable of the modulation frequency. This part is used for measuring thermal
diffusivity and optical absorption. On the other hand, a disagreement appears concerning the part which is independant

of frequency – the signal at saturation – which seems not to
be inversely proportional to the thermal effusivity, as was
predicted by the RG theory.

INTRODUCTION

When applied to thermally thick samples, the Rosencwaig-
Gersho theory (RG)[1] of the photoacoustic effect gives simple
expressions of the amplitude and phase of the photoacoustic
signal.[2] These expressions enable the optical absorption coefficient of the liquids to be measured, using either the
phase [2,3,4] or the amplitude [2,4,5] of the acoustic signal.

Furthermore, the amplitude must allow measurement of the
thermal diffusivity and effusivity, and therefore of the heat
capacity and thermal conductivity.

We shall demonstrate that the amplitude of the acoustic
signal is sufficient for measuring the thermal and optical properties. Simultaneous measurement of phase enables these properties to be examined at a single modulation frequency.

THEORETICAL CONSIDERATIONS

The expressions of the phase and amplitude of the photoacoustic signals produced by thermally thick samples are derived from the RG theory[2], and are given by :

$$\text{tg}\varphi = \beta\mu + 1 \qquad\qquad (1)$$

$$\Delta P\omega = \frac{AE_0}{\sqrt{k\rho c}} \cdot \frac{\beta\mu}{\sqrt{(\beta\mu+1)^2+1}} \qquad\qquad (2)$$

The above parameters are defined by :
φ the phase, and ΔP the amplitude of the photoacoustic signal
ω the pulsation of light modulation, E_0 the energy of the incident light, β the optical absorption coefficient of the sample, $\mu = (2\alpha/\omega)^{1/2}$ the thermal diffusion length of the sample,
$\alpha = k/\rho c$ the thermal diffusivity, k the thermal conductivity,
ρ the density and c the heat capacity of the sample.

A is a constant which is independent of the sample, and
a dependent variable of the geometry of the measuring cell
and the gas used. As the value of A depends on the volume of
gas, it will vary with the volume and shape of the sample.

Relations (1) and (2) above show that the two parameters
to be measured are $\beta\mu$ and $\sqrt{k\rho c}$. The first of these ($\beta\mu$) can be

measured either by direct use of phase, or by studying variations in amplitude in relation to the modulation frequency.

The second parameter ($\sqrt{k\rho c}$) can either be measured directly by simultaneous use of phase and amplitude, or by studying
the variations in amplitude, the A coefficient having first
been measured with a reference sample of known thermal properties.

MATERIALS AND METHODS

The photoacoustic spectrometer was built in our own laboratory[6], and consists primarily of a xenon arc lamp, a mechanical chopper, a monochromator, an aluminium cell, a condenser microphone, and a two-phase lock-in amplifier. All the
samples studied are two-millimeter thick liquids, which are
thermally thick at the frequency used (20-200Hz).

The reference sample used consisted of black ink, and
produced a saturated photoacoustic signal (the product
$\Delta P\omega = A\ Eo/\sqrt{k\rho c}$ is independent of the frequency).

The measured phase of the photoacoustic signal is the
sum of the theoretical phase given by (1), and of an experimental phase essentially due to the finite dimensions of the
gas[7,8], and which is supposed to be independent of the sample's properties[2,3,4,8].

This experimental phase shift is measured by using the
reference sample (black ink) for which the theoretical phase
is equal to $-\pi/2$.

It must be pointed out that the experimental phase is
not totally independent of wavelenght of the incident light.
It must be measured throughout the whole spectral range of
interest.

RESULTS AND DISCUSSION

Two experimental situations were taken into consideration :

- measurement of the optical absorption spectrum of samples of known thermal properties
- measurement of the thermal properties of samples of
  known absorption coefficient at a given wavelenght.

The first case is illustrated by figure 1, which deals with the optical absorption spectrum of an aqueous solution of potassium dichromate. The thermal properties of this solution are considered to be those of water. The spectra presented are:

a) the absorption spectrum obtained by measuring the phase, after correction of the experimental phase at each wavelenght.

b) the absorption spectrum obtained from the amplitude. This spectrum is obtained by dividing the amplitude of the photoacoustic signal produced by the potassium dichromate solution $\Delta P_{(s)}$ at each wavelength by the amplitude of the photoacoustic signal produced by the reference sample (black ink) $\Delta P_{(R)}$. If the two samples have the same thermal properties, then the ratio is equal to $\frac{\Delta P_{(s)}}{\Delta P_{(R)}} = \frac{\beta\mu}{\sqrt{(\beta\mu+1)^2+1}}$. This was shown to be true by studying the variations of $\Delta P_{(s)} \cdot \omega$ versus ω : the measured limit of $\Delta P_{(s)} \cdot \omega$ when $\beta\mu$ becomes infinite (saturation value) is close to the product of $\Delta P_{(R)} \cdot \omega$ (39 and 42 mVs^{-1} respectively). This result demonstrates that black ink and potassium dichromate solution both have the same thermal effusivity $\sqrt{k\rho c}$.

c) the optical absorption spectrum measured with a Cary 118 spectrophotometer, using a cell 0.1 millimeter thick.

Comparaison of the three spectra shows good concordance of the results, but it must be pointed out that the spectrum obtained with phase is less accurate than the spectrum obtained with amplitude, especially at high absorption coefficients $(\varphi \to -\pi/2$ and $d(tg\ \varphi)/d\ \varphi \to \infty)$. Furthermore, the phase does not tend towards $-\pi/4$ as β tends towards zero, as predicted by equation (1). The photoacoustic signal produced by the surface vibrations of the sample itself[9] or slight illumination of the microphone, can produce this effect.

In addition to the existence of an experimental phase which is dependent on the wavelength used, these two remarks reduce the interest of the use of phase, which resides essentially in the fact that there is no need to study the frequency-dependence of the photoacoustic signal.

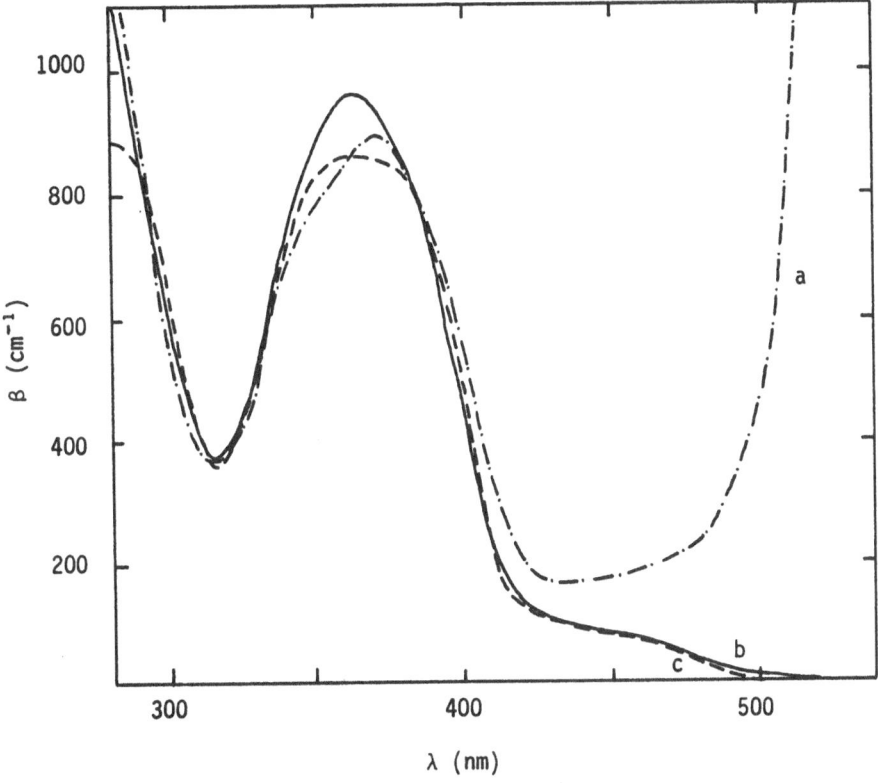

<u>Figure 1</u>

Optical absorption spectra of potassium dichromate in water, measured by photoacoustic spectroscopy, phase (a), amplitude (b), and by transmission spectroscopy (c).

In order to illustrate the second case (sample with known optical absorption at a given wavelength), a solution of methylene blue in methyl alcohol was used. The optical absorption coefficient of this solution, measured by conventional methods (Cary 118), is equal to 620 cm^{-1} at 600 nm.

A study of the variations of $\Delta P\omega$ in relation to the modulation frequency at 600 nm gives the following results :

- the saturation value of $\Delta P_{(s)}\omega$ = 119 mVs^{-1} is 4.1 times greater than the product of $\Delta P_{(R)}\omega$ at 600 nm – 29 mVs^{-1}.

- the frequency at which $\beta\mu$ = 1 is equal to 164 Hz.

The ratio of the saturation values is:

$$\frac{\Delta P_{(s)} \cdot \omega}{\Delta P_{(R)} \cdot \omega} = 4.1 = \frac{(\sqrt{k\rho c})R}{(\sqrt{k\rho c})S}$$

If the thermal effusivity of black ink is taken to be that of water, as shown with the aqueous solution of potassium dichromate : $\sqrt{k\rho c}$ = 37.9 x 10^{-3} cal/cm^{2}s$^{1/2}$ $^\circ$K, then the thermal effusivity of methyl alcohol must therefore be equal to 9.2 x 10^{-3} cal/cm^{2}s$^{1/2}$$^\circ$K (theoretical value 15.7 x 10^{-3}).

The thermal diffusivity of methyl alcohol is equal to 1.3 x 10^{-3} cm^2/s (theoretical value 1.1 x 10^{-3} cm^2/s).

The thermal conductivity k, calculated from these results, is 3.3 x 10^{-4} cal/cm s $^\circ$K (theoretical value 5.2 x 10^{-4}), and the heat capacity per unit volume ρc is 0.25 cal/cm^3 $^\circ$K (theoretical value 0.47).

These results call for some comments :

- the measured value of the thermal diffusivity is consistent with the known thermal diffusivity of methyl alcohol
- the measured value of the thermal effusivity is much lower than the predicted value. This is due to a saturation point which is higher than expected. The saturation value measured was 4.1 times that of an aqueous solution, instead of 2.4 times this saturation value, as predicted by the RG theory
- measurement of the diffusivity α and of the saturation value of $\Delta P_{(s)}\omega$ enables the optical absorption spectrum of a sample of unknown thermal properties to be calculated, as shown in figure 2.

The absorption spectrum measured by photoacoustic spectroscopy (a) of methylene blue in methyl alcohol is compared with the absorption spectrum measured by transmission spectroscopy (b). Spectrum (a) is obtained by dividing the signal produced by methylene blue by the signal produced by black ink at each wavelenght. This ratio is equal to :

$$\frac{\Delta P(s)}{\Delta P(R)} = 4.1 \ . \ \frac{\beta\mu}{\sqrt{(\beta\mu+1)^2+1}}$$

4.1 being the ratio of the saturation values as measured previously.

The concordance of the two spectra demonstrates the feasibility of measuring the optical absorption spectrum of a sample of unknown thermal properties by the use of photoacoustic spectroscopy.

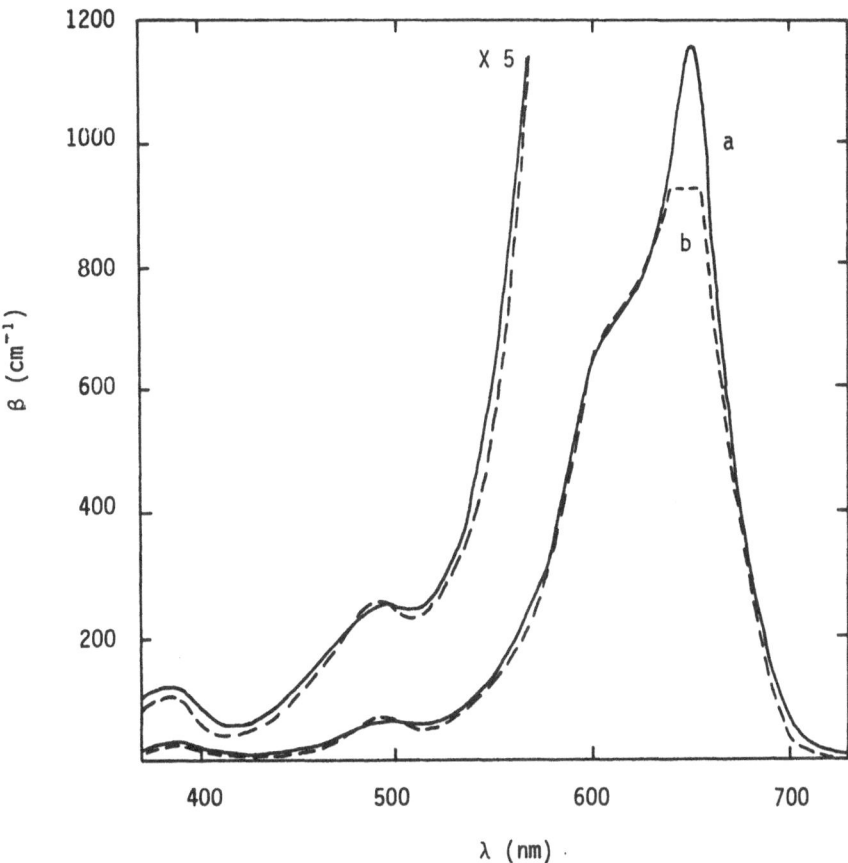

Figure 2

Optical absorption spectra of methylene blue in methyl alcohol, measured by photoacoustic spectroscopy, amplitude (a), and by transmission spectroscopy (b).

CONCLUSION

Rosencwaig and Gersho's theory can be used for measuring the optical absorption coefficient and the thermal diffusivity of thermally thick samples. The use of either the phase or the amplitude of the photoacoustic signal gives experimental results which are consistent with the expected values of β or α. Phase is less accurate than amplitude, and just as difficult to use, the latter appearing to be the best experimental parameter for measuring the optical and thermal properties of liquids.

It is essential to distinguish two separate parts in the expression of the photoacoustic signal obtained by the RG theory :

- The first part is frequency-dependent : $\beta\mu/\sqrt{(\beta\mu+1)^2+1}$.
  The concordance of the photoacoustic and transmission
  spectra and the measured value of the thermal diffusivity demonstrate that this part of the RG theory is
  verified by the experimental results obtained.
- The second part is independent of frequency :
  A $Eo/\sqrt{k\rho c}$, and corresponds to the saturation value of
  the photoacoustic signal.

The discrepancy observed between the measured and theoretical values of the thermal effusivity suggests that the photoacoustic signal is not inversely proportional to thermal effusivity.

REFERENCES

1) A. ROSENCWAIG, A. GERSHO,
   Theory of the photoacoustic effect with solids.
   J. Appl. Phys., 47, 64, 1976
2) P. POULET, J. CHAMBRON, R. UNTERREINER,
   Quantitative photoacoustic spectroscopy applied to ther-
   mally thick samples.
   J. Appl. Phys., 51, 1738, 1980
3) J.C. ROARK, R.A. PALMER, J.S. HUTCHISON,
   Quantitative absorption spectra via photoacoustic phase
   angle spectroscopy (ØAS)
   Chem. Phys. Lett., 60, 112, 1978
4) Y.C. TENG, B.S.H ROYCE,
   Absolute optical absorption coefficient measurements
   using photoacoustic spectroscopy amplitude and phase information.
   J. Opt. Soc. Am., 70, 557, 1980
5) S. MALKIN, D. CAHEN,
   Dependence of photoacoustic signal on optical absorption
   coefficients in liquids. (To be published).
6) P. POULET,
   Spectroscopie photoacoustique des milieux condensés
   opaques : théorie-expérimentations-applications.
   Thesis, Strasbourg, 1980.
7) L.C. AAMODT, J.C. MURPHY, J.G. PARKER,
   Size considerations in the design of cells for photoacoustic spectroscopy.
   J. Appl. Phys., 48, 927, 1977
8) R.S. QUIMBY, W.M. YEN,
   Three dimensional heat-flow effects in photoacoustic
   spectroscopy of solids.
   Appl. Phys. Lett., 35, 43, 1979
9) F.A. MAC DONALD, G.C. WETSEL Jr,
   Generalized theory of the photoacoustic effect.
   J. Appl. Phys., 49, 2313, 1978

THE PHOTOACOUSTIC EFFECT AT PHASE TRANSITIONS

P. Korpiun

Physik-Department, Technische Universität München

D-8046 Garching

ABSTRACT

The pressure response of the gas to the absorbed light is
determined essentially by the thermal properties of the
sample. At a first order phase transition, the latent heat
influences strongly the pressure signal. Recently, first
experimental observations of the PA at phase transitions
in various substances have been published. It has been
found that the amplitude of the pressure of the gas as a
function of temperature runs through a minimum in the transition region, whereas the phase angle shows different
patterns. For a formal description of the PAE in the transition region a "model of oscillating interface" has been
proposed. It is based on the assumption that the periodic
illumination of the sample creates a steady-state temperature gradient on which an oscillation of temperature is
superimposed. Therefore, there is a temperature region
around the transition temperature where the interface between both the thermodynamic phases oscillates around an
average position. It was found theoretically that the amplitude of the acoustic signal for samples of arbitrary
thickness becomes minimal when the mean temperature of the
sample surface is equal to the transition temperature. However, the phase-angle of the signal as a function of the
temperature depends strongly on the thermal thickness of the
sample. Experimental results of measurements on Ga, In, H_2O
K_2SnCl_6, $BaTiO_3$, VO_2, the metal hydrogen interstitial alloys
Ta $H_{0.5}$, Nb $H_{0.8}$, V $H_{0.517}$ and on DPPC vesicles with chlorophyll a are presented. They are qualitatively in fairly good
agreement with theory.

1. INTRODUCTION

The pressure signal measured in a gas-microphone cell is
determined by the optical absorption, the conversion to heat
and all quantities that govern the transport of heat from
the sample to the gas. These are the thermal conductivity
λ_k, the density ρ_k, the specific heat capacity c_{pk} of the
complete sample cell assembly, and the latent heat L of the
sample, Fig. 1a. The suffix k = 0 indicates the backing ma-

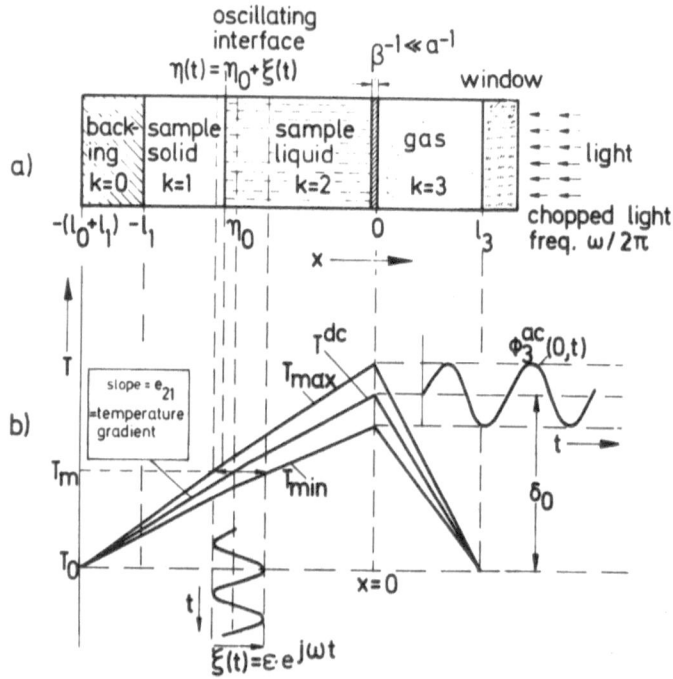

FIGURE 1   a) Schematic diagram of a gas-microphone PA-cell
           with a sample in two different thermo-dynamic
           states (solid and liquid).
           b) Temperature distribution in the cell sample
           assembly for optically opaque samples. Tempera-
           ture $T(x,t)$ oscillates between T_{min} and T_{max}
           around T^{dc}. T_m: transition temperature (melting
           temperature). η: position of the interface be-
           tween the two phases. $\xi(t)$: oscillation of the
           phase boundary (interface).

terial, $k = 1$ the low temperature phase, $k = 2$ the high temperature phase of the sample and $k = 3$ the gas. The measurement of the photoacoustic effect (PAE) at a first order phase
transition means the measurement of the pressure oscillation
of the gas as a function of the average temperature of the
photoacoustic cell. Obviously, the pressure signal should
decrease in the temperature range around the transition temperature T_m. Experimental observations of that effect have
been published first by Pelzl and coworkers[1]. Publications
of experimental results of other groups followed[2-6]. Some of
them will be presented in sec. 3.

Recently, we have presented a model of the PAE at first order phase transitions[7,8]. The basic ideas can be applied also to the description of higher order phase transitions[9]. According to that model the PAE at first order phase transitions should be observed passing the phase transition from lower to higher temperature as well as vice versa. We could confirm that experimentally. This fact is in contradiction to the conclusions of Pelzl and coworkers[1].

There are various models to describe the relation between the heat, created by the absorption of light and the pressure response of the gas[10-13]. The model most applied to interprete PA measurements in gas-microphone cells is the model of Rosencwaig and Gersho[10]. It is based on the assumption that it suffices to know the temperature distribution in the whole sample cell assembly to determine the pressure of the gas. Rosencwaig and Gersho obtained the temperature distribution from solutions of the differential equation of conduction of heat. To describe the PA at phase transitions we also start from that basic equation but we take into account the latent heat. Following Rosencwaig and Gersho we relate the variation in pressure δP to an average temperature variation $<\delta T>$ of the gas by the equation of state

$$\delta P(t) \sim <\delta T(t)>. \tag{1}$$

Here $<\delta T>$ is the temperature variation averaged along one wavelength of the temperature wave in the gas. It is proportional to the temperature at the sample to gas boundary with the amplitude V_g. Therefore the pressure oscillation in the gas is

$$\delta P(t) \sim V_g \, e^{j\omega t}. \tag{2}$$

Generally, the temperature amplitude V_g is complex and it is more convenient to write it in the form

$$V_g = |V_g| \, e^{j\varphi_g}, \tag{3}$$

where $|V_g|$ and φ_g are real valued quantities. The temperature amplitude $|V_g|$ is measured via the pressure oscillation. φ_g is the phase angle of the oscillation of the temperature at the sample to gas boundary relative to the modulation of the light intensity.

2. MODEL OF OSCILLATING INTERFACE

An optically opaque sample with length l_1 in a gas microphone cell, Fig.1a, is illuminated periodically. Since the light is absorbed essentially at the sample surface, one expects a spatial temperature distribution as sketched in Fig. 1b. The temperature of the phase transition is T_m.

The periodic heating causes a stationary temperature distribution $T_k^{dc}(x,T_o)$ that depends linearly on x and to which a temperature wave $\phi_k^{ac}(x,t)$ is superposed. For an appropriate interval of the temperature of the PA cell or the ambient temperature T_o, ref. 10, around the transition temperature T_m, there are two different thermal phases in the sample. They are seperated by an interface located at a position η where $T(\eta,t) = T_m$. A periodic variation of the intensity of light leads to an oscillation of that interface $\xi(t)$ around a mean position $\eta_o(T_o)$, Fig. 1b, and it holds

$$\eta(T_o,t) = \eta_o(T_o) + \xi(T_o,t) \ . \tag{4}$$

It becomes obvious from Fig. 2 that the mean position of the oscillating interface proceeds more and more into the sample with increasing abient temperature and vice versa. The amplitude will be proportional to the intensity of light.

2.1 Equation of Heat Conduction and Some Boundary Conditions

The geometry of the sample cell assembly is sketched schematically in Fig. 1a. We assume the extension of the cell and the sample and its illuminated area perpendicularely to the incident light to be large compared to the thermal diffusion length. In that case we have a one dimensional problem.

The temperature distribution in the sample cell assembly
can be obtained from appropriate solutions of the differential equation of conduction of heat without heat sources

$$\frac{\partial^2 \phi_k}{\partial x^2} - \frac{1}{\alpha_k} \frac{\partial \phi_k}{\partial t} = 0 \quad ; \tag{5}$$

$$\alpha_k = \frac{\lambda_k}{\rho_k c_{Pk}} \tag{6}$$

is the thermal diffusivity of medium k. We introduced

$$\phi_k(x,t) = T_k(x,t) - T_o \tag{7}$$

as the relevant quantity. It is the difference between the
temperature at x and t and the temperature T_o of the outer
cell wall and the window, Fig. 1b. A source term in Eq. (5)
can be omitted for the most experimental situations as
outlined in the following.

We consider opticallv opaque materials only. That means,
the light is absorbed appreciately within a layer of the
sample that is very thin compared to the length of the
sample and the thermal diffusion length, or the conditions

$$\beta l_1 >> 1 \quad , \quad \beta/a_1 >> 1 \tag{8}$$

should be fullfilled. Here, ß is the optical absorption
coefficient and $a_1 = (\omega/2\alpha_1)^{1/2}$ is the thermal diffusion
coefficient of the sample. Generally

$$a_k^{-1} = (2\alpha_k/\omega)^{1/2} \tag{9}$$

is the thermal diffusion length of medium k. If condition
(8) is valid and the intensity of light varies sinusio-
dally with frequency $\omega/2\pi$, the power per unit area converted from the energy of the absorbed light to heat at
the sample to gas boundary is

$$Q(t) = Q_o(1+e^{j\omega t}) . \tag{10}$$

We assume further that the extension of the interface, the
transition region from phase 1 to phase 2 in the sample, is
small compared to the thermal diffusion length. In that
case the process of phase transition determines a boundary
condition for the flux of heat at the interface η that
contains the latent heat L.

2.2 Temperature Distribution

Solutions of the equation of heat conduction (5) that describe the temperature distribution in the various regions
of the sample cell assembly are[7] for the backing (k = 0):

$$\varphi_o = e_{10} + e_{20}x + U_o e^{\sigma_o x + j\omega t} \tag{11}$$

for $-1_o - 1_1 \leq x \leq -1_1$; the sample, low temperature
phase (k = 1):

$$\varphi_1 = e_{11} + e_{21}x + U_1 e^{\sigma_1 x + j\omega t} + V_1 e^{-\sigma_1 x + j\omega t} \tag{12}$$

for $-1_1 \leq x \leq \eta$;

the sample, high temperature phase (k = 2):

$$\varphi_2 = e_{12} + e_{22}x + U_2 e^{\sigma_2 x + j\omega t} + V_2 e^{-\sigma_2 x + j\omega t} \tag{13}$$

for $\eta \leq x \leq 0$ and the gas (k = 3):

$$\varphi_3 = e_{13} + e_{23}x + V_3 e^{-\sigma_3 x + j\omega t} \tag{14}$$

for $0 \leq x \leq 1_3$, with the complex thermal diffusion
coefficient

$$\sigma_k = (1+j) a_k \tag{15}$$

where a_k is defined by eq. (9).

The first and second terms in Eqs. (11) to (14) represent
the stationary temperature distribution, Fig. 1b

$$\phi_k^{dc} = e_{1k} + e_{2k} x .$$
(16)

The constants e_{1k} and e_{2k} are real valued. Solutions for
the complete temperature distributions including the
stationary terms (16) have been given first by Rosencwaig
and Gersho[10]. Generally those terms play no role in the
description of the PA-signal. But as we shall see below
it is the temperature gradient e_{21} or e_{22} that determines
the dependence of the amplitude and phase angle of the
temperature oscillation in the gas on ambient temperature.

The third and forth terms represent temperature waves
travelling to the left, U_k, and to the right, V_k, respectively. The backing and the gas column are assumed to be
thermally very thick, $a_0 l_0 \gg 1$, $a_3 l_3 \gg 1$. Therefore, the
reflected temperature waves are damped out. The amplitudes
U_k and V_k are complex valued. They are determined together
with e_{1k} and e_{2k} by the boundary conditions for temperatures and heat fluxes.

2.3 Boundary Conditions

The flux of heat from the surface of the sample, $x = 0$, to
the sample and the gas is equal to the production of heat
expressed by the equation (10):

$$\lambda_2 \frac{\partial \phi_2(0,t)}{\partial x} - \lambda_3 \frac{\partial \phi_3(0,t)}{\partial x} = Q_0(1+e^{j\omega t}) .$$
(17)

From the continuity of the temperature follows

$$\phi_2(0,t) = \phi_3(0,t) .$$
(18)

At the boundary of the backing, $x = -(l_0+l_1)$

$$\phi_0(-l_0-l_1,t) = 0 ,$$
(19)

and at the window, $x = l_3$,

$$\phi_3(l_3,t) = 0 . \tag{20}$$

Similar one obtains at $x = -l_1$

$$\phi_0(-l_1,t) = \phi_1(-l_1,t) \tag{21}$$

and

$$\frac{\partial\phi_0(-l_1,t)}{\partial x} - \frac{\partial\phi_1(-l_1,t)}{\partial x} = 0 . \tag{22}$$

Finally, there are two boundary conditions at
$\eta(t) = \eta_0 + \xi(t)$, Eq. (4), where the phase transition
occurs. The temperature at η is the transition temperature
T_m. We define the difference between T_m and the ambient
temperature T_0 as

$$\delta = T_m - T_0 = \phi_1(\eta,t) = \phi_2(\eta,t) . \tag{23}$$

The flux of heat through the sample causes a phase transition along a volume per unit time that is equal to the
cross section of the illuminated area times dη/dt. There
holds the energy balance[14,15]

$$\rho_1 L \frac{d\eta}{dt} = \lambda_1 \frac{\partial\phi_1(\eta,t)}{\partial x} - \lambda_2 \frac{\partial\phi_2(\eta,t)}{\partial x} \tag{24}$$

at the boundary η if the thickness of the interface is
small compared to the thermal diffusion length. dη/dt can
be interpreted as the velocity of the moving boundary between the low and the high temperature state of the sample.
L is the latent heat.

2.4 Temperature Gradient and Oscillation

From the stationary parts of the relations (17) to (24)
one obtains the real valued coefficients e_{ik} and related
quantities. It is appropriate to introduce a thermal re-

sistance in the various part of the cell sample assembly
defined as

$$R_k = \frac{l_k}{\lambda_k} , \qquad\qquad (25)$$

with k defined above.

The temperature gradient in the low temperature state of the
sample is

$$e_{21} = \frac{\lambda_2 [Q_o R_3 + \delta(\lambda_1/\lambda_2 - 1)]}{\lambda_1^2 (R_o + R_1 + R_3 \lambda_2/\lambda_1)} \qquad\qquad (26)$$

and in the high temperature state

$$e_{22} = e_{21} \lambda_1/\lambda_2 \quad . \qquad\qquad (27)$$

For metallic samples and cells, $R_3 \gg R_o$, R_1; $\lambda_2 \sim \lambda_1$, the
temperature gradient is approximately

$$e_{21} \sim Q_o/\lambda_1 \quad . \qquad\qquad (28)$$

The stationary temperature at the sample to gas boundary
is

$$e_{12} = R_3 (Q_o - \lambda_1 e_{21}) = \frac{R_3 [Q_o (R_o + R_1) + \delta(\lambda_2/\lambda_1 - 1)]}{R_o + R_1 + \frac{\lambda_2}{\lambda_1} R_3} \quad .(29)$$

The average position of the boundary where the phase transition occurs as a function of temperature difference δ,
Eq. (23), is

$$\frac{\eta_o(\delta)}{l_1} = \frac{\delta}{l_1 e_{21}} - (1 + \frac{R_o}{R_1}) \quad . \qquad\qquad (30)$$

If the sample to gas boundary is at transition temperature
T_m, i.e. $\eta_o = 0$, we define the ambient temperature to be

$$\delta_o = \frac{Q_o R_3 (R_o + R_1)}{R_o + R_1 + R_3} \quad . \qquad\qquad (31)$$

δ_o is proportional to the heat Q_o. For $R_3 \gg R_o$, R_1

$$\delta_o \sim Q_o(R_o + R_1) \quad . \tag{32}$$

With the boundary condition given in sec. 2.3 we obtain from the periodic components of the temperatures and heat fluxes the following relations:

$$U_o e^{-\sigma_o l_1} - U_1 e^{-\sigma_1 l_1} - V_1 e^{\sigma_1 l_1} = 0, \tag{33}$$

$$U_2 + V_2 = V_3, \tag{34}$$

$$\lambda_o \sigma_o U_o e^{\sigma_o l_1} - \lambda_1 \sigma_1 (U_1 e^{-\sigma_1 l_1} - V_1 e^{\sigma_1 l_1}) = 0, \tag{35}$$

$$\lambda_2 \sigma_2 (U_2 - V_2) + \lambda_3 \sigma_3 V_3 = Q_o, \tag{36}$$

$$e_{21} \xi + U_1 e^{\sigma_1 (\eta_o + \xi) + j\omega t} + V_1 e^{-\sigma_1 (\eta_o + \xi) + j\omega t} = 0, \tag{37}$$

$$e_{22} \xi + U_2 e^{\sigma_2 (\eta_o + \xi) + j\omega t} + V_2 e^{-\sigma_2 (\eta_o + \xi) + j\omega t} = 0, \tag{38}$$

$$\rho L \frac{d\xi}{dt} - \lambda_1 \sigma_1 [U_1 e^{\sigma_1 (\eta_o + \xi) + j\omega t} - V_1 e^{-\sigma_1 (\eta_o + \xi) + j\omega t}]$$
$$+ \lambda_2 \sigma_2 [U_2 e^{\sigma_2 (\eta_o + \xi) + j\omega t} - V_2 e^{-\sigma_2 (\eta_o + \xi) + j\omega t}]. \tag{39}$$

The oscillation of the boundary can be expressed by[10]

$$\xi = \varepsilon \cdot e^{j\omega t} \quad . \tag{40}$$

Before we evaluate from the set of equations (33) to (39) the temperature V_3 at the sample gas boundary we make an assumption to simplify the calculation.

The amplitude ε of the oscillation of the interface should be much smaller than the thermal diffusion length:

$$a_{1,2} |\varepsilon| \ll 1 \quad . \tag{41}$$

This condition can be satisfied experimentally relatively easy.

Using the difinition (40) and taking into account the
condition (41) the equations (37) to (39) become

$$e_{21}\varepsilon + U_1 e^{\sigma_1 \eta_0} + V_1 e^{-\sigma_1 \eta_0} = 0 , \qquad (42)$$

$$e_{22}\varepsilon + U_2 e^{\sigma_2 \eta_0} + V_2 e^{-\sigma_2 \eta_0} = 0 , \qquad (43)$$

$$j\omega\rho L\varepsilon - \lambda_1 \sigma_1 (U_1 e^{\sigma_1 \eta_0} - V_1 e^{-\sigma_1 \eta_0})$$
$$+ \lambda_2 \sigma_2 (U_2 e^{\sigma_2 \eta_0} - V_2 e^{-\sigma_2 \eta_0}) = 0 . \qquad (44)$$

To get an idea of the magnitude of the amplitude ε of the
boundary oscillation we express its greatest value in
terms of the temperature amplitude V_3. Obviously, the amplitude ε should have a maximum if the oscillating interface is at the surface of the sample, i.e. $\eta_0 \to 0$. Therefore one obtains from eq. (43) for $a_1 l_1 \to 0$ and eq. (34)
for the amplitude

$$\varepsilon = - \frac{\lambda_1}{\lambda_2} \frac{V_3 (\eta_0 = 0)}{e_{21}} \qquad (45)$$

V_3 is complex numbered. The minus sign in eq. (45) indicates that the interface moves with increasing temperature
towards the inner of the sample. The magnitude of $|\varepsilon|$ de-
creases with increasing temperature gradient e_{21}. For
metallic samples and backing material $|V_3| \propto Q_0^2$, eq. (70),
as we shall see later, and $e_{21} \propto Q_0$, eq. (28). Therefore,
the amplitude $|\varepsilon|$ is proportional to the light intensity
and Q_0 respectively.

2.5 PA-Signal for Thermally Thick and Thin Samples

The pressure variation δP in a gas surrounding the sample
is according to eq. (2) proportional to the temperature V_3
at the sample to gas boundary. The set of equations (33)
to (36) and (42) to (44) can be solved relatively easy for
the limiting cases of thermally thick and thermally very
thin samples.

i) Thermally Thick Samples, $a_1 l_1 \gg 1$

Neglecting all the terms with $e^{-\sigma_1 l_1}$ one obtains

$$V_3 = \frac{Q_o}{\lambda_2 \sigma_2} \frac{\sigma_2(e^{-\sigma_2 \eta_o} + e^{\sigma_2 \eta_o}) + (e^{-\sigma_2 \eta_o} - e^{\sigma_2 \eta_o})(\sigma_1 + w)}{\sigma_2(e^{-\sigma_2 \eta_o} - e^{\sigma_2 \eta_o}) + (e^{-\sigma_2 \eta_o} + e^{\sigma_2 \eta_o})(\sigma_1 + w)} \tag{46}$$

with

$$w = j \frac{\omega \rho_1 L}{\lambda_1 e_{21}} . \tag{47}$$

The amplitude of the temperature oscillation defined by eq. (3) is in this case

$$|V_g| = |V_3| = K \left[(A^2 + B^2)/(C^2 + D^2) \right]^{1/2} \tag{48}$$

and the phase angle

$$\varphi_g = \varphi_3 = \tan^{-1} \left[(BC - AD)/(AC + BD) \right] \tag{49}$$

with the abbreviations

$$K = \frac{Q_o}{2\lambda_2 a_2 (1 + \lambda_3 a_3/\lambda_2 a_2)} , \tag{50}$$

$$A = \{ [f(\gamma + 1) + 1] \cos u + \sin u \} e^{-u} + f(\gamma - 1) - 1, \tag{51}$$

$$B = \{ \cos u - [f(\gamma + 1) + 1] \sin u \} e^{-u} - 1, \tag{52}$$

$$C = \{ [f(\gamma + 1)/2] \cos u + [f(\gamma + 1)/2 + 1] \sin u \} e^{-u} - p \, f(\gamma - 1)/2 , \tag{53}$$

$$D = \{ [f(\gamma + 1)/2 + 1] \cos u - [f(\gamma + 1)/2] \sin u \} e^{-u} - p[f(\gamma - 1)/2 - 1], \tag{54}$$

where

$$u = 2a_2 \eta_o , \tag{55}$$

$$\gamma = a_2/a_1 , \tag{56}$$

$$f = \frac{2 \cdot \delta \, c_{P1}}{a_1 L (1 + \rho_2/\rho_1)(l_1 + \eta_o)} , \tag{57}$$

and

$$p = \frac{1-\lambda_3 a_3/\lambda_2 a_2}{1+\lambda_3 a_3/\lambda_2 a_2} \quad .$$ (58)

ii) Thermally Very Thin Samples, $a_2 l_1 << 1$

With the approximation $e^{+\sigma_2 \eta_o} = 1$ in the equations (33) to (36) and (42) to (44) the complex amplitude becomes

$$V_3 = \frac{Q_o}{\lambda_2(\frac{\lambda_o}{\lambda_1} \sigma_o + \frac{\lambda_3}{\lambda_2} \sigma_3 + w)} \quad .$$ (59)

It is also assumed that the amplitude ε of oscillation is small compared to the length l_1 of the sample,

$$|\varepsilon|/l_1 << 1 \quad .$$ (60)

That condition limits the heat Q_o for a metallic sample to

$$Q_o << \omega \rho_1 L \, l_1 \lambda_2/\lambda_1 \quad ,$$ (61)

with the relations (45) and (70). Amplitude $|V_3|$ and phase angle Ψ_3 for $a_1 l_1 << 1$ are estimated in sec. 2.6.

2.6 Dependence on Ambient Temperature

In all experiments performed the ambient temperature T_o was varied. In the formalism developed in sec. 2.1 to 2.5 it corresponds to a variation of δ defined by eq. (23). It is appropriate to introduce a reduced temperature defined as

$$\theta = 1- \delta/\delta_o$$ (62)

with δ_o given by eq. (31). Then, $\theta = 0$ means that the ambient temperature T_o is by the amount δ_o below the transition temperature T_m and the average position of the interface should be at $\eta_o = 0$. If the ambient temperature is equal to the transition temperature there holds $\theta = 1$.

Discussing the dependence of amplitude and phase angle of
the PA signal on temperature, different temperature ranges
should be distinguished. The characteristic temperatures
boundering them are defined in sec. 2.7.

At ambient temperatures far away from T_m the temperature
distribution in the sample cell assembly is described by
the solutions given by Rosencwaig and Gersho[10] for opaque
samples. The amount of amplitude is denoted as $|v_{30}^{(k)}|$,
the phase angle as $\varphi_{30}^{(k)}$. Expressions for those quantities
are presented assuming $\lambda_o a_o \sim \lambda_1 a_1 \sim \lambda_2 a_2$.

i) Thermally Thick Samples, $a_1 l_1 > 1$

At temperatures below a characteristic reduced temperature
θ_b defined below, there is only the low temperature phase
of the sample (k = 1) and it holds

$$|v_g| = |v_{30}^{(1)}| \sim Q_o/\sqrt{2}(\lambda_1 a_1 + \lambda_3 a_3), \quad \varphi_3^{(1)} = -\pi/4 . \quad (63)$$

On the other hand, at temperatures above a characteristic
value θ_f there exists only the high temperature phase
k = 2 of the sample and it is

$$|v_g| = |v_{30}^{(2)}| \sim Q_o/\sqrt{2}(\lambda_2 a_2 + \lambda_3 a_3), \quad \varphi_3^{(2)} = -\pi/4 . \quad (64)$$

ii) Thermally Very Thin Samples, $a_1 l_1 << 1$

At temperatures far away from the transition temperature
$T \gtrless T_m$, the PA signal of thermally very thin samples should
be determined essentially by the thermal properties of the
backing material:

$$|v_g| = |v_{30}^{(1)}| \sim |v_{30}^{(2)}| \sim Q_o/\sqrt{2}(\lambda_o a_o + \lambda_3 a_3) \quad \text{for } \theta \begin{matrix} <\theta_b \\ >\theta_f \end{matrix} \quad (65)$$

and $\varphi_g = -\pi/4$.

In the formalism developed in the preceeding sections the
amplitude and phase angle as a function of temperature
have a minimum in the vicinity of $\theta = 0$ where $\eta_o = 0$, ref.
7,8. Both quantities will be determined now for $\eta_o \rightarrow 0$,

though their values expected experimentally are not described correctly because the influence of ξ is neglected[9].
One obtains, however, some important informations on those
quantities that determine essentially the PA signal in the
vicinity of a first order phase transition.

We start from the complex expressions (46) and (59) for
V_3 respectively. For the following discussion we assume
the sample to be a metal, e.g. indium. The backing material is a metal also, e.g. copper, the gas is air. The
relevant thermal data of these materials are listed in
Table 1. For such a sample cell assembly in eqs. (46) and
(59) there holds

$$|w| = \omega \rho_1 L / \lambda_1 e_{21} \gg a_1 \quad , \tag{66}$$

and further

$$a_1 \lambda_1 \sim a_2 \lambda_2 \sim a_o \lambda_o \gg a_3 \lambda_3 \quad . \tag{67}$$

For thermally thick as well as for thermally thin samples
for $\delta \to \delta_o$ of $\eta_o \to 0$ the complex amplitude is

$$V_3(\eta_o \to 0) \sim -j \frac{\lambda_1}{\lambda_2} \frac{Q_o e_{21}}{\omega \rho_1 L} \quad . \tag{68}$$

Comparing this expression with eq. (3) one obtains the
phase angle

$$\varphi_g = \varphi_3(\eta_o \to 0) \to -\frac{\pi}{2} \quad \text{for } a_1 l_1 \begin{matrix} \gg 1 \\ \ll 1 \end{matrix} \quad , \tag{69}$$

and for $\lambda_1 \sim \lambda_2$

$$|V_g| = |V_3(\eta_o \to 0)| \sim \frac{Q_o e_{21}}{\omega \rho_1 L} = \frac{Q_o^2}{\omega \lambda_1 \rho_1 L} \tag{70}$$

taking into account eq. (26) for e_{21}. The amplitude for
$\eta_o \to 0$ is proportional to the square of Q_o, the energy of
light absorbed in the sample. It arises from the tempera-

ture gradient $e_{21} \sim Q_0$. The normalized amplitude is finally
with relation (63)

$$|V_3(\eta_0 \to 0)|/|V_{30}^{(1)}| \sim Q_0/\sqrt{\omega \alpha_1} \; \rho_1 L \; , \tag{71}$$

where $\alpha_1 = \lambda_1/\rho_1 c_{p1}$ is the thermal diffusivity of the
sample. The minimum of the plot of the amplitude versus
temperature should be the lower the higher the frequency,
thermal diffusivity and heat capacity $\rho_1 L_1$ of the sample
and the lower the heat production Q_0.

In Fig. 2 the normalized amplitude $|V_g|/|V_{30}^{(1)}|$ and phase
angle φ_g expressed by the eqs (48), (49) and the limiting
values (63) and (64) for various values of $a_1 l_1 > 1$ are
plotted versus the reduced temperature. The dependence on
latent heat is shown. The corresponding plots of thermally
very thin samples, $a_1 l_1 \ll 1$, are shown in Fig. 4. They
are explained in sec. 2.7 after introduction of some
characteristic temperatures.

2.7 Characteristic Temperatures

We shall define now the characteristic temperatures θ_b, θ_f
and others.

Starting at an ambient temperature far below T_m and increasing it one approaches a temperature θ_b where the
maximum of the temperature oscillation is equal to the
transition temperature T_m, Fig. 3a. One can show using the
eqs. (63) and (31) that

$$\theta_b = -|V_{30}^{(1)}|/\delta_0 \approx \frac{R_0 + R_1 + R_3}{\sqrt{2} R_3 (R_0 + R_1)(\lambda_1 a_1 + \lambda_3 a_3)} \; . \tag{72}$$

The expressions (46) and (59) describe the temperature V_3
at temperatures $\theta \gtrsim \theta_{os}$ where along the entire period
$2\pi\omega^{-1}$ an interface between two phases exists, Fig. 3b. The
reduced characteristic temperature θ_{os} is defined by the
ambient temperature where the amount of the amplitude

$$\varepsilon = \eta_0(\theta_{os}) \tag{73}$$

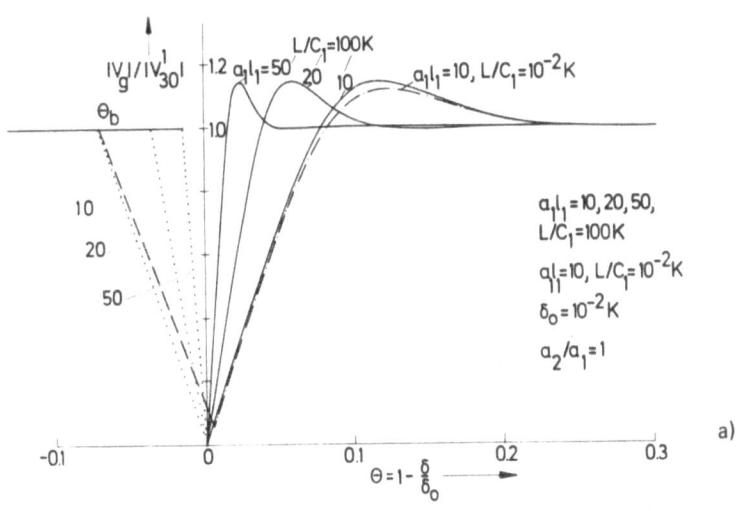

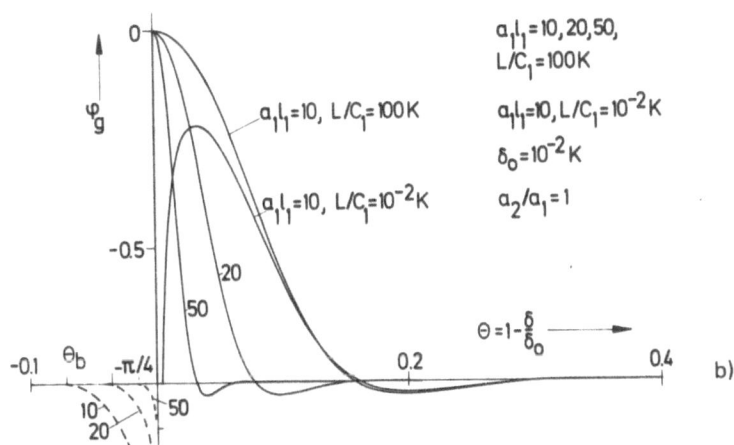

FIGURE 2   Temperature dependence of the amplitude (a) and
            the phase angle (b) of the PA signal at first
            order phase transition according to eqs (48) and
            (49) for different values of $a_1 l_1$ and L/C_1 with
            $\delta_0 = 10^{-2}$ K, $\lambda_3/\lambda_2 = 10^{-3}$, $\theta_b \sim 1/\sqrt{2}\ a_1 l_1$, and
            $|\xi|_0/l_1 = \delta_0 C_1/L(a_1 l_1)^2$. Dotted lines: possible
            form of interpolation between $\theta_b \leq 1-\delta/\delta_0 \lesssim \theta_{os}$
            for $L/C_1 = 100$ K (a), dashed lines for $L/C_1 =$
            10^{-2} K (a,b). $C_1 \equiv c_{p1}$.

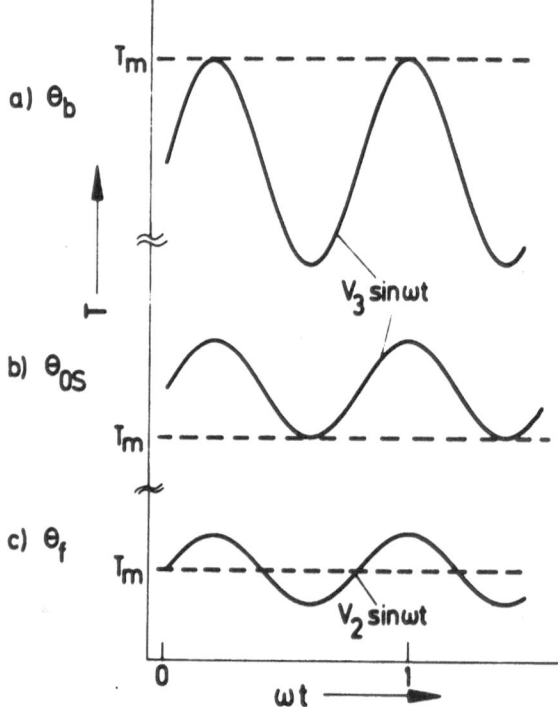

FIGURE 3   Temperature oscillation $V_3(t)$ and $V_2(t)$ respectively at the characteristic reduced temperatures
a) θ_b, b) θ_{os} and c) θ_F. The transition temperature T_m is indicated by the dashed line.

of the interface is equal to the distance η_o of its average position from the surface of the sample $x = 0$. With
the approximation $V_3(\eta_o = \varepsilon) \sim V_3(\eta_o = 0)$ one obtains with
eqs. (70) and (31)

$$\theta_{os} = \frac{\delta_o - \delta_{12}}{\delta_o} = \frac{V_3(\eta_o = 0)}{\delta_o} \sim \frac{Q_o}{\omega \lambda_1 \rho_1 L (R_o + R_1)} \qquad (74)$$

In the temperature range $\theta_b \lesssim 0 \lesssim \theta_{os}$ none of the expressions (46) or (63) alone describe the PA signal.
Therefore, no precise description can be given for amplitude and phase angle. It is known from the discussion in
sec. 2.6, however, that the amplitude for samples of all
thermal thicknesses is expected to run through a minimum.
Its hypothetical value has been evaluated, eq. (70).

With rising temperature $\theta > \theta_{os}$ the distance between the average position η_o of the interface and the sample surface $x = 0$ increases. The behaviour of the PA signal with respect to temperature for thermally thick sample differs from that of thermally thin samples. The phase angle for thermally thick samples runs through a maximum, Fig. 2, approaching finally together with the amplitude the values $\varphi_{30}^{(2)}$ and $V_{30}^{(2)}$ determined by the properties of the high temperature phase $k = 2$ of the sample. For $\theta \rightarrow 1$ eqs. (48) and (49) approach eq. (64), ref. 7.

The formal description of V_3 for thermally very thin samples as a function of temperature for $\theta \gtrsim \theta_{os}$ is more complicated. It will be presented in a forthcomming paper[9]. At this moment, however, some qualitative statements can be made. Measuring very thin samples, $a_1 l_1 \ll 1$, one will notice that the interface approaches to the backside of the sample, i.e. $\eta = -l_1$. The ambient temperature associated with the fictive average position $\eta_o = -l_1$ is expressed by

$$\theta_f \approx 1/(1+R_o/R_1) \quad ; \tag{75}$$

see also Fig. 3c. Therefore, amplitude $|V_3|$ and phase angle φ_3 after passing a minimum in the vicinity of $\theta = 0$ will approach the values $|V_3^{(o)}|$ and $\varphi_3^{(o)}$ respectively that are determined by the thermal properties of the backing material and the optical absorption coefficient of phase 2. In Fig. 4 is shown the qualitative plot of amplitude and phase angle with ambient temperature expressed by the reduced temperature θ.

3. EXPERIMENTS

3.1 PA-Cell

The scheme of the gas-microphone cell that we used for measurements in the temperature range between 290 and 400 K is shown[4,8] in Fig. 5. The temperature was controlled

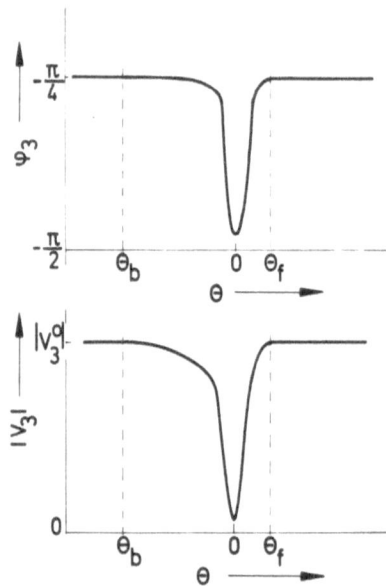

FIGURE 4   Schematic plot of amplitude $|V_3|$ and phase angle
φ_3 versus reduced temperature θ for a thermally
thin sample, $a_1 l_1 \ll 1$.

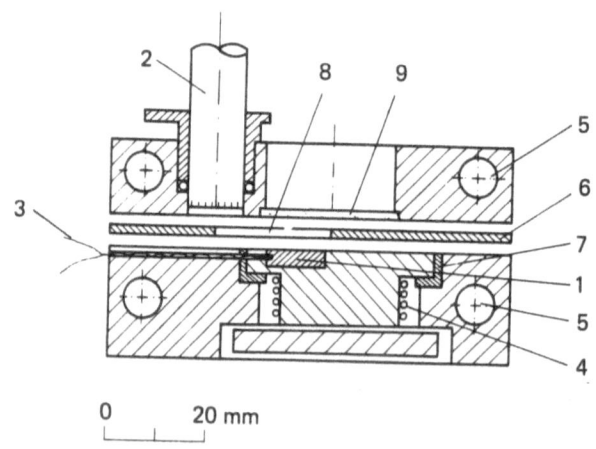

0   20 mm

FIGURE 5   Schematic diagram of a gas-microphone PA cell for
measurements in the temperature range 290 to 450
K. 1: sample, 2: microphone, 3: thermocouple,
4. heating coil, 5: holes of the cooling system
(water), 6: spacer and gasket (silicon rubber),
7: thermal insolation (Teflon), 8: gas, 9: optical
window.

by a heating coil wound on the sample holder and a water
cooling system. The sample holder was mounted thermally
isolated in the cell. Therefore, the temperature at the
condensor microphone (Bruel & Kjaer, Type 4166) never ex-
ceeded 330 K. The temperature was measured with an iron-
constantan thermocouple. The difference between the temperature of the sample holder and the water cooled part of
the cell leads to a temperature gradient parallel to the
plane of the sample gas boundary. It will smear out the
ideal plots shown in Fig. 2. The samples were illuminated
by the light of a 250 Watt halogen lamp. Its intensity was
modulated by a mechanical chopper (PAR 192). The chopping
frequency used varied between 28 and 86 Hz. Amplitude and
phase angle were detected using the lock-in technique
(Ithaco-Dyatrac 393 with phase option).

3.2 Melting of Thermally Thick and Thin Indium Samples

The essential results of the model of oscillating interface presented in sec. 2 could be confirmed by measurements
at the solid-liquid transition (melting) of indium[8]. The
photoacoustically relevant parameters are given in Table 1.
We investigated two samples of extremely different thermal
thickness: sample A was thermally thin with a thickness of
l_1 = 5 μm and $a_1 l_1$ = 0.01 at ν = 52 Hz; sample B with the
parameters l_1 = 2 mm and $a_1 l_1$ = 4.1 at 52 Hz was thermally
thick. The experimental results for the PA-amplitude
S $\propto |V_3|$ and the phase angle $\Delta \varphi_{exp}$ are plotted in Fig. 6.
The behaviour of the amplitude with temperature is the
same for both samples, it shows a minimum in the transition region. The plot of the phase angle in the transition
region is characterized by a minimum for the thermally very
thin and a maximum for the thermally thick sample. The
plots agree qualitatively very well with those expected
theoretically, Figs. 2 and 4.

3.3 Measurements on Various Materials

From the results of experimental investigations by various
groups[1-6] we present only those where beside the amplitude
also the phase angle of the PA signal was measured.

TABLE 1

Thermal Data of In, $BaTiO_3$ and VO_2 near transition temperature and of Air and Cu

	Transition temperature T_m K	ρ_k g cm^{-3}	c_{pk} Jg^{-1}k^{-1}	λ_k Watt cm^{-1}k^{-1}	L Jg^{-1}
In solid	429.76 [a]	7.31 [b]	0.266 [c]	0.729 [a]	28.6 [b]
In liquid	$\widehat{=}$156.6 °C	7.13	0.267	0.360	
$BaTiO_3$ +0.45 wt% Fe_2O_3	371 [d]	6.02 [e]	0.473 [c]	0.027 [a] (powdered)	0.46 [d]
VO_2	340 [f]	3.357 [d]	0.32 [f]		25.76 [f]
Cu (300 K)		8.93 [b]	0.38 [c]	4.0 [a]	
Air (300 K)		$1.18 \cdot 10^{-3}$ [e]	1.0 [e]	$2.6 \cdot 10^{-4}$ [e]	

a) Ref.18, b) Ref.19, c) Ref.20, d) Ref.21, e) Ref.22, f) Ref.23

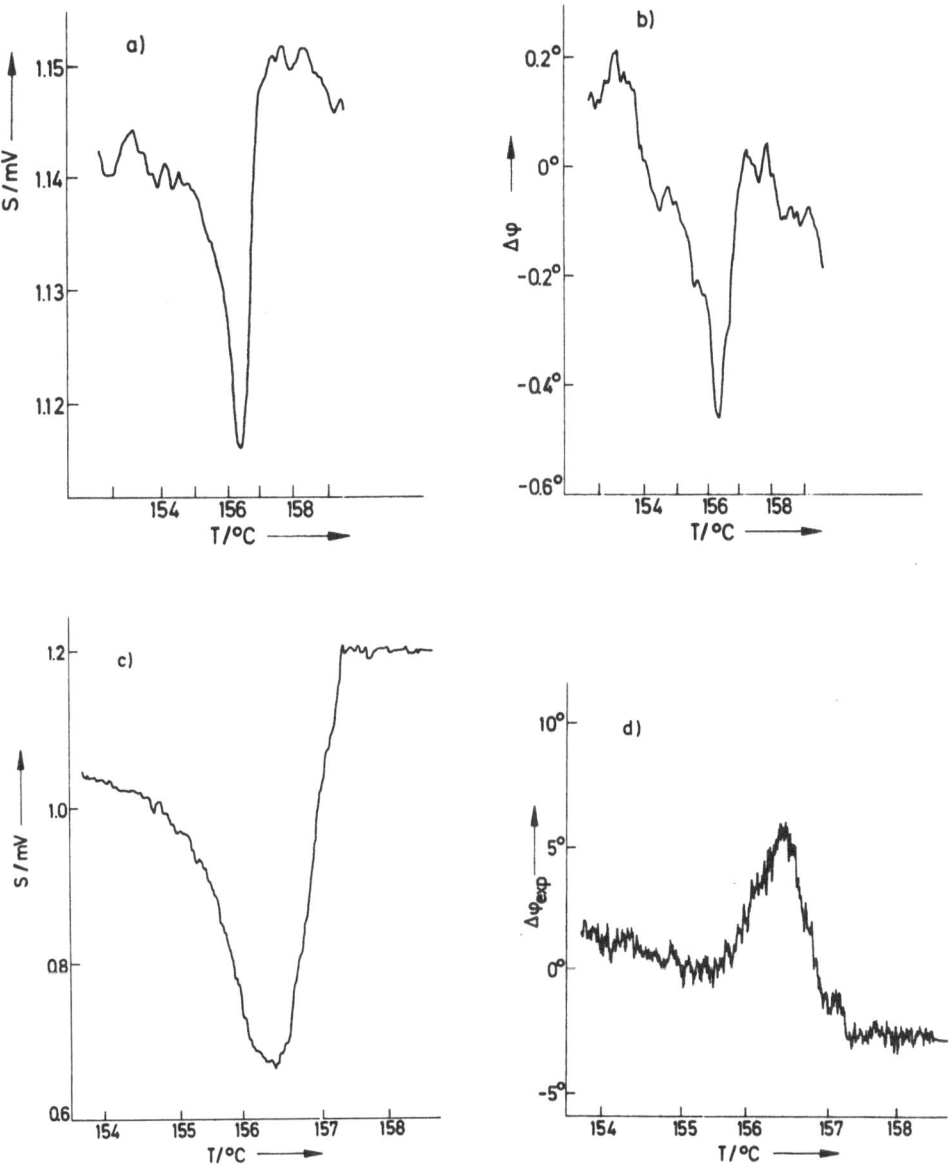

FIGURE 6   Amplitude S and phase shift $\Delta\varphi_{exp}$ of the PA-signal measured around the melting temperature of
indium, T_m = 156.4 °C. a),b): thermally thin
sample, $a_1 l_1$ = 0.01, l_1 = 5 μm; c),d): thermally
thick sample, $a_1 l_1$ = 4.1, l_1 = 2 mm. Modulation
frequency 52 Hz (Ref. 8).

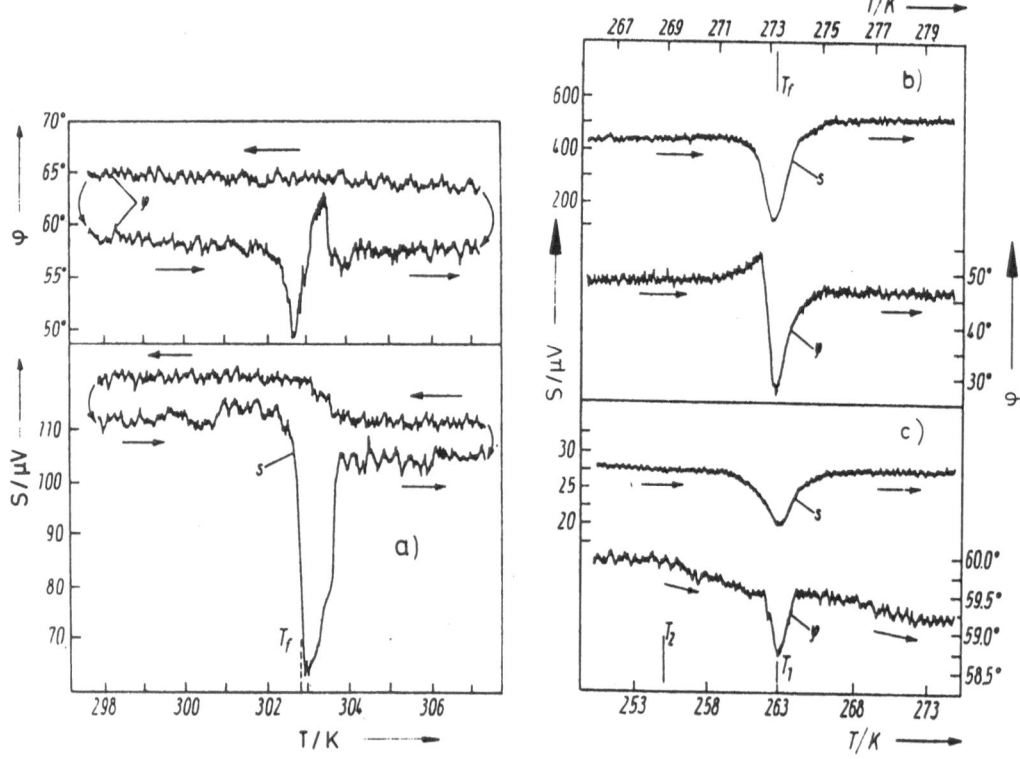

FIGURE 7   Amplitude S and phase angle φ_{exp} of the PA signal
            a) at the melting point of gallium T_m = 302.9 K
            for both increasing and decreasing temperature
            (arrow); b) at the melting point of ice; c) at the
            structural phase transition of K_2SnCl_6, $T_m \sim$ 263 K.
            Modulation frequency 36 Hz (Ref. 1).

We begin with the first photoacoustic detection of phase
transitions by Florian et al.[1]. In the Figs. 7a-c are
shown the results of measurements at the solid to liquid
transition of water and of gallium (T_m = 302.9 K) and at
a structural phase transition of potassium chlorostannate
(K_2SnCl_6, $T_m \sim$ 263 K). Florian et al. did not discuss
their results in detail. One essential statement of those
authors, however, should be mentioned. They could observe
the effect by passing the phase transition from lower to
higher temperatures. They reported to observe no essential
variation of the PA signal in the transition region measuring at decreasing temperature. Therefrom they concluded

that the processes determining the PA effect at the phase
transition is irreversible with respect to temperature.
The thermodynamic explanation they gave of the phenomenon
at increasing temperature is not consistent with that at
decreasing temperature. We suppose that their explanation
is a consequence of a misinterpretation of the measurements on gallium shown in Fig. 7. It is well known that
gallium may be supercooled easily by an amount up to 30 K
below the melting temperature. We have found experimentally
the PA effect to be reversible with respect to temperature
if supercooling is prevented[4]. We could observe the characteristic behaviour of amplitude and phase angle of the PA
signal with temperature in the transition region of all
samples at increasing as well as at decreasing temperature.
That behaviour is consistent with the model of oscillating
boundary. In the Figures 8a and 8b are shown the results
of measurements on the semiconductor to metal transition
of VO_2 at 340 K [16] and at the ferroelectric to paraelectric
transition of $BaTiO_3$ with 0.45 wt % Fe_2O_3 at about 371 K [4]
The samples were powders. Therefore, the effective thermal
thickness is unknown. The thermal data for bulk samples
are listed in Table 1. The plots show the microphone sig-
nal $\Delta S = S - S^{(2)}$ and $\Delta \varphi = \varphi_3 - \varphi_3^{(2)}$ as a function of ambient
temperature T. There hold $S \propto |V_3|$ and $S^{(2)} \propto |V_{30}^{(2)}|$.
Figs. 8a and 8c illustrate the dependence on the latent
heat, eq. (70), Table 1. Bechthold et al.[5] have measured
the PA effect at phase transitions in the metal hydrogen
interstitial alloys $TaH_{0.5}$, $NbH_{0.8}$ and $VH_{0.517}$. The
$\varepsilon \rightarrow \alpha$ order-disorder transition in $TaH_{0.}$ at 335 K, the
$\beta \rightarrow \alpha$ oder-disorder transition in $NbH_{0.8}$ at 386.8 $\pm$ 0.5 K
and the $\beta \rightarrow \varepsilon$ transition in $VH_{0.517}$ at 466 K are first
order phase transitions[16,17], whereas the $\beta \rightarrow \varepsilon$ transition
at 445 K in $VH_{0.517}$ is a λ-shaped second order phase transition[17]. Amplitude of the PA signal and shift of its
phase angle measured at increasing and decreasing temperatures are shown in the Figures 9a to 9c. In all examples
the amplitudes show a minimum and the phase angle φ a
maximum in the transition region. The position of the
extrema on the temperature scale in the plots are shifted

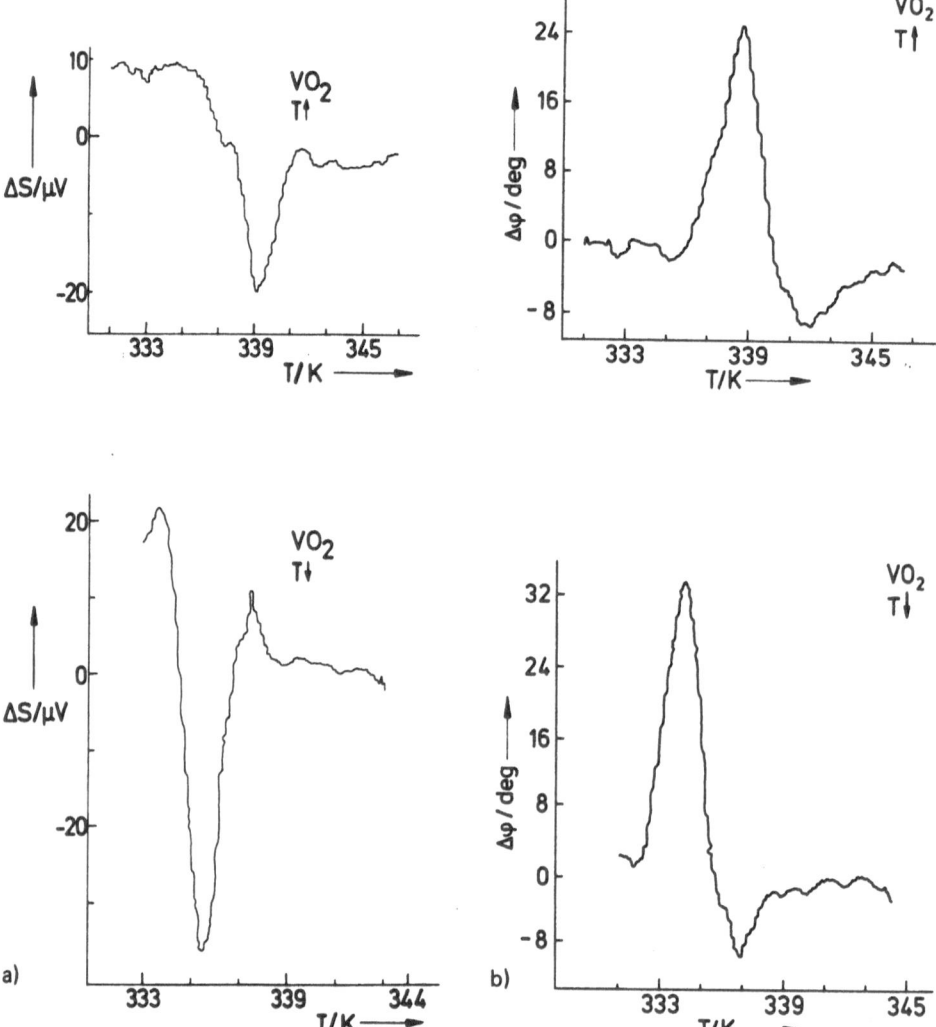

FIGURE 8   PA-amplitude S and phase shift $\Delta\varphi_{exp} = \varphi - \varphi^{(2)}$
           versus temperature measured at increasing (arrow
           up) and decreasing (arrow down) temperature at
           structural phase transitions. a,b: VO_2; c,d:
           $BaTiO_3 + 0.45$ wt % Fe_2O_3. Modulation frequency
           of light: 28 Hz; heating and cooling rate:
           2 K min^{-1}.

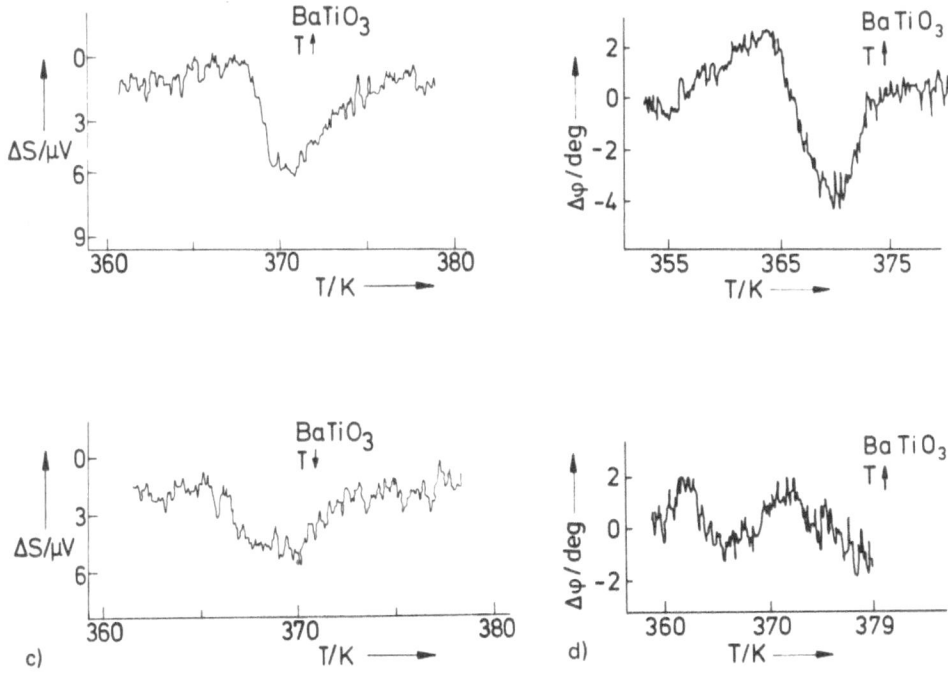

c)   d)

to lower temperatures passing the phase transition from
higher to lower temperatures. This shift can be explained
by supercooling that may occur at first order phase transitions[5]. All experimental results shown in the Fig. 9 are
qualitatively consistent with the model of oscillating
interface if all samples are assumed to be thermally thick.

Studies at a first order phase transition also in a bimole-
cular system were performed by Knoll et al.[6]. The PAE signal was measured on DPPC (dipalmitoyllecithin) containing
different concentrations of chlorophyll a, Fig. 10. DPPC
has a phase transition at about 43 °C, Fig. 10A. Chlorophyll
a, incorporated into such artificial membrane, can aggregate
depending on its concentration and temperature. The influence of this aggregation on PA amplitude and phase angle is
seen from Fig. 10B and C.

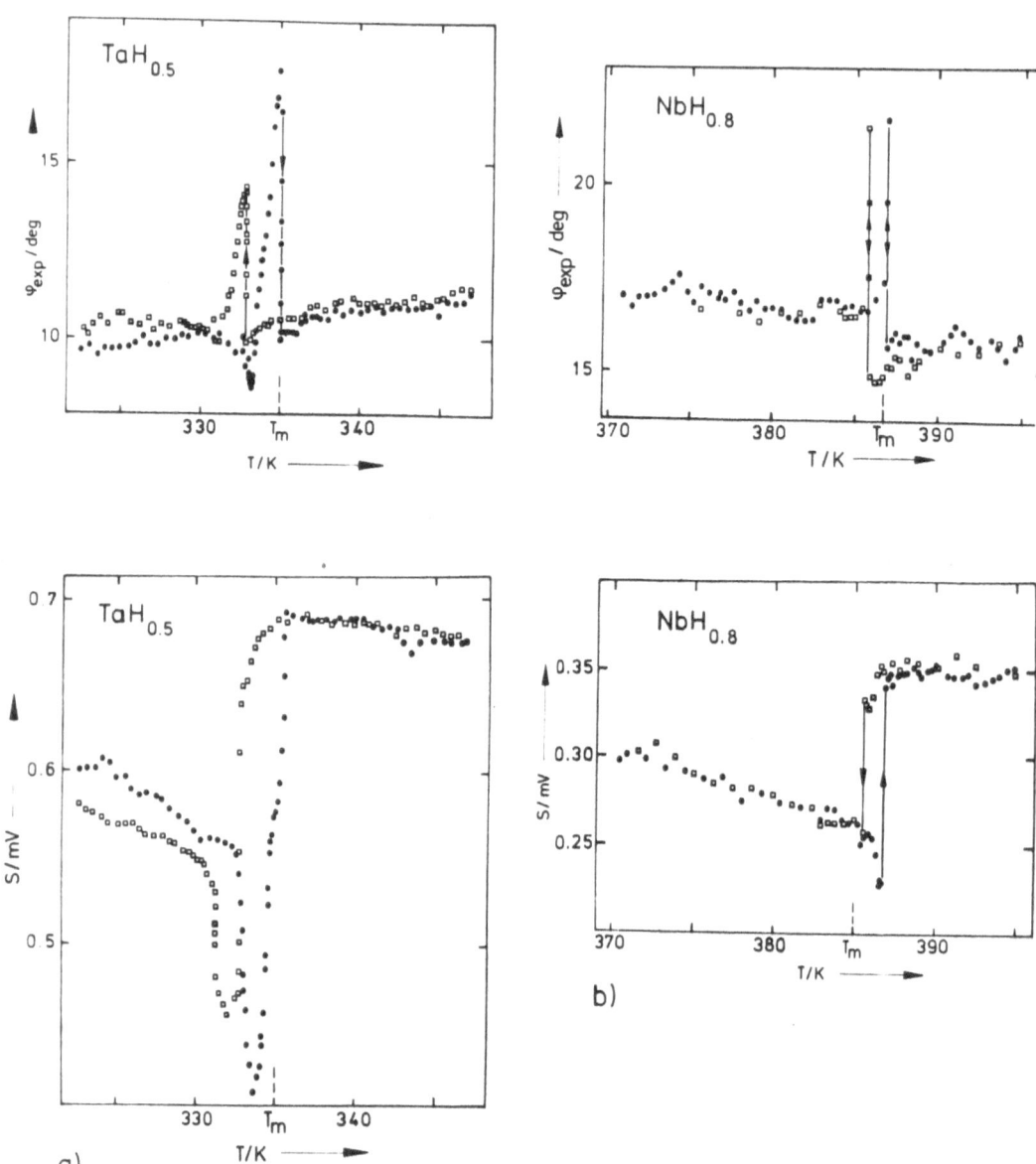

FIGURE 9   PA-amplitude S and phase angle φ_{exp} at first order
            phase transitions in metal hydrogen alloys.
            a) $TaH_{0.5}$ (T_m = 335 K), b) Nb $H_{0.8}$ (386.8 K),
            c) V $H_{0.517}$ (486 K) and at a second order phase
            transition of $VH_{0.517}$ (445 K). Full circles:
            measured at increasing, open squares: measured
            at decreasing temperature. Modulation frequency:
            40 Hz (Ref. 5).

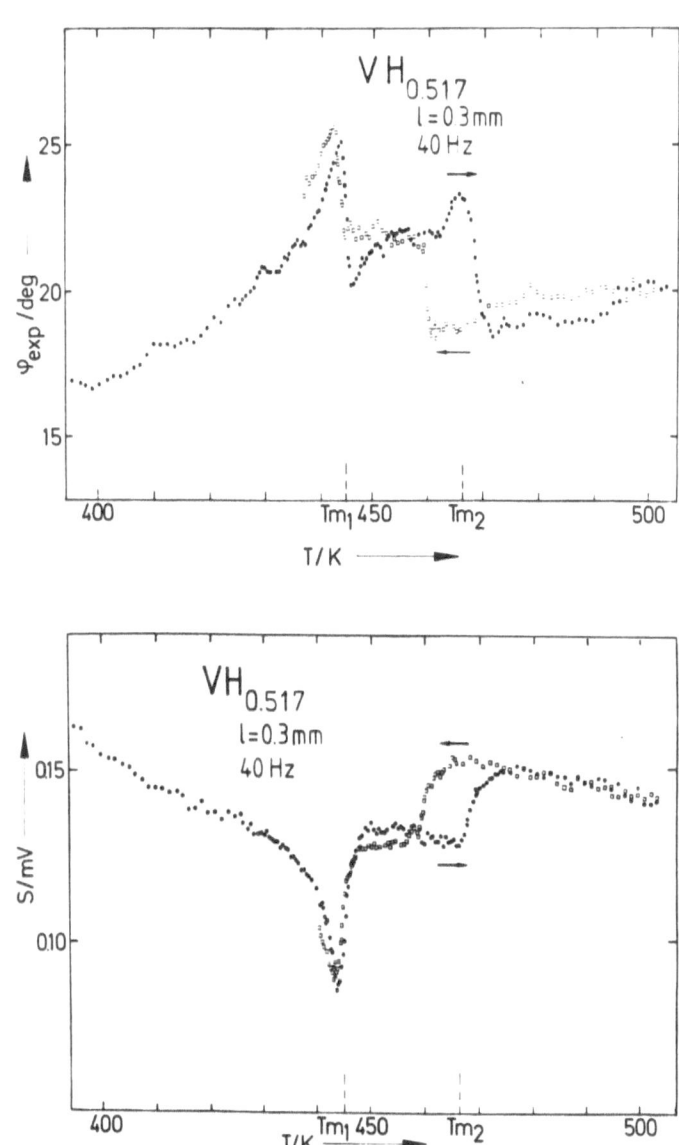

c)

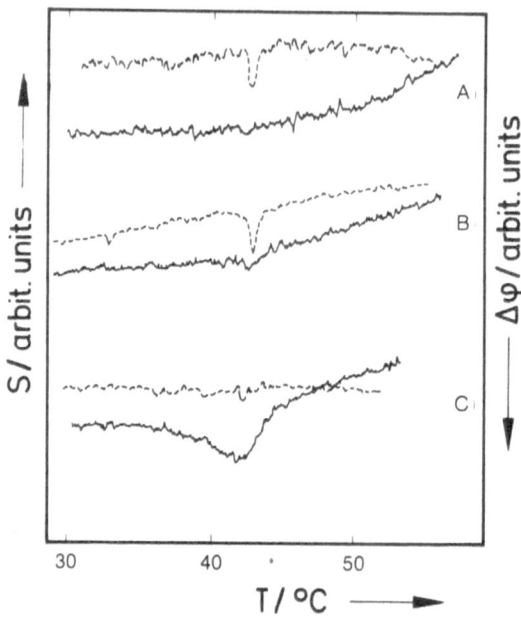

FIGURE 10 PA-amplitude S (full line) and negative shift in
            phase angle Δφ (dashed line) of DPPC dispersion
            (15 % wt/wt) without (A) with 0.33 mole % (B),
            and with 10 mole % (C) chlorophyll a.

4. CONCLUSION

The PAE is the base of a nonstationary method to investigate
phase transitions. The quantities measured are the amplitude
and the phase angle of a temperature wave. Both quantities
depend on the thermal conductivity, the specific heat or latent heat that determine refraction, reflexion and damping of
temperature waves, and on the sample thickness. Because of a
temperature gradient always present in the sample, different
parts of the sample are involved in the phase transition
process at different ambient temperatures. Thus, the experimental results reflect the geometrical structure of the
sample. The phase boundary oscillates. Therefore, time dependent phase transitions with relaxation times in the frequency range of the acoustical detector might be investigated.

REFERENCES

/1/ R. Florian, J. Pelzl, M. Rosenberg, H. Vargas, R. Wernhardt, phys.stat.sol. (a) $\underline{48}$, K35 (1978)

/2/ C. Pichon, M. Le Liboux, D. Fournier, A.C. Boccara Appl.Phys.Lett. $\underline{35}$, 435 (1979)

/3/ M.A. Siqueira, C.C. Ghizoni, J.I. Vargas, E.A. Menezes, H. Vargas, L.C.M. Miranda, J.Appl.Phys. $\underline{51}$, 1403 (1980)

/4/ P. Korpiun, J. Baumann, E. Lüscher, E. Papamokos, R. Tilgner, phys.stat.sol.(a) $\underline{58}$, K13 (1980)

/5/ P.S. Bechthold, M. Campagna, T. Schober, Sol.State Comm. $\underline{36}$, 225 (1980)

/6/ W. Knoll, J. Baumann, P. Korpiun, V. Theilen, Biochem.Biophys.Res.Comm. $\underline{96}$, 968 (1980)

/7/ P. Korpiun, R. Tilgner, J.Appl.Phys. $\underline{51}$, 6115 (1980)

/8/ P. Korpiun, R. Tilgner, phys.stat.sol.(a) $\underline{67}$, 201(1981)

/9/ P. Korpiun, R. Tilgner (to be published)

/10/ A. Rosencwaig, A. Gersho, J.Appl.Phys. $\underline{47}$, 64 (1976)

/11/ J.G. Parker, Appl.Opt. $\underline{12}$, 2974 (1973)

/12/ L.C. Aamodt, J.C. Murphy, J.G. Parker, J.Appl.Phys. $\underline{48}$, 927 (1977)

/13/ F.A. McDonald, G.C. Wetsel, Jr., J.Appl.Phys. $\underline{49}$, 2313 (1978)

/14/ H.S. Carslaw, J.C. Jaeger, Conduction of Heat in Solids, 2nd ed., chapt.11 (Clarendon, Oxford, 1978)

/15/ U. Grigull, H. Sandner: Wärmeleitung, chapt. 11, Springer, Berlin, 1979

/16/ T. Schober, H. Wenzl in: Hydrogen in Metals II, eds. G. Alefeld, J. Völkl, Springer, Berlin 1978

/17/ T. Schober, A. Carl, phys.stat.sol.(a), $\underline{43}$, 443 (1977)

/18/ Thermophysical Properties of Matter, Vol.1 and 2, eds. Y.S. Touloukian, R.W. Powell, C.Y. Ho, P.G. Klemens, IFI/Plenum, New York, 1970

/19/ Metals Handbook, Vol.1, eds. T. Lyman et al., ASM Metals Park, Ohio, 1966

/20/ Thermophysical Properties of Matter, Vol.4 and 5, eds. Y.S. Touloukian, E.H. Buyco, IFI/Plenum, New York, 1970

/21/ R. Tilgner, Private Communication

/22/ Handbook of Chemistry and Physics, ed. R.C. Weast, CRC, Cleveland, Ohio, 1971

/23/ G.V. Chandrashekhar, H.L.C. Barres, J.M. Honig, Mat.Res.Bull. Vol.8, 369 (1973)

PHOTOCALORIMETRIC INVESTIGATIONS OF ENERGY CONVERSION PROCESSES
USING PHOTOACOUSTIC DETECTION

David Cahen, Shmuel Malkin and Haim Garty
The Weizmann Institute of Science, Rehovot, Israel, 76100

ABSTRACT

The use of photoacoustic (PA) methods to
obtain calorimetric data on photoactive
systems is considered. With this application
in mind the theoretical basis of the
technique is outlined and some examples of
its use on chemical, physical and biological
systems are presented. Pitfalls that can be
encountered in chosing references, especially
for modulation frequency spectra as well as
in making proper background corrections are
considered and illustrated. Various methods
for calibrating the PA signal in terms of the
absolute fraction of radiation that is
absorbed or in absolute energy units, for
calorimetric purposes, are given. Absolute
calorimetric measurements on samples that are
photoacoustically saturated, are shown to be
feasible.

INTRODUCTION

In any process where radiant energy is converted partially

into chemical energy, electrical energy or radiation of a

different wavelength some heat production will occur. Therefore

calorimetry can help us in investigating such processes by

monitoring the heat generated as a function of experimental

variables. More accurately, for such studies we need to use

photocalorimetry, i.e. study the conversion of radiation

(especially UV-visible-IR) into heat. The history of

photocalorimetry goes back to the early experiments of Magee et

al.[1], who studied the quantum efficiency of photosynthesis in

Chlorella algae and the photobromination of several organic

materials. Bell[2] used similar instrumentation to derive

photosynthetic conversion efficiency from plots of ΔH vs. light intensity. In the last few years Jones[3], Adamson[4] and Olmsted[5] have revived photocalorimetry to study the enthalpies stored in various photochemical reactions. Also photocalorimetric studies on rhodopsin have yielded valuable information on the energetics of the visual process[6].

The energetics and the kinetics of photopolymerization processes have been studied by photocalorimetry as well[7]. Also "Photothermal Spectroscopy", used by Fujishima et al.[8] to determine quantum-and energy-efficiencies of photoelectrochemical cells, can be considered a photocalorimetric method.

Normal calorimetric methods as such are well established, not only in physics and chemistry, but also in biology[9], and various commercial instruments are available. More closely related is the, more sensitive, a.c. calorimetric method which has been applied recently to chemical and biological problems involving phase transitions[10].

The calorimetric use of photoacoustic (PA) methods was pioneered by Robin and colleagues[11] and by Hunter et al.[12], mainly for the study of gas-phase photoreactions. This last group also worked out a relaxation scheme for electronic excitation processes in terms of PA observables. Callis and coworkers[13] have developed a flash calorimetric technique for the study of photochemical and photobiological processes. This method is essentially the time domain equivalent of the photoacoustic one, and the relation between them is similar to that between flash photolysis and modulation excitation spectroscopy.

Photoacoustic and Photothermal methods.

The PA technique for calorimetry is essentially an indirect one, because the heat generated by the modulated exciting beam is converted into a pressure change, which is detected acoustically[14]. It would seem more straightforward, therefore, to measure this heat directly from the temperature change of the sample, or that of its direct surroundings. Brilmyer et al.[8,15] used a thermistor in direct contact with the sample to follow sample heating upon light absorption. Drawbacks of this, more direct, method are connected with the detection mechanism and its sensitivity. Use of modulated excitation, leading to modulated output, increases the sensitivity of the method considerably because much more noise and drift can be tolerated. However the response time of simple thermal detectors will limit the modulation frequency range that can be used. Furthermore, in this method, positioning the temperature sensor in, or very close to the sample is not always feasible and detection at a distance avoids this problem. Such detection can be accompished by sensitive IR detectors and this method has been put into practice by Nordal and Kanstad, and Busse (see their contributions in this volume). The detectors used have fast response times and thus modulation up to high frequencies is possible. At present the sensitivity of this technique (termed "Photothermal Radiometry") is such that rather high illumination intensities are needed. On the other hand no special sample chambre is necessary (as is the case for PA detection) and the technique may very well become a complimentary, or in some cases an alternative one to photoacoustics. As pointed out by Nordal and Kanstad it should be realized that sampling depths and signal saturation phenomena are quite different from those in photoacoustics.

Photoacoustic and a.c. calorimetric methods

If straight thermal detection of photoinduced energy changes is one link between normal calorimetry and (photoacoustic) modulation photocalorimetry, a.c. calorimetry[10] is the other one. In a.c. calorimetry a small oscillating voltage is applied across a heater contacting the sample, and the resulting temperature changes are measured as a function of temperature. Under certain conditions the heat capacity, C, can be calculated from the amplitude of the temperature oscillations. The method is used especially to detect and characterize phase transitions. Such phenomena have been studied recently by photoacoustics and a detailed theory for PA detection of them has been presented (Korpiun, this volume). It is therefore of some interest to briefly compare the two methods. Because in a.c. calorimetry, as in photoacoustics, modulated excitation is used, sensitivity is enhanced (over that of normal calorimetry). However, the upper frequency of the a.c. calorimetric method is much lower than that of PA, because no internal temperature gradients in the sample can be tolerated, i.e. bulk heating occurs, as opposed to thin surface film heating in many PA experiments (e.g. "thermally thick", non-transparent, or opaque "thermally thin" samples[14]). The limitations are similar to those encountered for the thermally thin sample case in the Rosencwaig-Gersho(RG)[14] theory in PA. Only thermally thin samples can be used for a.c. calorimetry, and this means that samples should be very thin and/or very low modulation frequencies should be used ($\leqslant$1mm at $\leqslant$1Hz for aqueous samples). This is so because the internal sample relaxation time, t_{in}, has to be shorter than the period of the exciting signal, 1/w. t_{in} is defined as $(C.\rho/k).d^2$ and $w.t_{in} \ll 1$

should be satisfied, i.e. $(w.C.\rho/k).d^2 \ll 1$. Here ρ is the sample density; k: sample thermal conductivity; w: angular modulation frequency; d; sample thickness. On the other hand, in the RG theory a sample is said to be thermally thin if $\mu_s > d$; here

$$\mu_s^2 = \frac{2k}{C.\rho} \cdot \frac{1}{w} \qquad \dots \dots \dots \dots \qquad (1)$$

where μ_s is the sample thermal diffusion length. By substituting eqn. 1 in $d^2/\mu_s^2 \ll 1$ we see that this is the same condition (except for a factor of 2) as found for a.c. calorimetry. Naturally the fact that the optical absorption length does not affect a.c. calorimetry limits the parallel and introduces other problems, such as PA saturation[14,16] (vide infra).

Photocalorimetric Uses of PA Effect

Up to now the widest calorimetric use of the PA effect has been in the investigation of luminescent samples, both liquids[17] and solids[18], mainly to determine quantum yields of radiative and non-radiative processes. The advantages over other methods are discussed at length in refs. 17 and 18, but they are valid only if correct calibration procedures, to be discussed below, are applied. From a PA study on the photopolymerization of diacetylenes the average number of polymerized molecules per absorbed photon could be derived[19] (cf. refs. 7). Photochemical and photobiological processes[20,21] have been studied by us, using i.a. the PA effect, to determine enthalpy changes and their chronology[22,23] in several systems (see also Prehn, this volume).

After a short theoretical introduction, we will give some examples of such calorimetric applications of the PA effect and use them to point out basic and practical problems that are encountered, when deriving quantum yields or enthalpy changes from raw PA data.

THEORETICAL BACKGROUND

Hunter and Stock[12] (HS) considered a simple photophysical process involving excitation to the first excited singlet state, subsequent radiationless and luminescent decay to the ground state, and intersystem crossing to the triplet state with subsequent radiationless decay to the ground state. They define two heat release parameters: H, the sum of heats that are released near-instantaneously, and L, the heat released in the slow step of the process (involving the triplet state), and derive the amplitude and phase angle of the PA signal as a function of modulation frequency (w), H, L and the triplet decay time, t_T.

Malkin and Cahen[20] (MC) considered inter alia a similar system, in terms of photochemical reactions and calculate the in-phase and quadrature PA signal as a function of w, incoming energy (Nhν), quantum yields (ϕ_i), and internal energy stored in intermediates (ΔE_i). They obtain for the general rate of heat production in a photoactive system (ref. 20, eq. 1):

$$H_{MC} = I \ (Nh\nu - \Sigma\phi_i\Delta Ep_i - \phi_L Nh\nu_L - \gamma Nh\nu); \qquad \ldots\ldots \qquad (2)$$

Here I is the absorbed light intensity, N is Avogadro's number, h is Planck's constant, ν and ν_L are the frequencies of incident and emitted radiation resp., the subscripts i and L indicate photochemical and luminescent processes resp., and γ is the

photovoltaic energy conversion efficiency. From eq. 2 we can
obtain the relative PA signal at wavelength λ and frequency w:

$$\rho(w,\lambda) = \alpha(\lambda) \cdot f(w)\{1-\Sigma \frac{\phi_i \Delta E_i}{Nh\nu} - \frac{\phi_L \nu_L}{\nu} - \gamma\} \quad \ldots \ldots \quad (3)$$

where $\alpha(\lambda)$ is the fraction of radiation absorbed by the sample at
λ, $f(w)$ is a correction factor depending on the sample, reference
and instrument, and the last three terms in the parentheses are
"photoactivity loss" terms, so called because they decrease the
maximum possible PA signal. If ρ is measured as a function of
modulation frequency, the life time of each photochemical
intermediate, i, and the luminescence decay time, will determine
at which frequency these intermediates and processes are "sensed"
photoacoustically. The energy content of the intermediate with
respect to the initial state and its formation efficiency will
determine the signal amplitude. This is illustrated schematically
in fig. 1. The simple photophysical process considered by HS is

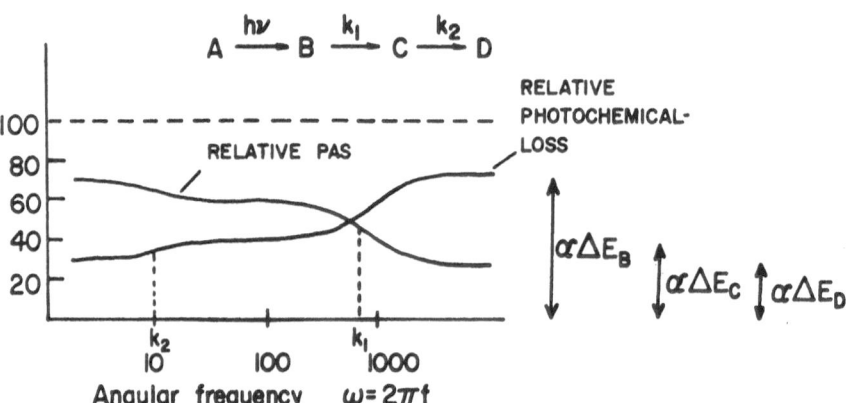

FIGURE 1.   Modulation frequency dependence of normalized in-
phase PA signal and relative "photochemical loss" for the process
indicated at the top. Formation of B is assumed to occur with
quantum yield φ, and intermediates C and D are formed from B and
C resp., with yield 1 (i.e. no branching). The maximal PA signal
(100 on l.h. scale) is obtained from a suitable reference. The
scheme is idealized for a case where the kinetic constants are
well-separated in magnitude, yielding a clear correlation between
the PA signal and the photochemically induced reaction sequence.

equivalent to the more general process $A + h\nu \rightarrow A^* \rightarrow B \xrightarrow{k_1} P$, if

the intermediate B is identified with the triplet state and the

final product $P \equiv A$. The rate of heat production for such a

process can be written as (ref. 20, eq. 18):

$$H_{MC} = (Nh\nu - \phi_L\ Nh\nu_L - \phi_B\ \Delta E_B + F) I_a sinwt - GI_a coswt \quad \ldots\ldots \quad (4)$$

where F and G are functions of ϕ_B, ϕ_p, ΔE_B, ΔE_p, k_1, and w, and

I_a is the amplitude of the modulated part of the absorbed light

intensity.

The results obtained for the PA signal are equivalent in the

HS and MC cases as can be seen by realizing that

$$H \propto Nh\nu - \phi_L\ Nh\nu_L - \phi_B\ \Delta E_B$$

and

$$L \propto \phi_B\Delta E_B - \phi_B\ \phi_p\ \Delta E_p$$

where H and L are defined as by HS, $Nh\nu$ corresponds to the energy

of incoming radiation, $\phi_L Nh\nu_L$ to that of luminescence, and $\phi_B \Delta E_B$

and $\phi_p \phi_B \Delta E_p$ to those stored in intermediate B and the final

product, P, resp. Noting finally that $t_T(HS) = 1/k_1(MC)$ we

arrive at the results shown in table I. From these it can be seen

that expressing the PA signal in terms of in-phase and quadrature

leads to more straightforward interpretations than use of the

amplitude/phase angle notation.

SOME EXAMPLES

Fig. 2 illustrates how measurement of the PA modulation

frequency spectrum allows one to estimate the lifetimes of

intermediates, in this case the triplet state of anthracene,

dissolved in ń-hexane. Both "uncorrected" and "corrected" data

TABLE 1

Expressions for the PA Signal, Obtained from a Simple Photophysical or Photochemical Process with One Intermediate.

Measured Quantity

Frequency (ω)	In-phase	Quadrature	Vector Amplitude	tan (phase angle)
0	$H+L$	0	$H+L$	0
$\to \infty$	H	0	H	0
k_1	$H+\dfrac{L}{2}$	$\dfrac{L}{2}$ (MAX)	$\sqrt{H^2+HL+L^2/2}$	$\dfrac{L}{2H+L}$
$k_1\sqrt{1+L/H}$	$H\left(1+\dfrac{L}{2H+L}\right);$	$\dfrac{L}{H+L}\cdot\sqrt{H(H+L)}$	$\sqrt{H^2+HL}$	$\dfrac{1}{2}\cdot\dfrac{1}{\sqrt{H(H+L)}}$ (MAX)

Inflection
point at $\omega=k_1$

are shown. The correction takes into account the variation of the detection system sensitivity, including the frequency dependent properties of the sample not connected with photoactivity (expressed by f(w) in eqn. 3) and the change in the relative absorbed light intensity ($\alpha(\lambda)$ in eqn. 3) within a thermal diffusion length, with frequency, because of the variation of this last parameter with frequency. The photoacoustic signal, ρ, given here by

$$\rho(w,\lambda) = \alpha(\lambda).f(w).\{1-\frac{\phi\cdot\Delta E}{Nhc}\cdot\lambda\} \qquad \ldots\ldots\ldots \qquad (5)$$

will, after correction for the first two factors, reflect the third factor, i.e. the photochemical loss. The corrected

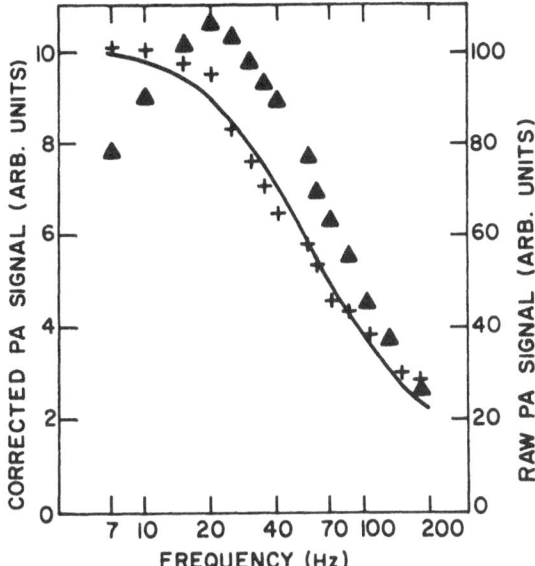

FIGURE 2.   Frequency spectrum of anthracene in hexane (1mM).
Measuring wavelength 340 nm. ▲ : raw data; +: data corrected for
variation of α (λ) and detection system sensitivity with
frequency. Correction was performed by dividing each signal by
that of the solvent to which black ink was added. Solid line:
simulated curve for first order decay at rate of 250 sec^{-1}. In
this case no stable products are formed and the PA signal at low
frequencies differs from the maximal one because of fluorescence.
At high frequencies the PA signal is decreased by energy storage
in the triplet state, as well.

frequency spectrum could be fitted to a simulated curve (solid

line) calculated according to ref. 20, for a first order decay

process at a rate constant of 250 sec^{-1}. Insufficient degassing

and, possibly, self-quenching of the triplet state at the high

anthracene concentrations used, cause a shift to faster decay, as

compared with literature values (60 sec^{-1})[24].

In fig. 3 results for the more complex photochemical cycle

of the bacteriorhodopsin - containing purple membranes from the

halophilic bacterium <u>Halobacterium</u> <u>halobium</u> are shown. The steps

in the lower, corrected, spectrum can be attributed to

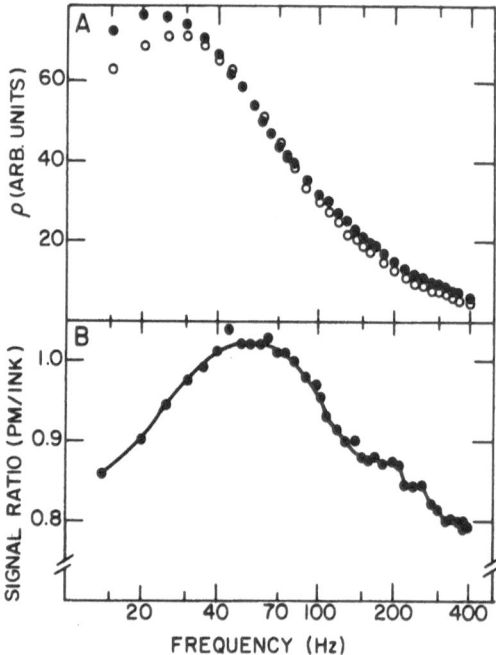

FIGURE 3.   Frequency spectrum of bacteriorhodopsin –containing
purple membrane fragments of Halobacterium halobium at pH 7. –A–
raw data for sample,   , and for reference, O, (of bleached
membranes + black ink)–B–corrected data. λ (excitation) = 565 nm.
(212kJ/mole); 1.0 on L.H. scale corresponds to 100% heat
dissipation. Only ∿85% of the 565 nm photon energy can be used
in the photoprocess (∿ 650 nm corresponds to excitation to the
lowest   vibrational   level   of   the   excited   state   of
bacteriorhodopsin). Taking into account a quantum yield of ∿ 0.7
for the very fast (psec.) primary process, at ∿ 400 Hz the PA
data show that some 80 kJ of the 180 kJ available (per utilized
photon) have been dissipated as heat. Using Table 1, approximate
decay times of several intermediates may be estimated from the
figure, e.g. the plateau at ∿ 150 Hz (∿1000 sec $^{-1}$), which may
correspond to the decay of the M_{412} photointermediate.

transitions   between   different   photochemical   or   photoinduced

chemical intermediates[22]. Because these measurements are done in

a gas-filled cell photo-induced volume changes will be of

secondary importance   here,   in   contrast   with   the   flash-

calorimetric studies of Ort and Parson,[25] where a completely

liquid-filled cell is used, connected to a very sensitive

capacitor   microphone.   (Flash   calorimetry[12]   is   essentially
equivalent to the pulsed optoacoustic method described by Patel
in this volume).

REFERENCING, CORRECTING AND CALIBRATING PHOTOACOUSTIC DATA

In order to be able to extract information from the raw PA
signals,   be   it   an   extinction   coefficient,   absolute   energy
values, or decay times, using either wavelength or modulation
frequency spectra, several questions have to be considered,
namely those concerning:

A - Saturation

B - Referencing (normalizing) with respect to λ - or w -
dependence, that is not related to the sample properties of
interest.

C - Background correction

D - Calibration

As can be seen from a comparison between raw and corrected data
in figs. 2 and 3, proper procedures to answer these questions are
imperative, if meaningful data are to be obtained.

A - Saturation

Photoacoustic   saturation[14,16]   limits   the   analytical
applications of PA spectroscopy, especially when data for more
than one wavelength are used. Total saturation may be recognized
readily, but cases of partial saturation leading to distorted
signal ratios for PA signals from several wavelengths, are less

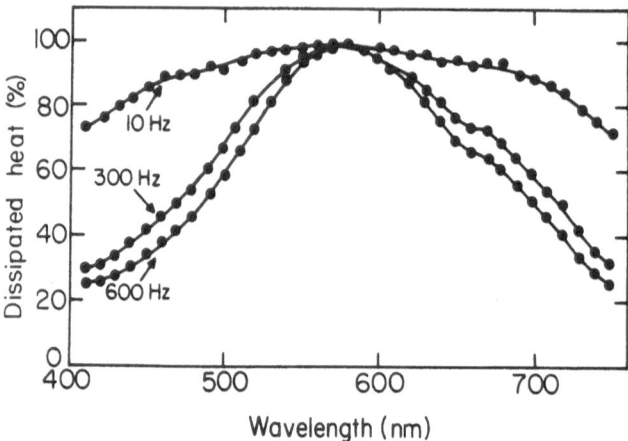

FIGURE 4. PA wavelength spectra of concentrated methylene blue
solution at several modulation frequencies, showing dimer
formation at these concentrations and illustrating PA saturation
effects.

easy to detect. The signal dependence on w may give a first

indication of saturation and in favourable cases, allows one to

avoid the problem. Fig. 4 illustrates the problem of PA

saturation for a very concentrated solution of methylene blue in

water. The spectra show increasing peak flattening at lower

modulation frequencies, where the thermal diffusion length (cf.

eqn. 1) will be greater. They also show another aspect of

performing measurements in highly concentrated solutions,

something that is often necessary because the PA method is in

many cases a thin film one and enough heat has to be generated in

a thin layer to enable PA detection. The unsaturated spectrum of

methylene blue differs markedly from the normal absorption

spectrum and rather resembles that of the dimer, which is usually

obtained at low temperatures. However, usually, PA saturation

should not be too serious a problem in aqueous solutions, because

99% absorption within ~50μm requires an optical density of 40 for an optical path of 1 cm, which is very high for normal chemical systems. Fig. 5 illustrates PA saturation for a solution of chlorophyllin in water, used by us as a model for studying the possibility of saturation in photosynthetic systems[23]. Both transmission (0.1 mm path length) and PA data are shown for two concentrations differing by a factor of ~10. From the differences between the PA and transmission data on the one hand, and between those for lower and higher concentrations on the other hand, the effect of PA saturation on absorption peak ratios can be seen clearly. However the situation in heterogeneous, finely dispersed systems may be quite different as the form and local concentration of the absorber can influence PA saturation considerably (Cahen, unpublished; compare also fig. 1A in ref. 28

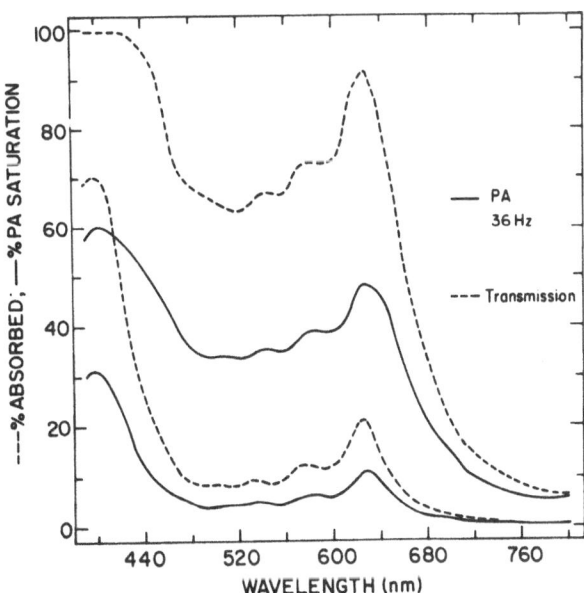

FIGURE 5. Wavelength spectra of chlorophyllin in water, measured photoacoustically and by transmission spectroscopy. Upper curves are for 10X more concentrated solution than lower ones.

with fig. 4 in ref. 29, spectra of tetraphenylporphyrin in different forms). Generally speaking the method of measuring the PA signal dependence on the optical absorption coefficient[26,27] is the most reliable one, especially as it may deviate from that derived from simplified RG theory[14,26,29] (see sub-section D.1).

B - References for λ - and w - Dependence.

Correction factors for the λ - dependence of the exciting beam intensity can be obtained from direct measurement of the incoming radiant power, or from the PA signal of carbon black or any other photoacoustically opaque sample. This last method can also correct for possible λ - dependence of the PA cell if the same or identical cells are used. If w - dependent data are sought for calorimetry, the PA signal should be corrected not only for the w - dependence of the PA cell and the electronics, but also for the normal w - dependence to be expected from a PA signal of a photoinactive sample (cf. figs. 2 and 3)[12,14,26,30]. Ideally the sample used for w - reference should have the same constitution (liquid, suspension, solid), thermal properties and optical absorption as the sample to be studied. Because carbon black and the like are PA opaque samples, they will not, in general, be satisfactory as w - reference. If, for example, the sample is a liquid suspension (such as most biological samples), carbon black will be a bad reference, undiluted black ink less so, while diluted ink may be satisfactory. A variation of this last method was used to obtain the data shown in fig. 3. Fig. 6 ilustrates the effect of the w - reference on the modulation frequency spectrum, for a heterogeneous biological system. For the inactive, _dry_, preparation of membranes dilute, dry carbon

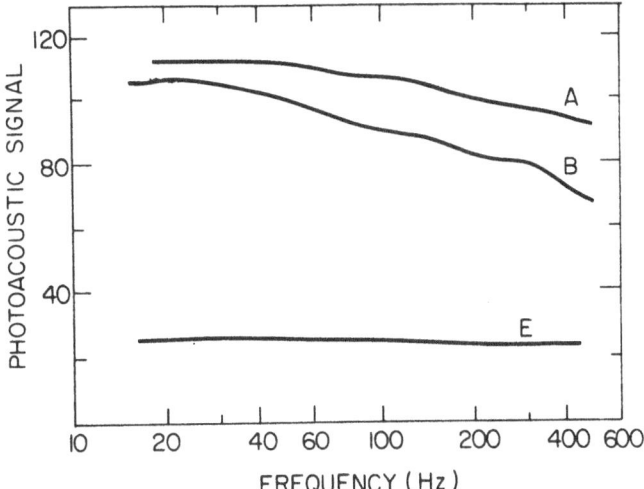

FIGURE 6. Effect of modulation frequency reference on normalization of PA signal for **H. halobium** purple membranes (=PM). Experimental conditions as in Fig. 3 -A- : PM in H_2O **vs**. Ink; -B- : PM in H_2O **vs**. carbon black; -E- dried PM (photochemically inactive down to 20 Hz) **vs**. dry carbon black. Neither of these two (A, B) methods of referencing is correct, but rather, a diluted suspension of bleached PM + ink has to be used (see text).

black is a reasonably appropriate reference. Fig. 7 shows the modulation frequency spectrum of a $Co(NH_3)_6^{3-}$ solution, which is presumably inert photochemically at the irradiation wavelength used (445 nm)[4]. Indeed there are only very small changes in the corrected signal **vs**. frequency[6]. The increase of the signal at low frequencies is due to background light absorption by the cell body and windows, which is not sufficiently corrected here and will be discussed below.

C - Background Correction.

Ideally data from the sample under investigation should be used for this correction e.g. if the sample is optically

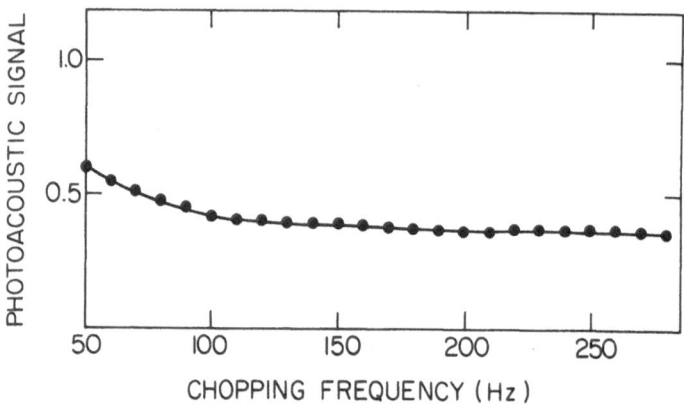

FIGURE 7. Modulation frequency spectrum of 200 mM $CO(NH_3)_6^{3-}$ in 0.01 M $HClO_4$ solution. Illumination at 445 nm; correction as in Fig. 2.

transparent at λ_1, the PA signal at λ_1 can be used to obtain the true background of cell and sample at the wavelengths of interest (by normalizing the λ- spectrum of a transparent sample at λ_1). The w - dependent background correction will depend on the thermal properties of the sample, sample holder and/or backing, and their optical ones at the wavelengths of interest, i.e. thermally thick or thin, optically transparent or opaque. If the signal of the empty PA cell is used for background correction, the decrease in gas volume because of the sample volume can lead to too low a value for the background. Finally, if a liquid sample is investigated with a gas phase around it, the effect of condensation on the cell walls should be considered[31].

D = Calibration

I. Optical absorption coefficient. Several methods can be used for this purpose, such as the use of phase angle and/or amplitude frequency dependence[32] or the use of an internal reference with known extinction coefficient.

Related to this is the method where a sample is used that absorbs all the incoming radiation, converts it to heat only, and has the same thermal transfer properties (especially: same surface area) as the sample under investigation, so that the ratio of the signal of this reference to that of the sample yields the true fraction of absorbed radiation. When liquid samples are used, and if the reproducibility of the measurement is good, this is probably the simplest method to use. By plotting the PA signal (normalized to the saturation value) against the optical absorption coefficient, a calibration curve can be constructed, which can be fitted theoretically, if the thermal diffusion length of the sample is known[26]. In this way knowledge of μ_s and ρ (saturated) allows determination of the optical absorption coefficient. Fig. 8 shows a test of this method, where

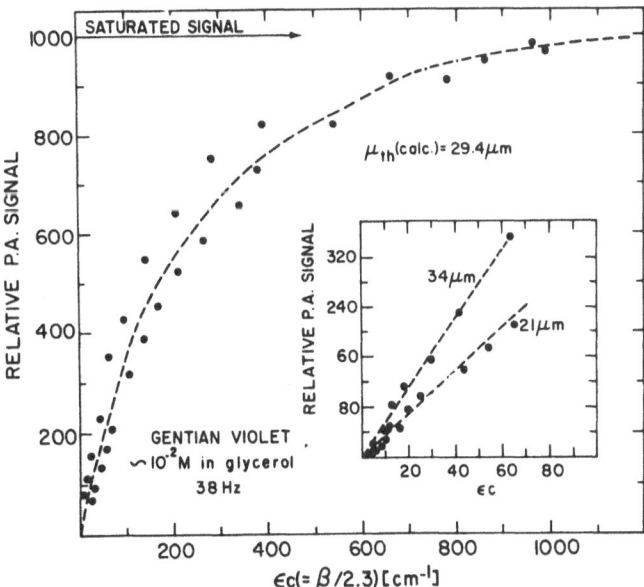

FIGURE 8. Dependence of PA signal on optical absorption coefficient (β: molar extinction coefficient, ε, times concentration). Insert: the linear part at low β values, expanded. Dashed lines are calculated ones for thermal diffusion lengths indicated next to them.

the optical absorption coefficient was determined separately and
μ_s is calculated, from the initial slope of the curve (compare
also Poulet et al. this volume).

II. PA signal in absolute energy units.

Electrical calibration, by the use of a resistance wire in the
cell[33], is wrought with many difficulties because of the need to
generate the heat under conditions that are identical to those
encountered in the PA signal generation (thermal properties,
distance from sample/gas interface). (We deal here only with PA
cells for the study of condensed phase samples, where a gas phase
is used to transfer the acoustic signal. Electrical calibration
is much more suitable for cells containing gas or condensed
phases only). The photothermophone[34] suffers from similar
problems, which are important chiefly when liquid or powdered
samples are used. If, without changing the thermal and optical
absorption properties of the sample, we can measure it in a state
where it undergoes thermal decay only, then the total heat
dissipation conditions defined in this way can serve to calibrate
the PA signal. For bacteriorhodopsin-containing purple membrane
fragments of H. halobium (cf. figs. 3, 6)[22] we found that the
only way to obtain such a photoinactive sample was to bleach the
purple membranes with hydroxylamine under illumination and
subsequently add black ink to them until the reference had the
same optical density at 565 nm (the irradiation wavelength, at
which the photoprocess is induced; absorption maximum of
bacteriorhodopsin) as the sample. In this way we could obtain a
scheme of enthalpy changes, calibrated in absolute energy units,
for those steps of the photoprocess that could be detected in our

set-up. Using the simple model discussed before[20], approximate
decay times for intermediates were determined as well (fig. 9).
Simpler ways to obtain proper calibration, are the use of
chemical inhibitors of photochemical reactions[21], the use of
strong background radiation to saturate photoactivity[23], and the
use of quenchers in the study of fluorescent samples[17,18]. Fig.
10 illustrates this last possibility for basic aqueous solutions
of sodium difluorescein (Na_2Fl), to which increasing
concentrations of quencher (KI) are added. The strongly quenched
sample (F) shows a spectrum closely similar to the transmission
and fluorescence excitation spectrum. The increase in PA signal
with increasing quencher concentration is seen clearly. We can
now use the PA - λ spectrum to calibrate the PA signal, because,
all other factors being equal, shorter wavelength excitation will
cause more heat generation than that at longer wavelengths,
because of vibrational decay in the excited state. Although, in
principle it is possible to use readings at two wavelengths only,
more reliable results can be obtained by using data over a wide
wavelength region. Then the wavelength dependence of heat
dissipation ("dissipation spectrum")[21] can be obtained from eqn.
5, where ΔE is equivalent to the integrated wavelength of
fluorescence emission. Use of eq. 5 is simplified if α can be
measured independently. The λ - dependence of the heat
dissipation is now given by:

$$\rho(\lambda)/\alpha(\lambda) = f \left\{ 1- \frac{\phi(\lambda) \cdot \Delta E(\lambda)}{Nhc} \cdot \lambda \right\} \quad \ldots\ldots \quad (6)$$

If we use values for α, measured by PA methods in the same set-up
as ρ (and thus expressed in the same units as ρ), the factor f is
eliminated, and the true dissipation spectrum is obtained. Such

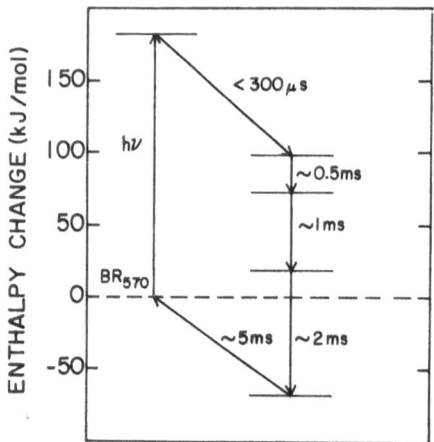

FIGURE 9.   Scheme of enthalpy changes during photocycle of H. halobium as derived from corrected modulation frequency dependent data (cf. Fig. 3) on purple membrane fragments; pH 7.

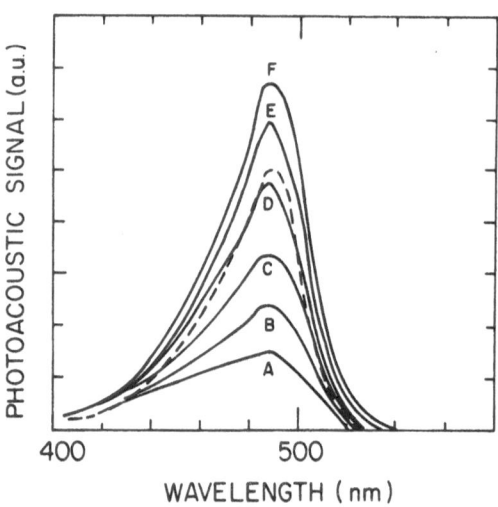

FIGURE 10.   PA spectra of 10^{-3} Na_2Fl in 10^{-2} M NaOH at 40 Hz. -A- no quencher added; B; C; D; E; F = 0.1; 0.2; 0.3; 0.4; 0.5 M KI added. Dotted line: transmission spectrum.

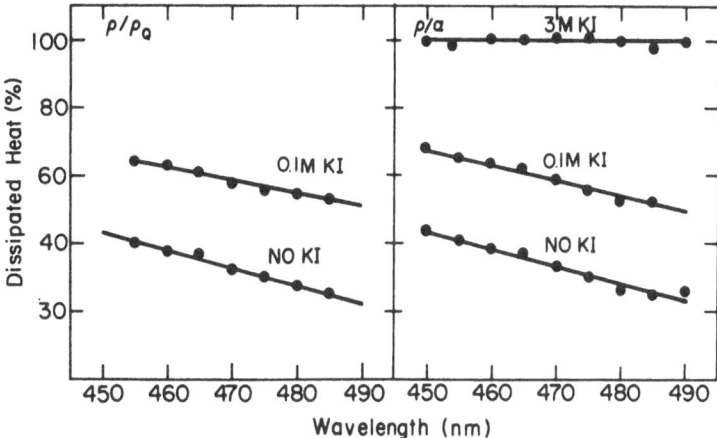

FIGURE 11.   Dissipation spectra for NA_2Fl, using PA data from a totally quenched sample (ρ_Q, left hand side), or data from transmission spectroscopy, taking into account μ_s as effective optical pathlength (α, right hand side), for correction. Knowledge of ΔE, from $\bar{\lambda}$(emission) = 532 nm; yields ϕ(fluor.) = 0.93 ± 0.3.   Reprinted with permission from J. Phys. Chem. 84, 3384 (1980). Copyright (1980) American Chemical Society.

in situ determinations of   are possible when conditions of 100% heat dissipation can be defined, as is the case here, at high (3M) KI concentrations. Fig. 11 shows such dissipation spectra (left hand side). Alternatively, if a cyclic photochemical process is studied it is possible, in some cases to obtain α from low frequency PA data, if the process can be assumed to be complete at such frequencies. Especially in this case, but also in the preceding and following ones the dependence of α on the modulation frequency should be considered. As α represents that fraction of the absorbed light that can contribute to the PA signal, it will depend on the thermal diffusion length, which represents, in the case of samples that are not photoacoustically opaque, the effective optical path length and varies with w.

When in situ measurement of α is impossible, knowledge of the thermal properties of the sample, combined with independent determination of α (by optical transmission, for clear solids or solutions, for example, or by diffuse reflectance, for turbid solutions or solids) will enable calculation of α with reasonable accuracy (fig. 11, right hand side).

The "photochemical loss" term, $(\phi.\Delta E/Nhc).$, in eq. 6 can be obtained from the dissipation spectrum, if the product $\phi.\Delta E$ is constant over a wavelength region, by extrapolating the straight line obtained in that region to $\rho/\alpha=o$. The wavelength where this line crosses the λ - axis will yield $\phi.\Delta E$ directly in nm. Alternatively the photochemical loss term can be obtained from the ratio of the slope of the straight part of the dissipation spectrum and the intercept of this line with the ρ/α axis.

PHOTOACOUSTIC SATURATION IN PHOTOCALORIMETRY

An interesting case arises when PA saturation is encountered in photocalorimetric measurements[17e,18d,19,35]. If the sample is photoinactive the wavelength spectrum that is obtained after correcting for the intensity variation of λ, will be a straight line parallel to the λ - axis. If, on the other hand, photoactivity is present over a certain wavelength region, the corrected spectrum will yield the dissipation spectrum directly, as illustrated in fig. 12. Here the spectral response of a p/n GaAs solar cell is shown together with its low frequency PA signal under conditions of optimal energy conversion. Because of the different electrical loads the spectra are not strictly comparable and S/N problems limit the accuracy of the measurement, especially at wavelengths longer than that

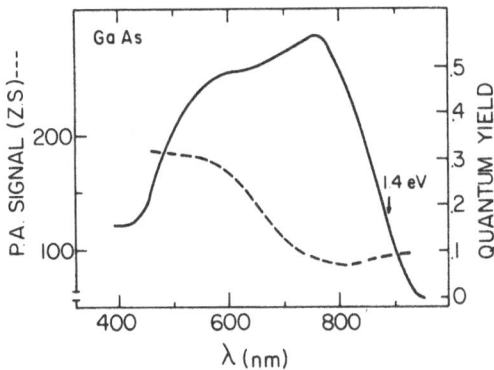

FIGURE 12. Spectral response (quantum yield under short circuit conditions) and zero-suppressed (z.s) PA signal for p/n GaAs solar cell. Modulation frequency: 22 Hz; quantum efficiencies: 33%, 22%; 50% and 13% at 560, 620, 820 and 900 nm, resp. Conversion efficiencies: 20, 17, 23 and 0% at these wavelengths, resp. Optimal load for PA measurements varied between 7,000 and 850 Ohms, depending on incident light intensity.

corresponding to the band edge. However, a dip in the PA spectrum around the region of optimal quantum yield (under short circuit conditions) can be seen clearly, and the shape of the

PA spectrum definitely does not resemble the reflection spectrum of GaAs (obtained on a Cary 17D spectrophotometer with reflectance attachment, to approximate the conditions under which the PA data are obtained as closely as possible), which shows a band edge around 900 nm ($\sim$1.4eV) with increasing reflectivity from 700 nm downwards. As shown in fig. 13 this saturation does not hinder us in any way obtain the power curve of this solar cell photoacoustically[35]. Here we encounter the simplest case of calibration because under open circuit and short circuit conditions no energy conversion into electricity occurs in a photovoltaic cell and all the absorbed radiation decays thermally (cf. also refs. 36, 37). The upper curve in fig. 13 represents a situation close to optimal for this solar cell assembly, and the

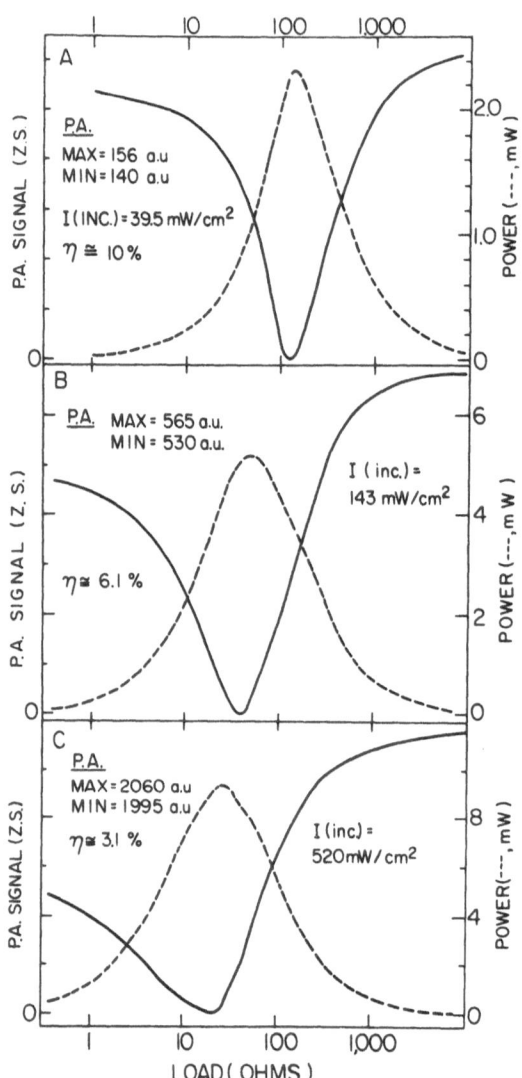

FIGURE 13. Power curves and zero-suppressed (z.s) PA signals for GaAs p/n solar cell. Illumination band between 480 and 750 nm. Modulation frequency: 22 Hz; 60 mm² active area; from top to bottom: open circuit voltage: 795, 835, 840 mV; short circuit current: 8.3, 28, 60 mA/cm²; fill factor: 58, 36, 32%; conversion efficiencies (PA): 10.2, 6.2, 3.15%; conversion efficiencies (from I-V curves): 9.8, 6.1, 3.0%.

PA power curve is the mirror image of the normal one. At higher light intensities power losses at the contacts apparently start to play a role leading to some heat generation outside this PA cell (described in ref. 31).

SUMMARY

Accurate, absolute calorimetric measurements by PA methods are subject to many experimental difficulties. Meaningful results can be obtained however, if the various corrections and calibrations that are needed, are carried out taking into account the processes leading to PA signal generation under the experimental conditions that are used. The difficulty of these correction and calibration procedures will vary from sample to sample and may be minor, as in the case of photovoltaic devices[35-37], or major as in the case of some biological materials[20-24]. The problem of saturation of the PA signal which, in some cases, rather limits analytical PA spectroscopy, can be overcome for calorimetric measurements[17e], and even turned into an advantage if completely saturated samples can be used[18d,19,35].

ACKNOWLEDGEMENTS

We thank S.R. Caplan and J. LeGrange for stimulating discussions, and Knowles Electronics Inc. for gifts of microphones. This work is supported, in part, by the U.S.-Israel Binational Science Foundation, Jerusalem, Israel.

REFERENCES

1. J.L. Magee, T. DeWitt, E.C. Smith, F. Daniels, J. Amer.
    Chem. Soc. 61, 3529-3533 (1939); J.L. Magee,
    F. Daniels, ibidem, 62, 2825-2833 (1940).

2. L.N. Bell in "Biofiz. Metody. Fizid. Rast." Yu. G.
    Molotkovskii ed., Nauka, Moscow (1971) pp. 106-128;
    Chem. Abstr. 76; R56153p.

3. G. Jones II, T.E. Reinhardt, W.R. Bergmark, Solar Energy,
    20, 241-248 (1978).

4. A.W. Adamson, A. Vogler, H. Kunkely, R. Watcher, J. Amer.
    Chem. Soc. 100, 1298-1300 (1978).

5. M. Mardelli, J. Olmsted, J. Photochem. 7, 277-285 (1977);
    J. Olmsted, J. Phys. Chem. 83, 2581-2584 (1979); J. Amer.
    Chem. Soc. 102, 66-71; 7619 (1980).

6. A. Cooper, C.A. Converse, Biochem. 15, 2970-2978
    (1976); A. Cooper, FEBS Lett. 100, 382-384 (1979);
    A. Cooper, Nature (London) 282, 531-533 (1979).

7. W.R. Hertler, P.J. McCartin, J.R. Merill, G.R. Nacci,
    W.J. Nebe, Photograph. Sci. Eng. 23, 297-301 (1979);
    R.W.Bush, A.D. Ketley , C.R. Morgan, D.G. Whitt,
    J. Radiat. Curing 7, 20-25 (1980).

8. A. Fujishima, Y. Maeda, K. Honda, G.H. Brilmyer, A.J. Bard,
    J. Electrochem. Soc. 127, 840-846 (1980).

9. See for example I. Wadsö, and others in "Applications of
    Calorimetry to Biochemistry," Biochem. Soc. Transact.
    44, 561-564 (1976).

10. O.S. Tanasijczuk, T. Oja, Rev. Sci. Instrum. 49, 1545-1548
    (1978).

11. M.B. Robin, in "Optoacoustic Spectroscopy and Detection"
    Y.H. Pao ed., Acad. Press, New York (1977) pp. 167-191,
    and references therein.

12. T.F. Hunter, M.G. Stock, J. Chem. Soc. Far. Trans. II, 70,
    1022-1028 (1974), and preceding and following articles.

13. J.B. Callis, J. Res. N.B.S.A., 80A, 413-419 (1976).

14. A. Rosencwaig, "Photoacoustics and Photoacoustic Detection",
    Wiley, N.Y. (1980).

15. G.H. Brilmeyer, A. Fujishima, K.S.V. Santhanam, A.J.
    Bard, Anal. Chem. 49, 2057-2062 (1977).

16. R.B. Somoano, Angew. Chem. Int'l. Ed. Engl., 17, 238-245
    (1978).

17. See for example a) E. Hey, K. Gollnick, Ber. Bunsenges. Phys.
    Chem. 72, 263 (1968); (Abstract only); Int. Conf.
    Photochemistry, Munchen, Sept. 6-9 (1967); Preprints,
    II. pp. 465-481. b) W. Lahmann, H.J. Ludewig, Chem. Phys.
    Lett. 45, 177-179 (1977); c) M.G. Rockley, K.M. Waugh,
    Chem. Phys. Lett. 54, 597-599 (1978); d) M.J. Adams,
    J.G. Highfield, G.F. Kirkbright, Analyt. Chem. 49,
    1850-1852 (1977); e) D. Cahen, H. Garty, R.S. Becker,
    J. Phys. Chem. 84, 3384-3389 (1980).

18a) R.G. Peterson, R.C. Powell, Chem. Phys. Lett. 53, 366-368
     (1978); R.C. Powell, D.P. Neikirk, J.M. Flaherty; J.G.
     Gualtieri, J. Phys. Chem. Solids, 41, 345-350 (1980).
  b) R.S. Quimby, W.M. Yen, Optics. Lett. 3, 181-183 (1978);
     J. Appl. Phys. 51, 1780 (1980).
  c) F. Auzel, D. Meichenin , J.C. Michel, J. Lumines .18/19,
     97-99 (1979).
  d) M.J. Adams, J.G. Highfield, G.F. Kirkbright, Anal. Chem. 52,
     1260-1264 (1980).

19. R.R. Chance, M.L. Shand, J. Chem. Phys. 72, 948-954 (1980).

20. S. Malkin, D. Cahen, Photochem. Photobiol. 29, 803-813
    (1979).

21a) D. Cahen, S. Malkin, E.I. Lerner, F.E.B.S. Lett. 91, 339-342
     (1978); b) D. Cahen, H. Garty, S.R. Caplan, F.E.B.S. Lett.
     91, 131-134 (1978); c) H. Garty, D. Cahen, S.R. Caplan,
     in "Energetics and Structure of Halophilic Microorganisms"
     (S.R. Caplan and M. Ginzburg eds.) Elsevier/North-Holland
     (1978), pp. 253-259.

22. H. Garty, D. Cahen, S.R. Caplan, Biochem. Biophys . Res.
    Commun. 97, 200-206 (1980).

23. N. Lasser-Ross, S. Malkin, D. Cahen, Biochim. Biophys. Acta.
    593, 330-341 (1980).

24. J.B. Birks, "Photophysics of Aromatic Molecules", Wiley N.Y.
    (1970).

25. D.R. Ort, W.W. Parson, J. Biol. Chem. 253, 6158-6164 (1978);
    Biophys. J. 25, 341-353; ibidem. pp. 355-364
    (1979).

26. J.F. McClelland, R.N. Kniseley, Appl. Opt. 15, 2658-2663;
    Y.C. Teng, B.S.H. Royce, J. Opt. Soc. Am. 70, 557-560
    (1980).

27. R. Poulet, R. Unterreiner, J. Chambron, J. Appl. Phys. 51,
    1738-1742 (1980) S. Malkin, D. Cahen, Anal. Chem. in
    press (1981).

28. W.H. Fuchsman, A.J. Silversmith, Anal. Chem. 51, 589-590
    (1979).

29. D. Cahen, E.I. Lerner, A. Auerbach, Rev. Sci. Instrum. 49,
    1206-1209 (1978).

30. F.A. McDonald and G.C. Wetsel Jr., J. Appl. Phys. 49,
     2313-2322 (1978).

31. D. Cahen, H. Garty, Anal. Chem., 51, (8), 1865-1877 (1979).

32. J.C. Roark, R.A. Palmer, J.S. Hutchison, Chem. Phys. Lett.
     60, 112-116 (1978).

33. J.M. McDavid, K.L. Lee, S.S. Yee, M.A. Aframowitz, J. Appl.
     Phys., 49, 6112-6117 (1978); D. Hursh, T. Kuwana, Anal.
     Chem. 52, 646-649 (1980).

34. J.C. Murphy, L.C. Aamodt, Appl. Phys. Lett. 31, 728-730
     (1977).

35. D. Cahen, Appl. Phys. Lett. 33, 810-811 (1978).

36. W. Thielemann, H. Neumann, Phys. Stat. Sol. A. 61, K123-125
     (1980).

37. A.C. Tam, Appl. Phys. Lett. 37, 978-981 (1980).

RELATIVE EFFICIENCY OF FLAT PLATE SOLAR COLLECTORS BY TRANSMISSION PHOTOACOUSTICS; IMPORTANCE OF PLATE VIBRATIONS IN SIGNAL GENERATION

David Cahen
Department of Structural Chemistry,
The Weizmann Institute of Science,
Rehovot, 76100, Israel

ABSTRACT

The photoacoustic effect can be used to obtain information on the performance of photothermal energy converters. The low frequency transmission signal can be correlated with over-all converter performance, while high frequency reflection data may become representative of absorber emittance. This is to be added to previously described relations between photoacoustic and absorptance data. A major complicating factor in photoacoustics on opaque flat plates is the possible occurrence of plate vibrations because of non-uniform sample heating. This has to be taken into account especially when corrections are to be made for variations in thickness beetween flat plate samples. Possible improvements, over the gas-phase microphone method used here, for quality control under actual operating conditions are suggested.

INTRODUCTION

The actual, and practical, efficiency of a photothermal converter is given by the ratio of useful heat that can be extracted from it Q_{out}, and the integrated incident energy, $I_{inc}[= \int W(\lambda)d\lambda$, where $W(\lambda)$ is the incident energy at wavelength λ]. This efficiency is determined mainly by the optical and thermal properties of the converter[1]. Ideally, such efficiencies should be determined under conditions that are close to those under which the converter will be used in practice. For selective surfaces the ratio of integrated solar absorptance,

$$\alpha = {}_0\int^{\infty} \alpha(\lambda,T).W(\lambda)d\lambda / {}_0\int^{\infty} W(\lambda)d\lambda$$

and integrated emittance

$$\varepsilon = {}_0\!\int^\infty \varepsilon(\lambda,T).W_B(\lambda,T)d\lambda / \int_0^\infty W_B(\lambda,T)d\lambda$$

where $W_B(\lambda,T)$ represents the black-body spectral density at λ and temperature T, at a specified temperature is often used as a figure of merit. However this method often overestimates the importance of one of these properties[1]. Therefore calorimetric methods are sometimes preferred, but in many cases they do not lend themselves to use under close to "practical" conditions, and often put severe restrictions on sample size and shape[2].

Use of the photoacoustic (PA) effect [3] for characterization of such converters has been suggested[4], and PA determinations of α have been reported[4]. Because the non-illuminated back-surface of the converter is in contact with the heat transfer fluid in most cases, the practical conversion efficiency will be determined by the temperature differential with the surroundings, ΔT, and the efficiency of heat transfer from the back-surface to the fluid. The PA signal, as derived in Ref.3 (in the "thermal piston" Rosencwaig-Gersho, RG, theory) depends on ΔT and thus monitoring the PA signal at the back-surface of the sample should yield a measure of the converter efficiency. While such a method will not yield absolute efficiency values, as can be obtained from normal calorimetry, it has the advantage of high sensitivity (as it is a differential method, using modulated excitation) and greater simplicity. It can be understood intuitively that the method will become more reliable at lower modulation frequencies, where sample heating will be more uniform and conditions will approach ultimately those used in normal transient calorimetry[2]. At higher frequencies the sample will become thermally thicker[3] and the acoustic signal detected at the non-illuminated, back-surface (transmission PA) will ultimately be generated by direct sample vibrations only. These vibrations, caused by non-uniform heating of the sample, which leads to sample buckling, especially with edge-clamped samples, also depend on ΔT, and through it on the sample temperature. At high frequencies, however, only the temperature differential across the first few thermal diffusion lengths (μ_{th}), is involved, making such measurements less useful for our purpose. Detection of the PA signal, generated at the illuminated, front-surface (reflection PA) at any frequency, is not satisfactory because then non-radiative heat losses and the nature of the back-surface are not properly taken into account. At high frequencies such reflection PA data may, however, constitute some measure of the sample emittance.

EXPERIMENTAL

A PA cell was used that allowed both reflection and transmission measurements, using the sample in the form of a flat plate, as one of the cell walls, (Fig. 1). The sample was pressed against an O-ring around this "open" cell, and illuminated through light-guides. For transmission measurements the side of the sample, which had the solar radiation absorbing coating, was facing away from the cell. By turning the sample around and illuminating from the other (left hand, in Fig. 1) side front-surface data could be obtained. This arrangement, described and evaluated in detail elsewhere[6], can give rise to measurable mechanical vibrations of the sample (buckling) as predicted first by Rayleigh[7], for the case of uneven sample

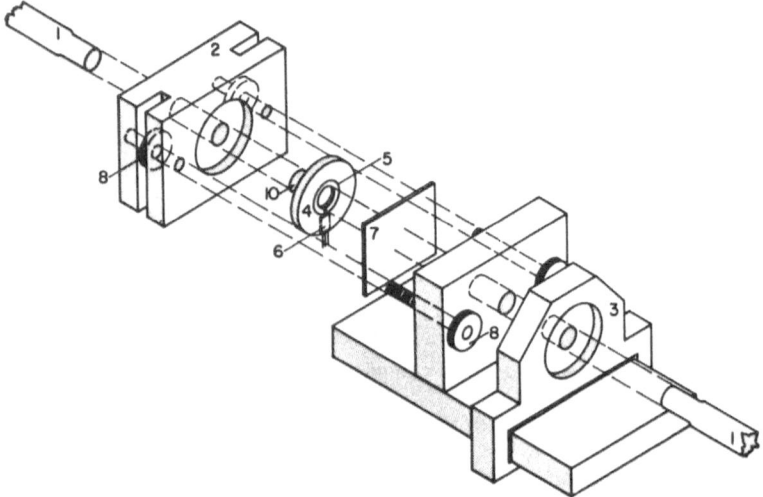

FIGURE 1 Exploded view of PA cell. Illumination reaches sample
[7] through flexible light guide [1] (left: reflection; right:
transmission). The PA cell [4] is clamped between holders [2] and
[9] by screws [8], and sealed by O-ring [5]. Rigid light guide
[10] seals the PA cell from the other side. [3]: light guide
support; [6]: Knowles BL 1685 microphone. Actual PA cell volume
is 85mm^3.

heating. Therefore, if samples of different thicknesses are
compared, the relative contributions of the thermal wave (which
car be severely damped in transmission PA) and of the direct
mechanical sample vibrations have to be taken into account. This
is discussed in detail below (see CALIBRATION).

Samples were illuminated by a 450W Xe-arc lamp, using a Schott
113 IR filter, somewhat simulting solar irradiation (2kW/m^2, i.e.
$\sim$ 2xAM1), and a mechanical chopper. A 0.9 cm diameter circle in
the centre of the 5x5 cm samples was in contact with the PA cell.
On all samples a 2x2cm central area on the back-surface was
sandblasted to remove oxidation layers and provide uniform
surface roughness (vide infra). Sample thickness varied between

0.38 and 1.00mm. Most plates were of copper, onto one side of which a copper oxide layer was plated[8] as a selective coating. These samples were part of a series with varying performance, the best of which are manufactuued commercially (by Miromit Ltd.). Some samples of black chrome on copper and of a proprietary tin oxide-based coating on stainless steel (from Electra Ltd.), were included as well. Integrated absorptances were measured on a CARY 17D Spectrometer equipped with a 1711 integrating sphere attachment, yielding near normal (6°) absorptance values. Emittance values were determined using a thermopile $(60^{\circ}$ angle), and for some samples by vacuum calorimetry (hemispherical) with the sample at $100^{\circ}C$.

CALIBRATION

Because samples with different thicknesses were used, the transmission (and to a lesser extent also the reflection) PA signals had to be corrected for the effect of the thickness t, on the signal. This was done empirically as the signal is composed of at least two components (thermal piston, sample buckling), each with its own dependence on t. For this purpose series of clean plates of copper and stainless steel (303) with varying t were used. These plates were sandblasted on both sides, to obtain uniform surface roughness (and remove oxidation layers). This treatment is important because increased surface roughness increases sample emittance. The extent of the increase is related to the surface micro-roughness and grain-size[9], and can be quite considerably[10] (a 20-40% increase is reported in ref.10). Also the PA signal is known to depend on surface

roughness, even after correction for reflection decrease (a ~ 10%
signal increase was found after sandblasting, for the
transmission PA signal, for these samples; Cahen, unpublished;
cf. also refs 11,12 for particle size effects). For all plates
reflection and transmission PA measurements were performed at
several frequencies. In this way data such as those shown in
Fig.2 (for transmission PA) were obtained. As can be seen
clearest from the low frequency data for the copper plates, the
behaviour is non-linear, and deviates strongly from that expected
for straightforward damping of the thermal wave, according to
$\exp(-t/\mu_{th})$. As is well-known now acoustic waves caused by local
expansion and contraction of the sample may also contribute to
the PA signal ("acoustic piston")[13]. While such a contribution
may become important at high frequencies for thick samples
(thermally thick), it is negligible for the thermally thin copper
plates showing the non-linear behaviour [13,14]   Another

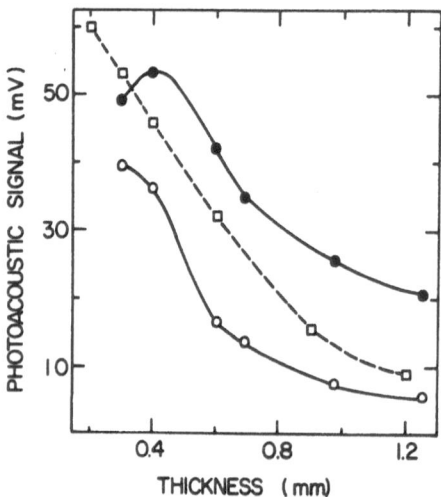

FIGURE 2 Transmission PA signal as function of sample thickness
for copper (o, •) and stainless steel (□) plates, at 31 Hz, (•,
□) and 1005 Hz (o).

possibility could be interference of reflected (and direct) thermal waves in the sample, but such interference is expected to be constructive, rather than destructive, in thermally thin metal samples in contact with air[13].

Apparently the most important contribution to the PA signal, besides the "thermal piston", comes from periodic sample vibrations, because of non-uniform sample heating. Rayleigh[7] felt that this was the main cause for any PA signal from plates, and recent work by Charpentier and Lepoutre[14], Pelz[1] (unpublished), and Jackson and Amer (on piezoelectric detection)[15] confirms its occurrence under certain conditions. In our PA cell the sample is clamped circularly at its edges and this provides the possibility of sample buckling because of the stress caused by uneven expansion under non-uniform sample heating. Such non-uniform heating will occur especially if a sample is thermally thick (t> μ_{th}) and optically dense (t >α^{-1}; α = optical absorption coefficient). Both conditions are fulfilled for most of our samples. The occurrence of these sample vibrations in our set-up could be demonstrated clearly by transmission PA on a 1.25 mm thick Cu plate at 1015Hz. Under these conditions the thermal wave is damped to 0.1% of its original amplitude, but a strong PA signal was obtained, more than 10% of that measured in reflection mode. The PA transmission signal decreased only slightly when the non-illuminated back-surface was covered with a layer of vacuum grease. Such a layer should effectively damp out any thermal waves, also at low frequencies, but a strong PA signal was obtained also at ~30Hz. Table 1 illustrates the above for some

stainless steel and aluminum (Dural) plates.

TABLE 1

Calculated Thermal Wave Damping and Measured PA
Transmission Signals for Some Metal Plates.

Material	t(mm)	1020Hz		10Hz	
		%T[a]	PA(mV)[b]	%T[a]	PA(mV)[b]
Steel	1.2	0	0.041	3.1	35
	0.2	0.28	5.69	56	136
Al	1.2	0.02	0.24	44	17.7
	0.2	25	1.57	87	56.4

[a] Calculated fraction of transmitted thermal wave.
[b] Measured PA transmission signal

If we assume the PA signal to be due only to the thermal

piston and sample buckling, then knowledge of the normal

reflection PA signal, on the one hand, and of the fraction of

heat transmitted through the sample, on the other hand, allows us

to calculate the relative contributions of the two effects. In

Table 2 results of such calculations are given, for the

reflection PA signal, as in this case, too, sample buckling will

occur, but with opposite sign compared to transmission PA.

Although such calculations are greatly

TABLE 2

Relative Contributions from Thermal Piston (TP)
and Sample Buckling (SB) to Reflection PA Signal
for Copper Plates

thickness (mm)	31Hz		680Hz	
	TP (mV)	SB (mV)	TP (mV)	SB (mV)
0.3	49.8	-2.4	7.0	-2.0
0.6	58.2	-0.7	8.1	-1.5
1.25	57.7	+1.1	5.9	-0.6

oversimplified (phase lags between the two contributions are not considered, for example) the results show that sample buckling may contribute significantly to the measured PA signal.

Such a conclusion can be put on more solid, and semiquantitative grounds, if we try to calculate the maximal actual displacement of the sample, Δx. This can be done in various, approximative ways.

1.-EXP-. We can use the experimental data from Table 2 and knowing the microphone sensitivity (~ 0.4 mV/μbar), taking into account a preamplification factor of 10, and using ideal gas laws, we calculate the volume change corresponding to the measured pressure change. Δx, in the centre of a symmetrically buckled plate, can be found using standard geometry.

2.-RG-. From simplified RG theory[3] ΔT can be estimated as the ratio of I_{inc} and the thermal mass of the sample (only that part that is contained between the PA cell and the illuminated surface). Using the linear expansion coefficient of copper, Δx can be found. Such a calculation holds if the sample is thermally thin, otherwise the thermal mass of the sample should be replaced by that of the volume up to one thermal diffusion length.

3.-R-. It it also possible to follow Rayleigh's original proposal exactly, which is meant for thermally thick samples, so that "as great a difference of temperature as possible"[7] will be established between the sample surfaces.

4.-TI: Later workers, especially Timoshenko, treated the mechanics of plates like those used here, in great detail. Timoshenko's analysis[16] of vibrations of circular plates with clamped edges, allows estimation of Δx if the force applied to the sample is known. In an admittedly circular argument that force can be estimated from Δx found by other methods. Flügge[17] proved that elastic plates can undergo forced vibrations with significant amplitudes at <u>any</u> frequency, not just the resonance ones (which are hundreds of kHz for the samples used here).

<u>5.-CL-</u>. Very recently Charpentier and Lepoutre, in their study on the use of the PA effect to determine thermal diffusivities, considered sample buckling. They assume a constant temperature gradient across the sample and, using ΔT values obtained from simplified RG theory, we can calculate Δx.

Table 3 gives some values for Δx as estimated by the methods described above. Although none of these is rigorous, the different assumptions used in them lend credibility to the conclusion that the buckling effect is not at all negligible. Jackson and Amer have given a quite complete treatment of this

TABLE 3

Estimated Maximal Sample Displacement upon Buckling, by Various Methods, for Cu Plates (in Å)

Method	31 Hz		680 Hz	
	t=1.25mm	t=0.6mm	t=1.25mm	t=0.6mm
EXP	8	5	4	11
RG(a)	4	8	1	1
R	11	47	0.5	2
TI(b)	14	230	3	28
CL(c)	3	12	1	1.5

(a) $\Delta T : 6 \times 10^{-4}$ °K (680 Hz).
  3×10^{-3} °K (31 Hz, 1.25mm).
  5×10^{-3} °K (31 Hz, 0.6mm).

(b) Forces calculated from ΔX values obtained by RG method.

(c) ΔT values, from RG method used.

kind of phenomenon (in piezoelectric transducers)[15] and a combination of their analysis with that of Flugge[17] may lead to a rigorous theory of sumple buckling. It should be noted that the reported excitation of mechanical vibrations in solids by light pulses was found at normal resonance modes of the samples [18] and, as such, is quite far removed from our experimental situation.

RESULTS FOR SOLAR COLLECTORS

Using data such as those shown in Fig. 2, the transmission PA results for the flat plate collector samples, which had different thickness, were put on a common scale. The effect of sample thickness on reflection PA data is much less, but was also taken into account, using the clean copper plates to correct for it. Figure 3 shows a plot of 680Hz reflection data vs. measured absorptance values for nine absorbers on copper substrates. Clearly for highly absorbing samples this PA signal cannot be used to differentiate properly between them. Figure 4 shows,

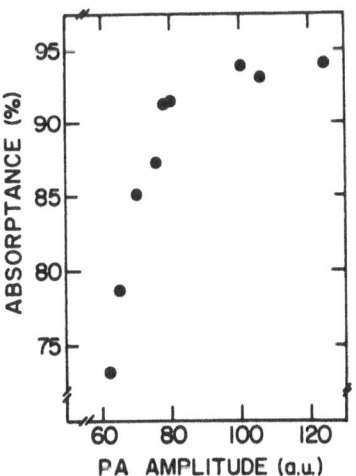

FIGURE 3 Reflection PA signal (680 Hz) vs. absorptance of nine copper oxide on copper flat plate absorbers. Sample thicknesses ranged from 0.49 to 1.00 mm.

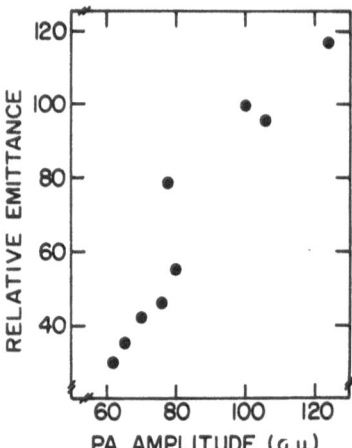

Reflection PA signal (680 Hz) <u>vs</u>. emittance . '100' ('30') on emittance scale corresponds to 20.3% (6%) absolute emittance. Same samples as in Fig.3.

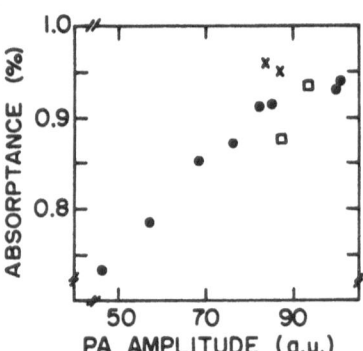

<u>FIGURE 5</u> Transmission PA signal (28 Hz) <u>vs</u>. absorptance. ●:copper oxide on copper;   x: black chrome on copper; □ :   tin oxide on stainless steel.

however, the reason for the spread in PA reflection signals between these samples, by plotting them <u>vs</u>. relative emittance. Except for two samples a rather good correlation is obtained in this case.

Figures 5 and 6 show plots of low frequency transmission PA signals (corrected for thickness variations) as a function of

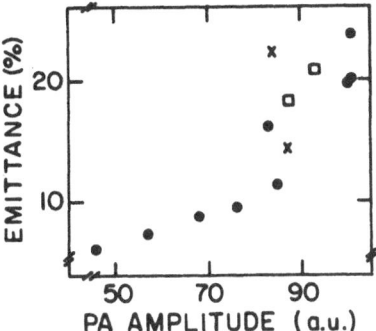

<u>FIGURE 6</u> Transmission PA signal (as Fig. 5) <u>vs</u>. emittance.

sample absorptance and emittance, respectively. In these plots
data for black chrome on copper, and for tin oxide on stainless
steel are included as well. The correlation between absorptance
and PA transmission values is quite good, but breaks down for the
black chrome samples. As is to be expected these transmission
data do not reflect properly the front-surface, emittance values
of the samples.

If conductive and convective losses are neglected, the
absorptance figure of merit [1] will be given by the surface
efficiency of the absorber, η_s[1,9]; $\eta_s = \alpha - \varepsilon \cdot (T_a^4 - T^4) \cdot \sigma / I_{inc}$
(T_a, T:temperature of absorber, and of surroundings;
σ:Boltzmann's constant). Table 4 gives some values for η_s for
several of the samples studied, together with their spectral
selectivity and the normalized transmission PA signals. Data on
the black chrome and tin oxide samples are not included because
their normalization is problematic (Cf. Fig. 5), as their
absorber structures (coating/substrate) differ from that of the
copper oxide/copper samples. The transmission PA data do,
however, reflect correctly differences between samples of the
same type, i.e. higher values are obtained for better converters.

TABLE 4

Absorptance (α), Emitance (ϵ), Spectral
Selectivity (α/ϵ), Surface Efficiency (η_s),
and Normalized Transmission PA Signals [PA(tr)]
for Several Copper Oxide/Copper Flat Plate Absorbers.

Sample	thickness (mm)	(α) (%)	(ϵ) (%)	(α/ϵ)	η_s(b)	PA(tr)(c)
1	0.99	73.2	6.2	11.8	0.72	45
2	1.00	78.7	7.3	10.8	0.78	57
3	0.49	85.1	8.8	9.7	0.84	68
4	0.49	91.6	16.2	5.7	0.89	82
5	0.71	94.1	23.8	4.0	0.91	100
Mir(a)	0.77	93.2	19.5	4.8	0.90	100

(a) MIR: Commercial product; best performance.

(b) Calculated for $T_a=340^{\circ}K$; $I_{inc}=2kW/m^2$.

(c) Normalized to MIR=100; 28Hz data.

The results in Table 4 show that use of α/ϵ only is misleading while that of η_s is more correct. However, the appreciable differences in performance between e.g. sample 4 and MIR are barely reflected in η_s, while they stand out clearly in the PA data, indicating at least that convective and conductive losses cannot always be neglected.

We noted above that the two main contributions to the PA signal depend on ΔT, which we can take to represent the thermal amplitude. As such it is correlated directly with the maximal temperature difference between the sample surface and the gas, and with the maximal temperature differential reached in the sample[3,7,14,15]. The dependence of ΔT on α is quite obvious. From a standard heat balance analysis it is found that if stagnation temperature, T_s, is reached, and if convective and conductive heat losses are excluded, then ΔT depends linearly on α/ε. No such dependence is found here, because T_s is not reached at our modulation frequencies. But the effect of ε will be felt, for constant α, through the maximal and minimal temperatures reached during the heating (light) and cooling (dark) cycle. This effect is seen clearest in the high frequency reflection PA data (Fig.4). It can be argued that, because most samples are very strongly absorbing anyhow, ε will influence mostly the minimal temperature reached during the dark cycle. Increasing ε will then increase ΔT. At increasing frequencies the influence of ε, compared to that of non-radiative convective and conductive losses, should increase. Front-surface, reflection, PA measurements do not correspond properly to surface efficiencies , because of this effect of ε during the dark, cooling cycle.

Using the data presented here we can try to formulate optimal conditions under which the PA method may be used to characterize photothermal converters. Because a normal converter will function under constant illumination, as low a frequency as possible should be used to obtain PA transmission data. While this factor was considered, the frequency used here was dictated mainly by signal to noise considerations. Detectors with better low frequency response should be used. Another variable, not used here, is the steady state temperature of the absorber. If only a small part of the sample is illuminated, the whole sample should be brought to a temperature higher than its surroundings ($\leqslant 90^\circ$C for most low temperature collectors, without concentration). This can be accomplished by DC heating or by simultaneous illumination with a continuous light source[18]. Also, no backing other than the heat transfer fluid, should be used. The reflection PA data at high frequency suggest that a more reliable measure of ε may be obtained by increasing the frequency further. However determination of ε can probably be done more conveniently by "photothermal radiometry"[20]. This same method has been used by Busse[21] for transmission probing of metal samples and, because no special cell is needed, it may be an alternative to the method described here. In all these methods detection is done at a distance, but the heat transfer fluid is not the one commonly employed in normal flat plate collectors, viz. water. While normal microphones cannot be used easily in water, PZT piezoelectric transducers can[15]. Figure 7 shows schematically how such detectors could be employed to characterize flat plate collectors. Because of the IR absorption by water photothermal radiometric detection is problematic in this case.

It should be stressed that the PA method used here, or related ones suggested above, are not proposed as alternatives to

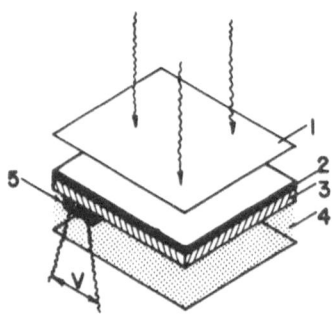 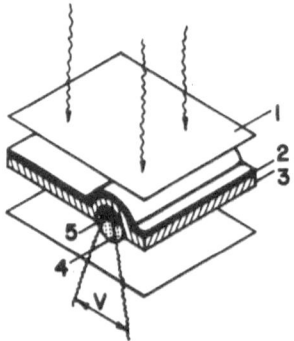

FIGURE 7 Schematic pictures of use of PA method to monitor flat plate collector performance. [1] window (could be tip of light guide; [2] solar radiation absorbing coating; [3] substrate; [4] working fluid; [5] transducer. [5] could be placed into [4], rather than in contact with [3], and [4] may be contained in the measuring cell. Right side drawing represents cheaper and simpler collector where the heat transfer fluid contacts only part of the collector back-surface. 'V': measured voltage.

accurate   normal   calorimetric   methods   for   thorough
characterization of solar collectors.   Rather, their use should
be considered for quality control and other comparative studies.

ACKNOWLEDGEMENTS

I thank the Photothermal Conversion Group at the Weizmann
Institute for samples and M. Wanna and V. Duval for assistance
with non-PA measurements.   I am grateful to A. Rosencwaig for
suggesting the importance of the sample buckling effect, and
bringing refs. 7 and 16 to my attention.   Microphones were kindly
donated by Knowles Electronics Inc.

REFERENCES

1. B.O. Seraphin, Topics in Applied Physics,
    31, 5 (1979).

2. H. Willrath, R.B. Gammon, Sol. En. 21, 193 (1978).

3. A. Rosencwaig, "Photoacoustics and Photoacoustic
    Spectroscopy", J. Wiley, N.Y. 1980.

4. J.F. McClelland, R.N. Kniseley, Proc. Semin. Test.
    Sol. En. Mater. Syst. 245 (1978);   Chem. Abstr. 90,
    22:171494.

5. S.I. Yun, J. Kor. Phys. Soc., 13, 74 (1980).

6. D. Cahen, Rev.Sci.Instr., in press.

7. J.W.S. Rayleigh, Nature (London) 23, 275 (1881).

8. M. Epstein, I. Dostrovsky, Isr. Pat. Appl.
    58214 (1979).

9. A.B.Meinel, M.P. Meinel, "Applied Solar Energy",
    Addison-Wesley, Reading, Mass. (1976) Ch.9.

10. P. Kokoropoulos, E. Salam, F. Daniels,
    Sol. En. 3 (4), 19 (1959).

11. S. Kawamato, Y. Yokoyama, S. Ikeda,
    Bull. Chem. Soc. Japan, 53, 391 (1980).

12. Y. Sugitami, K. Kato, Bull. Chem. Soc. Japan,
    52, 3499 (1959).

13. F.A. McDonald, Am. J. Phys. 48, 41 (1980).

14. P. Charpentier, F. Lepoutre, J. Appl. Phys.
    submitted.

15. W. Jackson, N.M. Amer, J. Appl. Phys. 51,
    3343 (1980).

16. S. Timoshenko, "Theory of Plates and Shells",
    McGraw-Hill, N.Y. (1940). p.519 cf.

17. W. Flügge, Z. Tech. Phys. 13, 199 (1932).

18. J.W. Hodby, A.V. Lewis, C. Alexandrou,
    C. Schwab, L.A. Boatner, Appl. Phy. Lett.
    36, 736 (1980).

19. Cf. D. Cahen, H. Garty, Anal. Chem. 51,
    1865 (1979).

20. P.E. Nordal, S. Kanstad, Phys.Scripta, 20, 659
    (1979).

21. G. Busse, Infrared Phys. 20, 419 (1980).

5. Photochemistry

THE PHOTOCYCLES OF BACTERIORHODOPSIN IN BUFFERED AQUEOUS
SUSPENSION STUDIED AT LOW TEMPERATURES BY MEANS OF PHOTO-
ACOUSTIC SPECTROSCOPY.

P.S.Bechthold

Institut für Festkörperforschung
Kernforschungsanlage Jülich GmbH
Postfach 1913, D-5170 Jülich,
W.Germany

K.-D.Kohl, W.Sperling

Institut für Neurobiologie
Kernforschungsanlage Jülich GmbH
Postfach 1913, D-5170 Jülich,
W.Germany

ABSTRACT

The photoacoustic spectra of light- and dark-adapted bacteriorhodopsin in aqueous and mixed water-glycerol (1:2)-suspensions
are studied in the temperature range of 90-300 K. Photochemical
transients are observed and interpreted in terms of the known
photocycles of trans and 13-cis bacteriorhodopsin. Differences
of the spectra with respect to those obtained by low temperature
optical absorption spectroscopy of glassy glycerol-water suspensions are discussed.

1. INTRODUCTION

Halobacterium halobium is an extremely halophilic bacterium
and requires a high concentration of NaCl ($c_{NaCl} > 2$ M) for growth
and maintenance of structure. It occurs in salt lakes (e.g. Dead
Sea, salt ponds) and survives even in crystalline salt. Under
conditions of limited oxygen supply, light-energy-converting
membrane patches, the so-called purple membranes, are synthe-

sized within the outer membrane of the bacteria. The purple
membrane contains practically only one protein called bacteriorhodopsin (BR) /1/, which is arranged in a two-dimensional
lattice /2/. BR is a chromoprotein and the pigment of the purple
membrane, which uses the energy of light to drive the transport
of protons across the cell membrane /3/. Like the visual pigment
rhodopsin it contains retinal as the chromophore (see ref. 4
for a review).

Optical absorption spectroscopy and flash photolysis have
been applied to understand the light induced proton pump and the
related photochemical reactions of BR. However, a detailed molecular model is still lacking.

PAS may, therefore, provide additional information. It has
already been used in room temperature measurements to identify
the enthalpy changes of various reaction products   /5-8/.
Enthalpy changes and the quantum yield of proton release of BR
were also measured with a capacitor microphone calorimeter /9-11/.

Chemical reactions, where short lived intermediates are involved, are easier to study when the reaction rates are decreased
by cooling the sample. To obtain low temperature spectra of biological suspensions by conventional transmission spectroscopy,
the material has to be frozen to a glassy state. For this
reason, the samples are generally suspended in glycerol-water
mixtures, which usually contain more than 50 % (weight) glycerol.
Aqueous suspensions cannot be investigated because the strong
light scattering due to the formation of micro-crystals and
cracks will prevent any useful results. The occurrence of
interference effects in the optical vessel often causes additional problems. However, both effects present no major problems
in photoacoustic spectroscopy.

In this paper we report low temperature photoacoustic
spectra of frozen aqueous and glycerol-water suspensions of BR
in the temperature range 90 to 300 K. We compare the results to
those obtained with glassy water-glycerol suspensions by low
temperature optical absorption measurements.

2. THE PHOTOCYCLES OF BACTERIORHODOPSIN

The purple membrane can be isolated by lysis of the cells
and following purification by different centrifugation steps.
The chromophore of BR, retinal, is alternatively either in the
all-trans or 13-cis configuration (fig. 1) /12/. BR which contains all-trans retinal is called trans BR (BR_{trans}^{568}), BR which
contains 13-cis retinal is called 13-cis BR   (BR_{13-cis}^{548}).
The two BR-isomers are interconvertible by means of several
pathways (fig. 2) /13/. One pathway does not require light
('dark adaption'); the others are initiated by light. Each
isomer forms its distinct primary photoproduct. In the case
of trans BR, the photoproduct returns via dark reactions through
a series of transients to its initial isomer, trans BR (trans
BR cycle). The light-energy-converting process, i.e. the active
transport of protons from one side of the purple membrane to
the other, can be correlated to the trans BR cycle. The function
of the 13-cis BR cycle is not yet known. In the case of 13-cis
BR, the dark pathway is split: most of the molecules return to
their initial isomer, 13-cis BR (13-cis BR cycle), but some,
on an alternative path, go to trans BR. Both photocycles are
connected by light-reaction pathways, which originate
from different transients /13/. The environmental light condi-

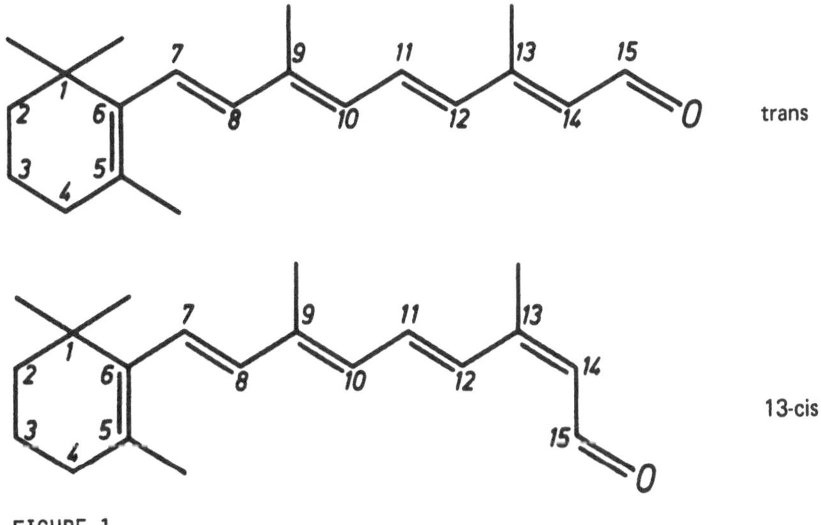

FIGURE 1
Trans retinal and 13-cis retinal

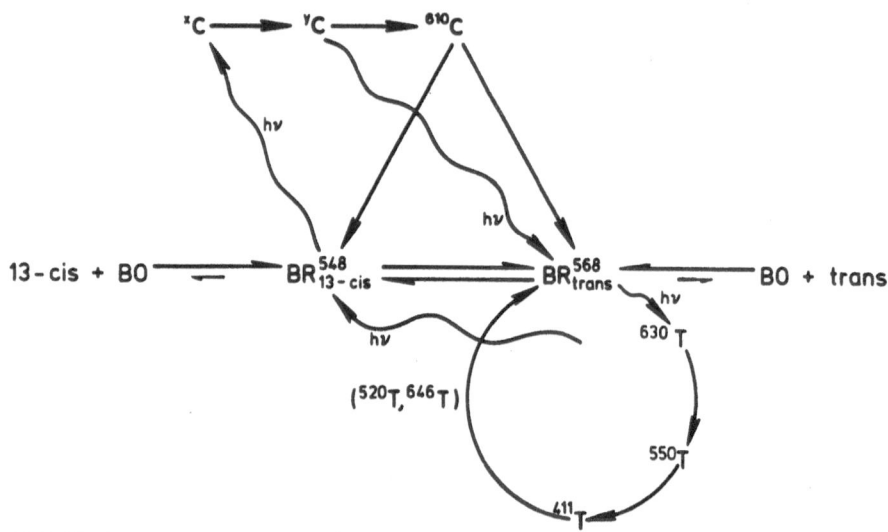

FIGURE 2

Simplified reaction scheme of the photochemistry and of the dark
reactions of 13-cis and trans bacteriorhodopsin (BR). Products
originating from 13-cis BR after light absorption are designated
as C, products originating from trans-BR are called T. A superscript on the right indicates the wavelength maximum of the absorption spectrum, a superscript on the left indicates the wavelength maximum of the difference absorption spectrum (spectrum of
C product minus spectrum of BR_{13-cis}^{548}; spectrum of T product minus
spectrum of BR_{trans}^{568}). Superscript x denotes that the exact maximum of the difference spectrum is presently unknown.

tions determine the isomeric composition of BR. In the dark-
adapted state, BR contains a mixture of 13-cis BR and trans BR
in a ratio of 1:1. On illumination of a dark-adapted sample
with moderate light (e.g. 1 mW/cm^2 white light) at room temperature, nearly all BR molecules are found in the trans BR state
within a few seconds.

3. METHODS AND MATERIALS

We used a Princeton Applied Research model 6001 photoacoustic spectrometer together with our previously described low
temperature photoacoustic cell, which is suitable to cover the
90 K to 320 K temperature range /14-16/.

Helium was used as the sound-transmitting gas. The 1 kW
Xenon arc-lamp of the photoacoustic spectrometer was sinusoi-

dally modulated at a frequency of 10 Hz. The spectra reported
were obtained by single runs scanned at a speed of 20 nm/min
with a spectral resolution of 8 nm. The low temperature spectra
are normalized with respect to the signal of a pyroelectric
detector, the spectral response of which differs only slightly
from that of a photoacoustic cell in the 250-800 nm spectral
range /16/. The shape of the spectra is therefore only insig-
nificantly distorted.

The purple membranes were isolated by lysing the halobac-
teria in water which breaks up the cell membrane into fragments.
Differential centrifugation and further centrifugation on a
sucrose density gradient yields pure purple membranes which
were dialysed against a buffer of pH 6.9. The purple membranes
were centrifuged to a pellet which was diluted to give a suspension of an optical density in the range of 25 to 30 at 570 nm.
For the measurements with the glycerol/water suspension a
2:1 (w:w) glycerol/water mixture was used as a solvent, resulting in an optical density of~10.

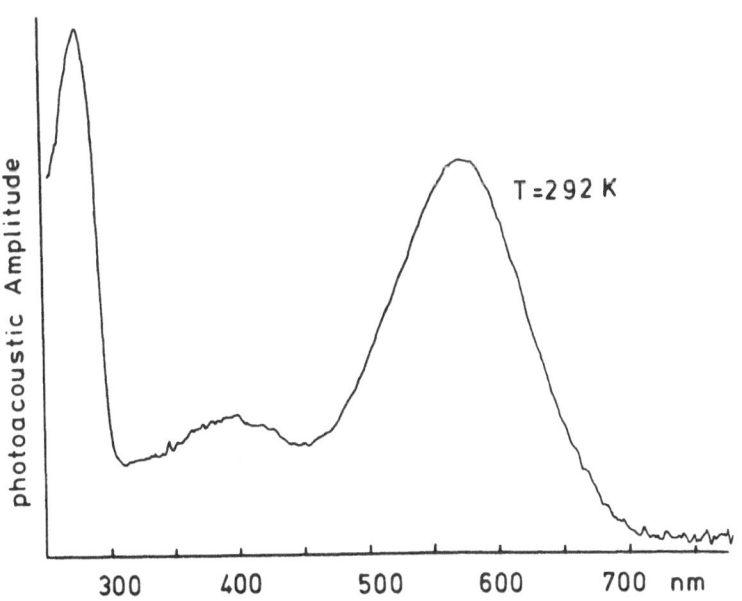

FIGURE 3
Photoacoustic spectrum of an aqueous suspension of bacteriorhodopsin at room temperature.

4. RESULTS AND DISCUSSION

Fig. 3 shows a room temperature photoacoustic spectrum of an aqueous BR suspension, which closely resembles the optical absorption spectrum with its major chromophore absorption peak near 570 nm, a minor chromophore peak near 400 nm and the protein peak at 280 nm.

Fig. 4 shows successively recorded photoacoustic spectra (27 min per run) of a dark adapted buffered aqueous suspension (pH 6.9) of BR, which had been cooled to 90 K in the dark. The scanning was started on the long wavelength side to avoid photochemical reactions in the beginning of each run. In the wavelength range from 800 to about 560 nm and below 470 nm the photoacoustic amplitude increases in successive runs whereas it de-

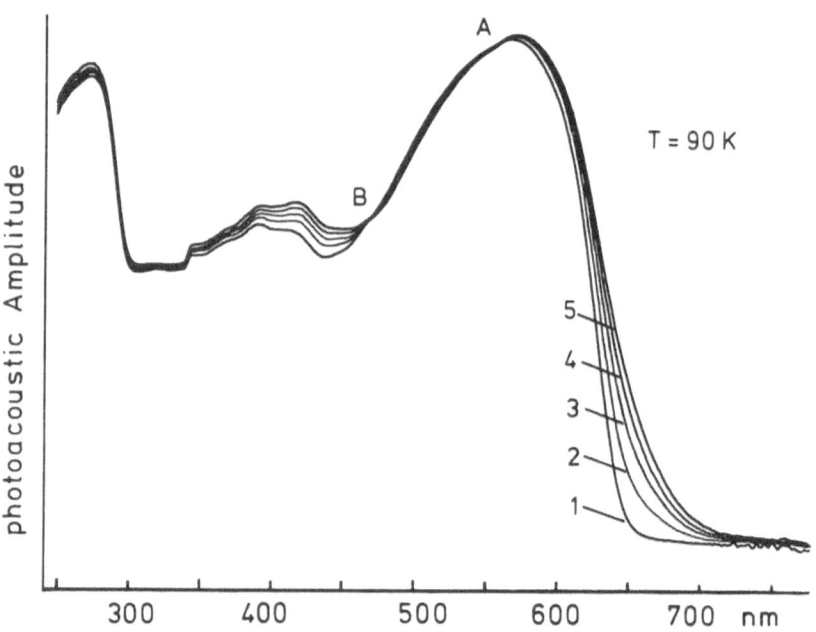

FIGURE 4

Successively recorded photoacoustic spectra of an originally dark adapted buffered aqueous suspension of bacteriorhodopsin measured at 90 K. Optical density 25 at 570 nm. The spectra are scanned from longer to shorter wavelengths. Spectral changes are due to photochemical reactions caused by the measuring light. A and B indicate points of equal heat production. The step at 340 nm is due to improper source compensation caused by a grating change in the monochromator.

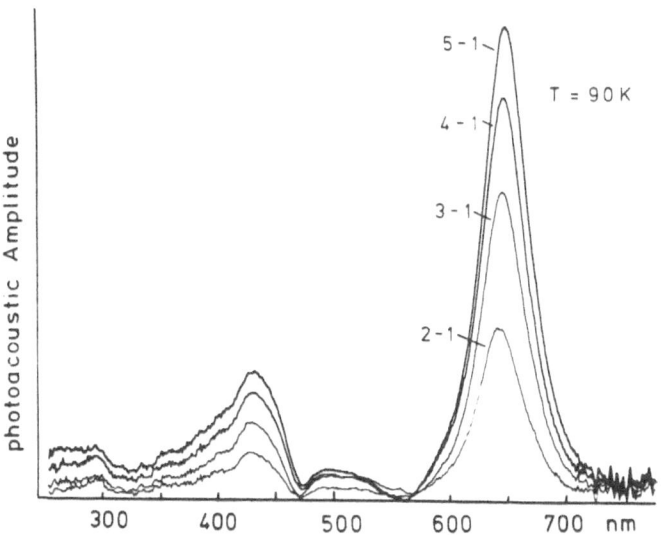

FIGURE 5
Photoacoustic difference spectra. Drawn is the magnitude of the
vector difference of the photoacoustic spectra of fig. 4. Curve 1
of fig. 4 is taken as the reference.

creases between 560 nm and 470 nm. Two distinct points of equal
heat production, A and B in fig. 4, can be seen at 560 nm and
470 nm. The spectral changes are produced by the measuring
light, i.e. the sample is exposed to the light of each wavelength in the range from 800 to 250 nm during one run. The
light intensity at each wavelength depends on the spectral characteristics of the xenon lamp and the optical system (see
fig. 9 of ref. 16). In fig. 5 the light effect is clearly
demonstrated by the difference spectra, where the spectrum of
the dark-adapted sample, curve 1 in fig. 4, was taken as the
reference for the following ones. Since the photoacoustic signal
is a vector characterized by magnitude and phase, the quantity
displayed in fig. 5 is the magnitude of the vector difference
and shows only positive values. The difference spectra have
maxima at 640 nm , 500 nm, and 430 nm. The heights of the peaks
at 640 nm and 430 nm increase by the same ratio. This indicates
that the two original BR isomers react to photoproducts at a
constant ratio. After 6 runs the points of crossover with the
initial curve 1, A and B, start to move. They are shifted to

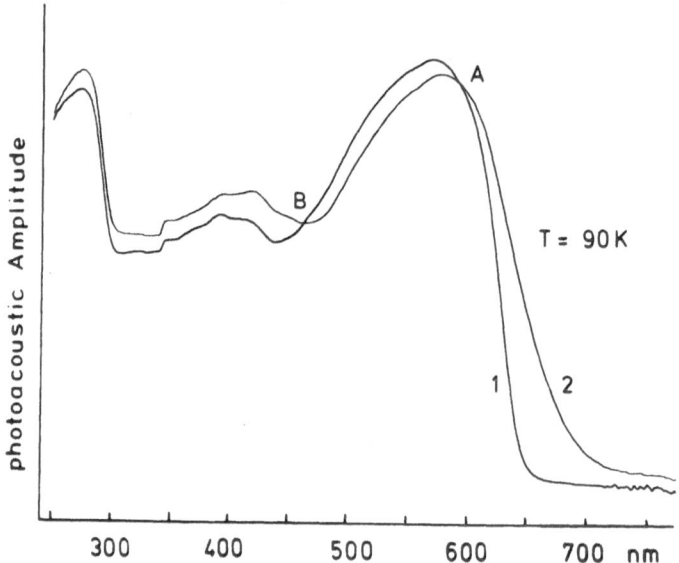

FIGURE 6
Initial and 'photosteady state' spectrum of originally dark adapted bacteriorhodopsin at 90 K. Spectrum 1 is identical with spectrum 1 of fig. 4.

higher and lower wavelengths, respectively. After further runs a
"photosteady state" is finally reached as demonstrated in fig. 6
(curve 2). The point A is shifted by 30 nm to a longer wavelength,
point B by 10 nm to a shorter wavelength.

    The question arises which intermediates the peaks can be
assigned to and which reaction pathways can produce such a "photosteady state". The composition of the BR-isomers and of their
intermediates reached after the illumination is the result of
different photoreactions (see fig. 2). From absorption spectroscopy of the trans BR and 13-cis BR cycle it is known that the
optical difference spectra of five intermediates of the two
photocycles have nearly the same maxima as the photoacoustic
difference spectra. These intermediates are ^{630}T and ^{411}T from
the trans BR cycle, ^{610}C and presumably ^{x}C and ^{y}C from the
13-cis BR cycle /13/. We assign the photoacoustic amplitude at 640 nm to the increasing amount of ^{x}C, ^{y}C, and
^{610}C and of ^{630}T. The apparent suggestion that the signal
increase at 430 nm represents the formation of ^{411}T disagrees

with a report that ^{411}T is not formed at such low temperatures
as 90 K /17/. Further experiments are necessary to prove whether
or not ^{411}T can be populated under our experimental conditions.

For a light adapted BR sample at 90 K we also obtained two
points of equal heat production in consecutive spectra, in this
case at 458 nm and 613 nm. In contrast to the dark adapted samples, these points don't shift with successive illumination. The
size of the peak at about 430 nm is smaller than that of the
dark adapted sample. This difference might result from the fact
that the original dark-adapted sample contains two BR isomers
while the original light adapted sample contains only trans BR.

To study the influence of glycerol on the behaviour of BR,
we measured dark adapted BR in glycerol/water suspensions at
90 K. Also in this case two points of equal heat production are
observed at 467 nm and 616 nm, which in contrast to the dark
adapted aqueous suspension remain constant during all consecutive
runs. A "photosteady state" is already reached after the fourth
run.   A strong increase of the 430 nm band is observed compared with the dark adapted aqueous suspension.

It should be noted that in all photoacoustic spectra the relative height of the protein peak at 280 nm is much smaller at
low temperatures than at room temperature (fig.3). This is in contrast to transmission spectra where no such difference exists.
It indicates the appearance of fluorescence of the aromatic amino
acids, and possibly photochemical activity.

From absorption spectroscopy of glycerol/water suspensions
with optical densities near 1 we know that the photoproducts
of 13-cis BR and trans BR are both photoreversible at 90 K /18,
19/, i.e. the photoproducts ^{630}T and ^{x}C can be photoconverted
to the original BR isomers, 13-cis BR and trans BR. Photorever-
sibility could not be seen in our photoacoustic measurements.
This is probably due to the different light conditions.

On illumination at 90 K only spectroscopic changes of BR due
to the photoproducts can be seen. All following dark reactions
in the BR cycles are practically stopped at this temperature. By
warming up the sample to higher temperatures, the dark reactions
are "thawn" corresponding to their rate constants. Fig. 7 is an

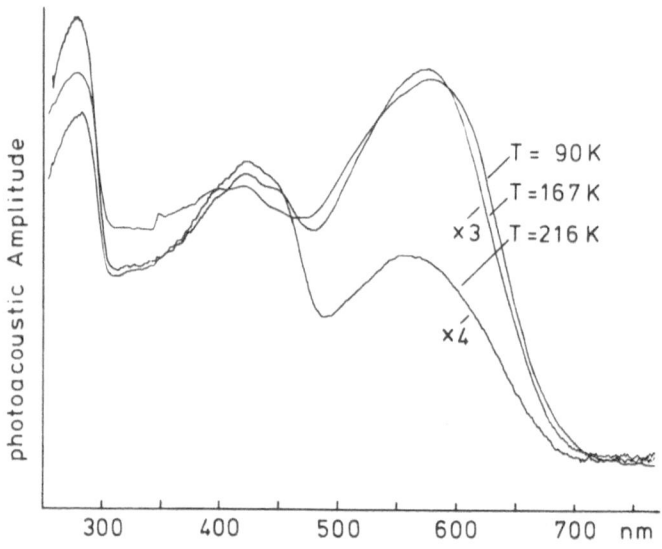

FIGURE 7

Photoacoustic spectra of originally dark adapted bacteriorhodopsin
at 90, 167 and 216 K. The spectra are plotted with different normalization factors because the photoacoustic amplitude decreases
as temperature increases /15,16/. The spectrum at 90 K is identical with spectrum 2 of fig. 6. The spectra at 167 K and 216 K
were recorded after stepwise warming up the sample. At 216 K the
^{411}T intermediate is strongly enriched and practically trapped.

example for such an experiment. The curve for T=90 K is identical with curve 2 of fig. 6. On warming the BR sample to 167 K
the maximum of the photoacoustic amplitude is slightly shifted
to shorter wavelengths. The relative photoacoustic amplitude decreases in the wavelength range around 630 nm and increases around
410 nm. This may be explained by a concentration decrease of
^{630}T and a concentration increase of ^{411}T. The effect is more
clearly demonstrated by the curve for T=216 K. Here the concentration of ^{411}T is higher than the concentration of trans BR or
of the rest of the intermediates.

    Our results demonstrate that PAS can be used for the study
of frozen aqueous suspensions. This information is not available
by conventional absorption spectroscopy. Good S/N ratios can be
achieved in spite of the boiling coolant. Further experiments
have to be done to clarify the new results on the BR cycles,
e.g. the formation of ^{411}T at 90 K.

ACKNOWLEDGEMENTS

This work was supported in part by the Deutsche Forschungsgemeinschaft (SFB 160) and by the Friedr. Krupp GmbH. We thank
Prof. M.Campagna for his support and Mrs. A.Böhme for her typing
of the manuscript.

REFERENCES

1. D.Oesterhelt, W.Stoeckenius, "Purple membrane - Rhodopsin-
    like protein from Halobacterium halobium", Nature New Biol.
    233, 149 (1971)

2. P.N.T.Unwin, R.Henderson, "Molecular structure determina-
    tion by 'non-destructive' electron microscopy of unstained
    specimens", J. Mol. Biol. 94, 425 (1975)

3. A.Danon, W.Stoeckenius, "Photophosphorylation in Halobac-
    terium halobium", Proc. Nat. Acad. Sci. USA 71, 1234 (1974)

4. R.Henderson, "The purple membrane of Halobacterium halobium",
    Ann. Rev. Biophys. Bioeng. 6, 87 (1977)
    W.Stoeckenius, R.H.Lozier, R.A.Bogomolni, "Bacteriorhodop-
    sin and the purple membrane of Halobacterium halobium",
    Biochim. Biophys. Acta 505, 215 (1979)

5. D.Cahen, H.Garty, S.R.Caplan, "Spectroscopy and energetics
    of the purple membrane of Halobacterium halobium: A photo-
    acoustic study", FEBS Lett. 91, 131 (1978)

6. H.Garty, D.Cahen, S.R.Caplan, "Photoacoustic Calorimetry
    of Halobacterium halobium Photocycle", Biochem. Biophys.
    Res. Comm. 97, 200 (1980)

7. H.Garty, S.R.Caplan, D.Cahen, "Photoacoustic photocalori-
    metry and spectroscopy of Halobacterium halobium purple
    membranes", (in press)

8. D.Cahen, G.Bults, H.Garty, S.Malkin, "Biological appli-
    cations of the photoacoustic effect", this conference

9. D.R.Ort, W.W.Parson, "Flash-induced volume changes of bac-
    teriorhodopsin - containing membrane fragments and their
    relationship to proton movements and absorbance transients",
    J. Biol. Chem. 253, 6158 (1978)

10. D.R.Ort, W.W.Parson, "The quantum yield of flash-induced
    proton release by bacteriorhodopsin - containing membrane
    fragments", Biophys. J. 25, 341 (1979)

11. D.R.Ort, W.W.Parson, "Enthalpy changes during the photo-
    chemical cycle of bacteriorhodopsin", Biophys. J. 25,
    355 (1979)

12. N.A.Dencher, C.N.Rafferty, W.Sperling, "13-cis and trans Bacteriorhodopsin: Photochemistry and Dark Equilibrium", Berichte der Kernforschungsanlage Jülich, Jül-1374 (1976)

13. W.Sperling, C.N.Rafferty, K.-D.Kohl, N.A.Dencher, "Isomeric composition of bacteriorhodopsin under different environmental light conditions", FEBS Lett. 97, 129 (1979)

14. P.S.Bechthold, M.Campagna, J.Chatzipetros, "Variable temperature photoacoustic spectroscopy, I. Instrumentation", Opt. Comm. 36, 369 (1981)

15. P.S.Bechthold, M.Campagna, "Variable temperature photoacoustic spectroscopy, II. Temperature characteristics and applications", Opt. Comm. 36, 373 (1981)

16. P.S.Bechthold, "Instrumentation for photoacoustic spectroscopy and calorimetry of liquids and solids", this conference

17. T.Iwasa, F.Tokunaga, T.Yoshizawa, "A new pathway in the photoreaction cycle of trans-bacteriorhodopsin and the absorption spectra of its intermediates", Biophys. Struct. Mech. 6, 253 (1980)

18. K.-D.Kohl, W.Sperling, unpublished

19. R.H.Lozier, R.A.Bogomolni, W.Stoeckenius, "Bacteriorhodopsin: a light driven proton pump in halobacterium halobium", Biophys. J. 15, 955 (1975)

PHOTOCHEMISTRY OF ADSORBED THIOINDIGO DYES

H. D. BREUER, H. JACOB, G. DÜSTER

FR 13.2, Physikalische Chemie,

Universität des Saarlandes

D-6600 Saarbrücken, West Germany

ABSTRACT

The photochemistry of two thioindigo dyes adsorbed on
alumina has been investigated. The results are compared to
those obtained in solution. Unsubstituted thioindigo can be
isomerized at the surface. In contrast to the findings in
solution the cis isomer is stabilized at the surface and
cannot be back converted. If thioindigo is substituted in
6,6'-positions only the trans form is found in the surface
spectrum and no trans → cis photoisomerization is observed.

INTRODUCTION

The mechanism of the photoisomerization of thioindigo in
solution has been investigated in recent papers [1-4]. It now
seems to be well established that the trans → cis and the
cis → trans isomerization proceeds via a common triplet state[1].
Since the photochemical behaviour of the first-excited singlet
state of trans-thioindigo, which is involved in the isomerization, is very sensitive to solvent properties, one should expect some influence of a solid surface onto which thioindigo
is adsorbed. If the molecule, which is known to be planar in
its trans form, is adsorbed as a whole parallel to the surface, the torsional degree of freedom about the central double
bond may be lost due to the fixation at the surface.

EXPERIMENTAL

For the adsorption experiments, thioindigo was dissolved
in benzene (Merck, Uvasol) and adsorbed on Al_2O_3 (Woelm,

200 m^2/g). In all experiments surface coverages were kept at
about 0.1 monolayer. When the adsorption was completed the
samples were dried and kept in the dark until they were used.
Spectra of the adsorbed thioindigo and of the photoproduct
were recorded in a photoacoustic spectrometer. The double-beam
spectrometer used for this work was built in our laboratory.
The instrumentation includes a 450 W xenon lamp, a 25 cm Ebert
monochromator and two identical non-resonant cells, with Senn-
heiser KE 13-227 microphones. Both sample and reference channels consist of a preamplifier driving a lock-in amplifier.
The outputs from the two lock-in amplifiers are then ratioed
to give a source compensated (I/I_o) signal to the chart recorder. Source modulation can either be done by a mechanical
chopper or by modulating the lamp current.

For the isomerization experiments the same lamp could be
used. Adsorbed dyes were irradiated in the photoacoustic cell.
Cis or trans solution were prepared in a water bath to eli-
mate heat effects.

RESULTS AND DISCUSSION

Curve (a) in fig. 1 is the absorption spectrum of thioindigo adsorbed on Al_2O_3. The peak near 470 nm was neither
observed in the spectrum of a freshly prepared solution from
which thioindigo was adsorbed, nor in a dark-adapted solution.
Since it coincides with the adsorption maximum of the cis isomer, it can be concluded that part of the trans-thioindigo is
converted into cis-thioindigo by adsorption.

If the sample of curve (a) is irradiated for 2 h at the
wave length of the trans absorption, spectrum (b) is obtained.
Here clearly the cis isomer is in excess of the trans configuration. At present we cannot determine exactly the quantum
yield for the trans→cis isomerization, since we do not know
the absorption coefficients of the adsorbed thioindigo. However, taking the absorption coefficients in benzene [4] as an
approximation, it is evident that in the photostationary state

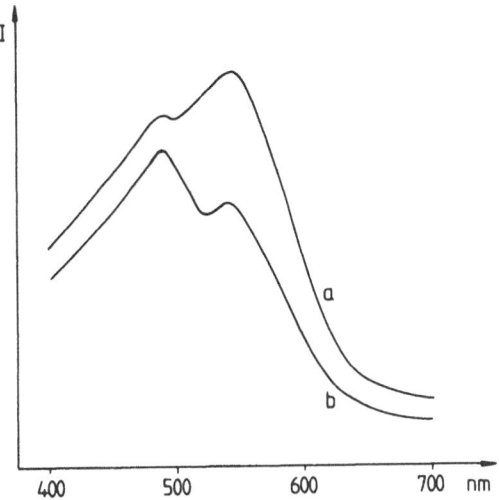

FIGURE 1.   Thioindigo adsorbed on alumina, (a) trans-thioindigo, (b) after irradiation.

more than 50 % of the adsorbed thioindigo is in the cis form.
Performing the same experiments at a surface coverage higher
than 0.5 monolayer, we did not observe photoisomerization.
The corresponding spectra, however, have the same shape as
curve (a) in fig. 1.

    In solution cis-thioindigo is not stable over a longer
period of time. If kept in the dark it reacts back to the
trans isomer. At the surface, however, no cis → trans isomerization could be observed. Even illumination at the wave length
of the cis-absorption or with the unfiltered xenon lamp did
not produce any changes in the spectrum. This indicates that
the cis isomer, which is known to be more polar than the trans
form, is energetically favoured at the surface. These findings
are supported by another observation. If thioindigo is adsorbed from a solution containing about equal amounts of trans-
and cis-thioindigo we obtain the spectrum shown in fig. 2.
Here only cis-thioindigo is adsorbed while the concentration
of the trans form in the solution remained unchanged. Thermal
conversion of adsorbed trans-thioindigo into cis can be ex-

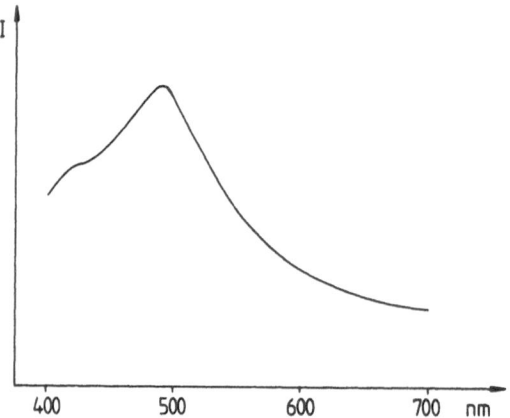

FIGURE 2. Thioindigo adsorbed from a solution containing trans and cis.

cluded since in a number of control experiments no thermal reaction in either direction has been observed in the adsorbed state.

Since the absorption intensity of the cis isomer in fig. 2 is comparable to that of curve (b) in fig. 1, the surface coverage is assumed to be about 0.05 to 0.1 monolayer.

As in the case of photoisomerization at the surface, the adsorbed cis-thioindigo is stable and cannot be converted photochemically or thermally into the trans configuration.

The fact that photoisomerization is observed in the adsorbed state indicates that only one part of the trans-thioindigo molecule can be coplanar with the surface. The other ring system can rotate about the central double bond. This is in agreement with the observation that at higher surface coverages no photoisomerization occurs. In this case the rotation is hindered by neighbouring molecules. Since the more polar cis isomer is stabilized at the surface and cannot be converted back into the trans form, both ring systems now seem to have very similar adsorption bonds and the molecule no longer has a torsinal degree of freedom.

On the other hand, the deformation of the molecule at the
surface can only be very small. The absorption maxima of both
isomers in the adsorbed state agree very well with those observed in solution. Any distortion or twisting of the C=C
double bond, however, would result in a significant change in
the absorption spectra. The attribution of the spectra to the
trans and cis isomer of thioindigo is further supported by
fluorescence measurements. In solution fluorescence is observed only in trans-rich mixtures. If thioindigo is adsorbed
from a dark-adapted solution and excited at the wave length
of the adsorption maximum we observe a fluorescence whose intensity increases with decreasing temperature. Samples which
are prepared from cis-rich solutions or preirradiated do not
exhibit fluorescence.

In solution the ratio of the quantum yields for cis → trans
and trans → cis isomerization of thioindigo, $\emptyset_c/\emptyset_t$, is 11,
while the same ratio for 6,6'-diethoxythioindigo is 2 [4]. There
is also a great difference in the fluorescence quantum yields[1]
(0.71 and 0.04 respectively). This should also be reflected
in the surface photochemistry of both dyes. Fig. 3 shows both
the spectrum of trans-6,6'-diethoxythioindigo in benzene (a)
and on alumina (b). The same surface spectra are obtained if
the dye is adsorbed from a photostationary state solution or
from a mainly cis solution. At the first glance this would
suggest a completely different mechanism than in the case of
unsubstituted thioindigo. However, following the spectrum of
the solution during the adsorption one observes a continous
decrease of the cis absorption while the trans absorption remains nearly unaffected. From this it can be concluded that
also in the case of 6,6'-diethoxythioindigo the more polar cis
form is adsorbed. In contrast to thioindigo the cis isomer is
not stabilized at the surface but immediately converted into
trans. Photoisomerization of the adsorbed dye is not possible
and no fluorescence is observed in the adsorbed state.

Comparing the two dyes, the following picture is obtained:
In both cases the more polar cis form is adsorbed preferentially. While for thioindigo the cis isomer is stabilized at

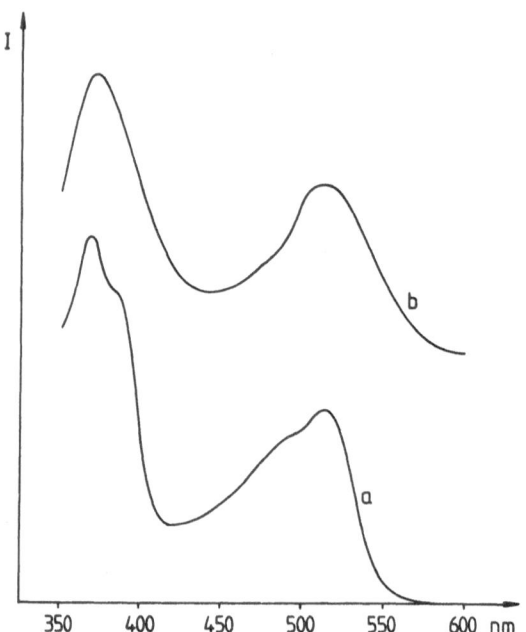

FIGURE 3. 6,6'-diethoxythioindigo (a) in benzene, (b) adsorbed
on alumina.

the surface 6,6'-diethoxythioindigo is converted thermally
or surface-catalyzed completely into the trans isomer. If we
assume that similar to the solution[1] a twisted triplet state
is involved in the photoisomerization intersystem crossing
can proceed for thioindigo from trans to cis. For the reverse
reaction there is either no intersystem crossing or only one
single route from the intermediate triplet to the S_o-state
of the cis isomer.

Since in the case of 6,6'-diethoxythioindigo no photoisomerization is observed the relatively long lived triplet may
be quenched by interaction with the surface.

Our results on thioindigo and 6,6'-diethoxythioindigo are
somewhat different from those obtained by Eggerton and Galil [5].
The authors investigated the spectra of thioindigo and some
derivatives as dyeings on cellophane sheet. In their spectra
thionindigo only occurs in the trans form, while for 6,6'-di-

ethoxythioindigo a stable mixture of trans and cis is observed. Photoisomerization of the dryed dye is considered to be impossible.

ACKNOWLEDGEMENT

The authors wish to thank Professor G. M. Wyman for helpful discussions. Financial support by the Fonds der Chemischen Industrie is greatfully acknowledged.

REFERENCES

1: K. H. Grellmann, P. Hentzschel,
   Chem. Phys. Letters 53, 545, 1978

2: H. Görner, D. Schulte-Frohlinde,
   Chem. Phys. Letters 66, 363, 1979

3: T. Karstens, K. Kobs, R. Memming,
   Ber. Bunsenges. Physik. Chem. 83, 504, 1979

4: G. M. Wyman, B. M Zarnegar,
   J. Phys. Chem. 77, 831, 1973

5: G. S. Egerton, F. Galil,
   J. S. D. C. 78, 167, 1962

PHOTOISOMERISATION OF DODCI STUDIED BY
PHOTOACOUSTIC SPECTROSCOPY

S. Schneider and U. Möller

Institut für Physikalische und Theoretische Chemie
Techn. Universität München, D8046 Garching, FRG

H. Coufal

Physik Department E13, Technische Universität München
D 8046 Garching, FRG

A B S T R A C T
DODCI (3,3'-diethyldicarbocyanine iodide), a widely
used mode-locking dye, was taken as an example to study
the influence of concentration, chopper frequency and
light intensity onto the photoacoustic spectra of this
compound in various solvent matrices. Due to a photoinduced isomerisation reaction reversible changes are
detected in the photoacoustic spectra which can be used
to gather information on this process.

INTRODUCTION

Passive mode-locking of a Rhodamine 6G-dye laser is

usually performed by means of a solution of 3,3'-di-

ethyldicarbocyanine iodide (DODCI) as a saturable ab-

sorber [1]. In order to understand the physics behind

this process, the excited state kinetics of this dye

has been studied extensively with strongly different re-

sults. Using laser flash photolysis as tool Dempster and

coworkers [2] resolved this puzzle finally by measuring

the uv-absorption spectra of the photoinduced isomeric

form of DODCI (fig. 1) whose absorption maximum is

shifted bathochromic by about 45 nm compared to the one

of the all-trans configuration.

    In ethanolic solution at room temperature the ex-

cited state lifetime of the all-trans isomere ("normal

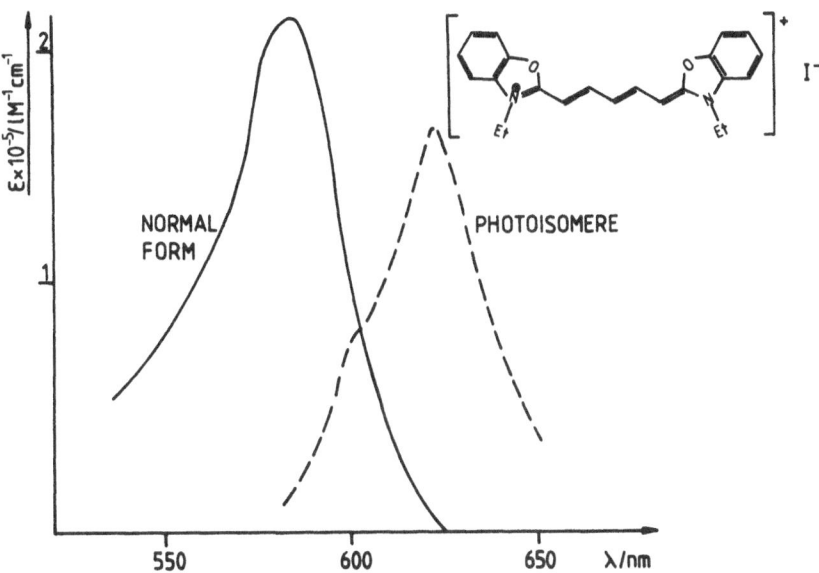

FIGURE 1   Uv-absorption spectra of DODCI in the
"normal" and photoinduced isomeric form,
respectively. Solvent is ethanol (after reference 2)

form") of DODCI has been measured by several authors to
be 1.3 $\pm$ 0.2 ns, while the lifetime of the isomere is
much shorter, namely 350 $\pm$ 100 ps[1]. As an important
consequence of this fact, one must conclude that the rate
for a radiationless deactivation of the first excited
singlet state is much higher in the isomeric form than
in the "normal" form. With respect to photoacoustic
spectroscopy one must infer from the lifetime data that a
photon absorbed by a molecule in the isomeric form gives
rise to a larger acoustic signal than one absorbed by
a molecule in the normal form. In analogy to the isosbestic point of uv absorbtion spectroscopy one should
therefore define an "iso-optoacoustic" point determined
by the wavelength of the photon which gives rise to a
photoacoustic signal of the same magnitude.

EXPERIMENTAL

Figure 2 displays a schematic of the experimental set up
used in this work. Light source is a synchronously
pumped dye laser normally used for time resolved spectroscopy. The laser beam is intensity modulated by an acousto-
optic modulator at frequency f_1. About 4 % of the beam
are directed onto a reference cell holding carbon black
as target. The remainder is used after proper attenuation
to excite the sample under study. The design of both cells
which make use of a Brüel and Kjaer microphone, is described in detail elsewhere in this volume[3]. A Bessel-
filter protects the lock-in amplifiers from running into
"overload", when a sudden low frequency noise spike is
generated (the experiment is run on a vibrationally isolated table but without a hood for sound isolation). The
outputs of the lock-in amplifiers (amplitude and phase)
are fed to a microprocessor, digitized in a 12 bit ADC
and stored after averaging over a predetermined number of
data points. The microprocessor also controls the wavelength tuning of the dye laser and the frequency setting
of the modulator drive unit, respectively, depending on
what kind of spectra are to be measured.

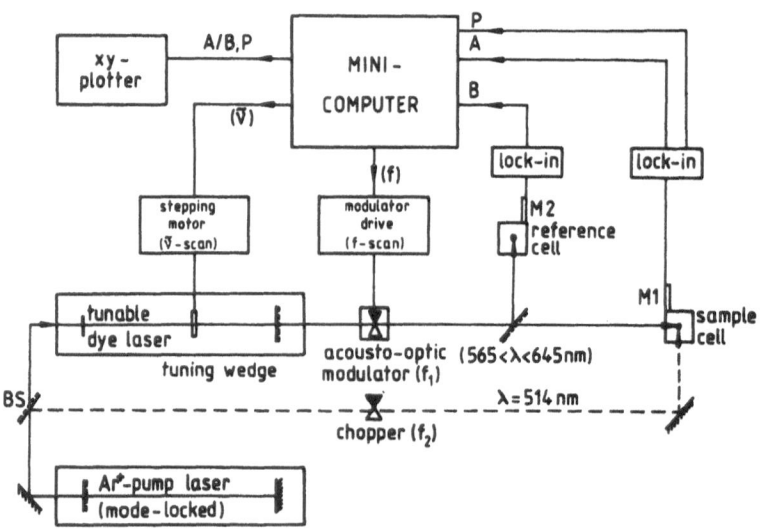

FIGURE 2   Experimental set up (for details see text)

In order to induce photochemical reactions provision
is taken to illuminate the sample with light of the laser
pumping frequency (e.g. λ = 514 nm) modulated by a mechanical chopper at a different frequency f_2. To monitor the
photoacoustic excitation spectra, a He-Ne laser has been
used as a probe laser because of the good wavelength
matching with the absorption maximum of the isomere.

RESULTS

In figure 3, the photoacoustic spectra of a $2 \cdot 10^{-3}$ M
solution of DODCI in ethylene-glycol are shown as a function of both the modulation frequency and the sample temperature. Ethyleneglycol was chosen as solvent for two
reasons. Firstly,the thermally induced back isomerisation
is temperature and viscosity dependent, therefore giving
rise to a higher quasi -stationary concentration of the
metastable photoproduct in highly viscous solvents.
Secondly,because of the low vapor pressure which prevents
a damage of the microphone by solvent vapor.

The room temperature spectra taken with low excitation
intensity show the well known saturation effects. Therefore

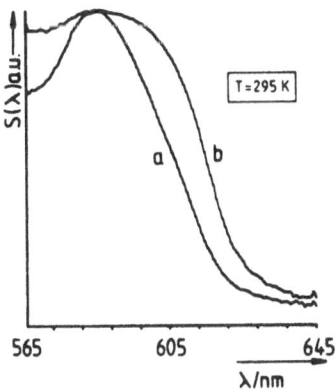
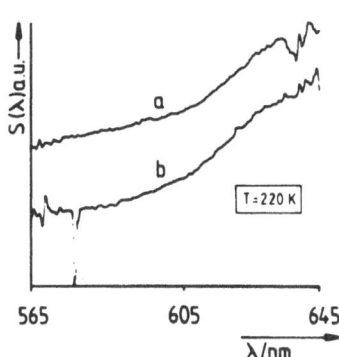

FIGURE 3   Normalized PA-spectra of DODCI in ethylene
                  glycol ($2 \cdot 10^{-3}$M) at T = 295 K (top) and T =
                  220 K (bottom). Modulation frequency is 20 Hz
                  (curve a) and 950 Hz (curve b), resp.

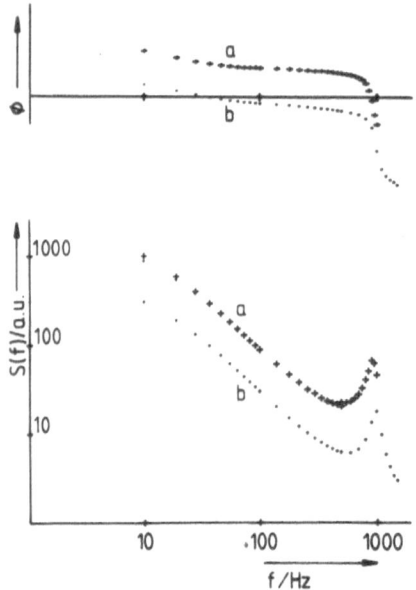

FIGURE 4

Dependence of phase (top) and amplitude (bottom) of the PA-signal on modulation frequency.

a) $T = 220$ K, $\lambda = 630$ nm
b) $T = 295$ K, $\lambda = 585$ nm

the modulation frequency dependence of the PA-signal was recorded (figure 4). The excitation wavelength coincided with the maximum in the PA-spectra, i.e. it was $\lambda = 585$ nm at room temperature and $\lambda = 630$ nm at low temperature. Both the phase and the slope of the amplitude/frequency plot show that at modulation frequencies below 100 Hz, problems with saturation effects occur. But even at higher frequencies the theoretically predicted dependence like $f^{-3/2}$ is not fullfilled, which may in part be caused by the cell resonance at $f_R \approx 1000$ Hz.

Figure 5 displays the PA-spectra of a $8 \cdot 10^{-4}$M solution of DODCI in ethyleneglycol at room temperature obtained with different intensities of the exciting laser beam. The change in the spectra is completely reversible upon switching between high (ca. 10 mW) and low (ca. 1mW) light intensity. For the sake of convenience, the difference in the normalised PA-spectra is drawn in an enlarged scale in figure 6 for both modulation frequencies. The zero crossing of the difference curves occurs at $\lambda \approx 575$ nm in contrast to the isosbestic point at $\lambda = 605$ nm observed by Dempster et al.

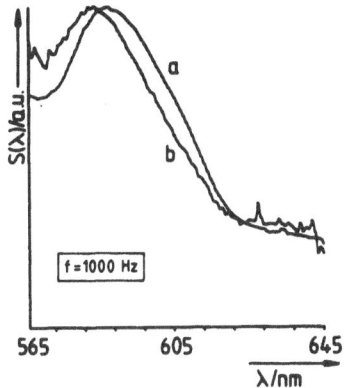 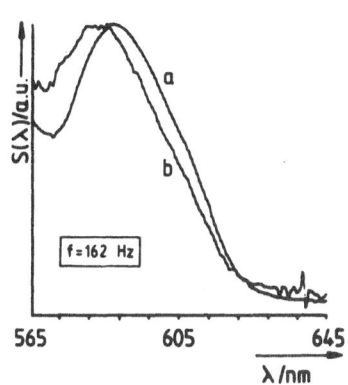

FIGURE 5   Normalized PA-spectra of DODCI in ethylene
             glycol ($8 \cdot 10^{-4}$M) at T = 295 K in dependence
             of modulation frequency and intensity of the
             exciting light a) high intensity; b) low
             intensity.

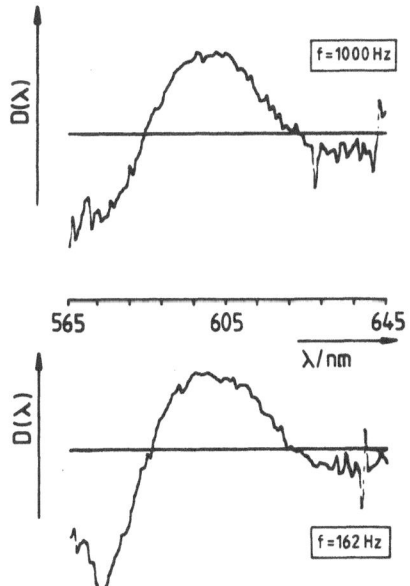

FIGURE 6

Difference of spectra
displayed in figure 4.

DISCUSSION

The PA-signal amplitude of a $8 \cdot 10^{-3}$ M solution of DODCI
shows up to 1000 Hz a clear f^{-1} dependence. As pointed out
above, a reduction in DODCI concentration by a factor of

4 causes the turning point between an f^{-1} and an $f^{-3/2}$-
dependence to fall into the 50-100 Hz region with the
effect that the shape of the PA-spectrum changes drastically with the modulation frequency (see fig. 3). Lowering
the concentration to $8 \cdot 10^{-4}$ M and thereby increasing the
value of $1/\mu_\beta$ to approximately $1/(2 \cdot 10^{+5} \cdot 8 \cdot 10^{-4}) =$
$6.25 \cdot 10^{-3}$ cm makes the turning point fall into the 10 Hz
region (as can be observed experimentally). As a consequence, the PA-spectra at this concentration are fairly
independent of modulation frequency.

Despite this fact one observes that the PA-signal
drops much less upon going from $\lambda \sim 585$ nm to $\lambda \sim 570$ nm than
what one would expect from inspection of the absorption
spectrum (fig. 1). Although very often ignored, this behaviour is a consequence of the mechanism by which the
photoacoustic signal is generated. To make this point
clear three molecular levels are picked out in figure 7:
the ground state $S_{o,o}$, the vibrational groundstate of the
first electronically excited state $S_{1,o}$ and an electronically and vibrationally excited state $S_{1,v}$. For the sake
of simplicity, we assume that the electronic excitation
energy corresponds to $\tilde{\nu} = 16000$ cm^{-1} and the vibrational
frequency to $\tilde{\nu} = 1600$ cm^{-1} which is typical for C = C-

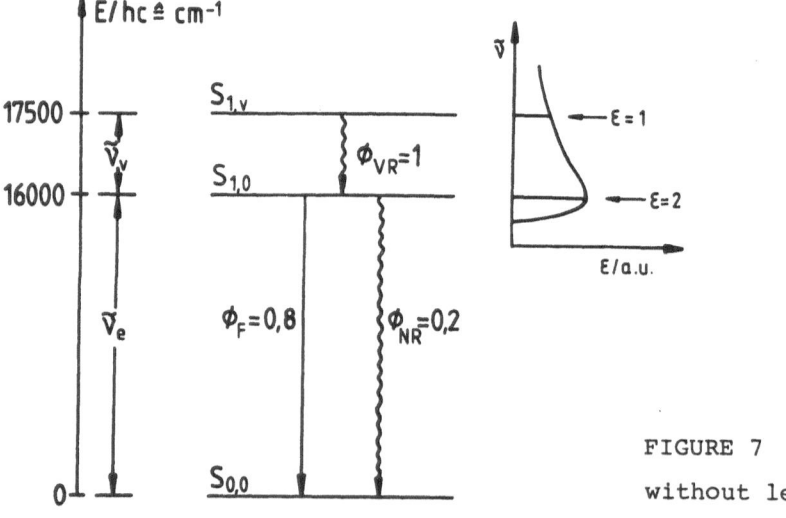

FIGURE 7

without legend

stretching vibrations. Furthermore, we assume that the ra-
tio of the molar extinction coefficients ε (16000) : ε
(17600) is 2 : 1. The fluorescence quantum yield is taken
as 0.8 to describe a situation as given in the case of
DODCI in the trans-configuration.

Suppose now, the photon flux is chosen such that 10
photons are absorbed per unit time to induce a transition
$S_{o,o} \rightarrow S_{1,o}$. The same flux would lead to an absorption of
5 photons/unit time when tuned to the transition $S_{o,o} \rightarrow S_{1,v}$.
The energy released as heat by nonradiative transitions
varies with the excessenergy, i.e. the vibrational energy
in the excited state. In case of zero excess energy, the
normalized PA-signal would be:

$$E_r/E_o = 2 \cdot \tilde{\nu}_e / 10 \cdot \tilde{\nu}_e = 0.2$$

In the case of the vibrationally excited molecule, we get:

$$E_r/E_o = (1 \cdot \tilde{\nu}_e + 5 \cdot \tilde{\nu}_v)/ 10 (\tilde{\nu}_e + \tilde{\nu}_v) =$$
$$(16000 + 5 \cdot 1600) / (10 \cdot 17600) = 0,136$$

Instead of ε(16000) : ε(17600) = 2 : 1 we observe a
ratio of 0.2 : 0.136 = 1.47 : 1 in the PA-spectrum of this
model molecule.

The conversion   efficiency of absorbed light energy
into sound is, as just demonstrated, a function both of the
excessenergy and the quantum yield of fluorescence. Furthermore, it will depend in principle on the quantum yield of
phosphorescence and the amount of vibrational energy created when a fluorescence photon with an energy less than
$\Delta E(S_{1,o} \rightarrow S_{o,o})$ is emitted (shape of fluorescence spectra!).

On the basis of these considerations it is easy to
understand why the isosbestic point observed in the flash
photolysis experiments coincides with a maximum in the PA-
difference spectra. At that wavelength (λ= 605nm) the molecular extinction coefficients of both the normal and the
photoinduced isomeric form are equal, i.e. the conversion
of a trans into a cis molecule does not result in a change

in optical density. The amount of heat released upon absorption of one photon, however, differs strongly for the
two isomeres; it is in average larger for the cis isomere
(shorter excited state lifetime, higher excess energy).
Upon excitation at λ = 575 nm, on the other hand, the
average amount of heat released is the same for both
species. To differentiate this wavelength from the isosbestic   point of absorption spectroscopy we propose to call
this the"iso-optoacoustic" point in the PA-spectrum.

As mentioned above, the photoinduced spectral changes
are reversible because, due to an thermally induced isomerisation on the groundstate potential surface, the equilibrium between cis- and trans- isomeres is established again
after a stop of the illumination. Since the rate of equilibration depends exponentially on temperature, comparably
low light intensities can be sufficient to maintain a high
quasistationary concentration of the metastable photoiso-
mere. The PA-spectra of DODCI at low temperature are therefore dominated by the absorption of the isomere (fig. 3,
bottom).

As a final proof, that the light induced changes in
the PA-spectra are due to photoisomerisation, we recorded
photoacoustic-excitation spectra. To this end we monitored
the PA-signal intensity generated by a He Ne-laserbeam in
dependence of the wavelength of the photolysis laser. The
result is essentially identical with the low intensity
response of the DODCI-solution. Blocking of either beam
assures that there is no "cross-talk" between the two
excitation sources.

CONCLUSION

The results of flash photolysis experiments can not directly be compared to those of photoacoustic studies;
the difference spectra generated by the latter technique
do not allow an easy conclusion with respect to the ab-

sorption spectra of the metastable species. Nevertheless, PA-spectroscopy may prove itself a valuable tool for photochemical studies for two reasons:

(i) the sample does not need to be transparent, a restriction, which is very severe in low temperature solution spectroscopy.

(ii) the problem of exact matching of excitation and probing volume to achieve good signal/noise ratio seems to be less severe in PAS.

ACKNOWLEDGEMENTS

Financial support by the "Deutsche Forschungsgemeinschaft" and the "Fonds der Chemischen Industrie" is gratefully acknowledged. Furthermore, we wish to thank the members of the Institute's electronic shop for their valuable help in computerizing the experiment.

REFERENCES

1/ For a review see:

S. Schneider
Flashlamp-pumped mode-locked dye lasers
Phil. Trans. R. Soc. London A 298, 233 (1980)

2/ D.N. Dempster, T. Morrow, R. Rankin and
G.F. Thompson
Photochemical Characteristics of Cyanine Dyes
J. Chem. Soc. Faraday Trans II 68, 1479 (1972)

3/ H. Coufal, U. Möller and S. Schneider
Photoacoustic cells for measurements at various temperatures and pressures-design and characterisation -. This volume

6. Nondestructive Testing

OPTOACOUSTIC AND PHOTOTHERMAL IMAGING AND MICROSCOPY

G. Busse, Zentrale Wiss. Einrichtung Physik,
Hochschule der Bundeswehr München, D-8014 Neubiberg, Germany F.R.

1. INTRODUCTION

Modulated radiation shining on an absorbing sample causes a temperature modulation which can be detected in different ways. The classical method is based on the gas pressure modulation when the sample is kept in a gas-filled cell [1]. One can also detect the modulated thermal expansion of the solid sample itself by using piezoceramic material or even a strain gauge attached to it [2-4]. While these are optoacoustic (also called photoacoustic) methods [5,6], in photothermal detection the temperature is monitored via the modulation of infrared thermal emission correlated with it [7], as is described by P.E. Nordal and S.O. Kanstad in this book. It is true remote sensing like the "mirage effect" where the change of sample temperature causes deflection of a monitoring beam of light [8] (see article by D. Fournier and A.C. Boccara in this book).

Irrespective of how the signal is detected, the temperature modulation of the sample indicates the thermal response to periodically absorbed power. Hence the temperature amplitude ΔT can be described in the Gaussian plane by a complex vector. In Cartesian coordinates, the signal components are the in-phase and the quadrature parts. For many applications, however, description in polar coordinates is more practical: The length of the complex vector ΔT (magnitude A) is the maximum temperature change, and the angle φ with the real axis is the phase delay between the maximum optical power and the maximum of temperature, therefore φ can be considered the normalized propagation time between the light and the resulting thermal effect. The signal ΔT depends on how much heat has been pro-

duced originally and on how well it is dissipated by heat transport into
the material. Different theoretical models describe the influence of optical and thermal properties [9-11].

Most activity in optoacoustic research has been concentrated on spectroscopy where the wavelength of the incident light beam is varied [5,6].
Experiments have been performed recently where the sample was scanned instead of the wavelength: The optoacoustic signal obtained locally is monitored as a function of the sample coordinates [12,13], and mapping of the
local signal results in an optoacoustic image which does not only show
optical but also thermal structures because it is based on thermal wave
propagation [14]. Signal phase φ and magnitude A are both suited for imaging, the information they provide is usually different. It is one aim of
this article to show examples where optoacoustic and photothermal images
give information that cannot be obtained by optical means.

2. EXPERIMENTAL ARRANGEMENT

The standard equipment in optoacoustic spectroscopy is a light source
with a beam modulator, the detecting device (microphone, piezoceramics, or
infrared detector), and a lockin amplifier for phase sensitive signal analysis [15]. All one needs additionally for imaging is relative motion of the
beam with respect to the sample, and a recorder that shows which signal
was obtained at which part of the sample. A schematical diagram is shown
in Fig. 1 [16]. A convenient light source is a cw laser (CO_2 or Ar-ion) since
both focusing (to generate a thermal wave locally) and modulation are
easily achieved.

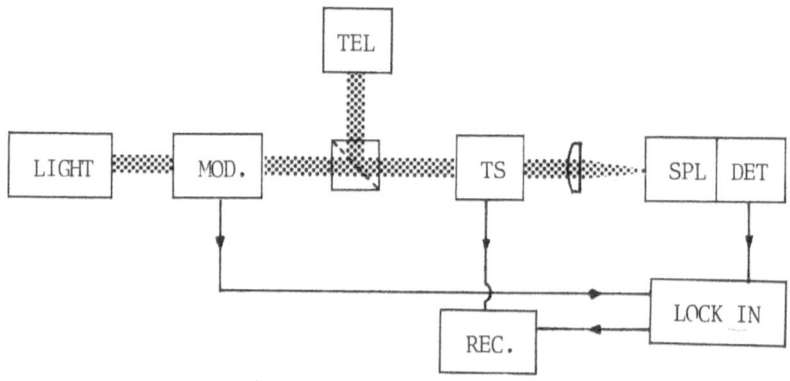

Figure 1: Experimental arrangement

The box indicated by TS symbolizes that part of the setup where the sample coordinates x and y are scanned. As only the relative position is of interest, one can use beam steering optics (e.g. rotatable mirrors, one for x- and one for y-deflection) or alternatively move the sample across the stationary laser beam, but this motion causes additional noise if detection is based on microphones or piezoceramics. On the other hand, there is no limitation in the field of view. Therefore sample motion is advantageous at least in photothermal detection of emitted modulated infrared radiation. Beam scanning may be faster, but off-axis beams can limit resolution and the field of view. Sometimes a combination of both techniques is reasonable. One application is local beam modulation which can be used instead of beam intensity modulation [16,17]: The faster superposed linear or circular beam oscillation at small amplitudes is well achieved with mirrors, while for the slow coordinate scan one can move the sample. A beam splitting cube connected to a telescope TEL and/or an optical detector allows for optical monitoring during the scan.

The box indicated by REC is the system which shows the collected data $A(x,y)$ or $\varphi(x,y)$ and allows for inspection and evaluation of the results. An x-y recorder can be used to show a projection of the three-dimensional signal surface [16]. This method is well suited for quantitative analysis, but visualization is difficult if the structures are complicated and if the signal is noisy [18]. In this case, halftone image representation makes inspection easier, and additional electronic filtering enhances structures [17]

3. EXAMPLES FOR OPTOACOUSTIC AND PHOTOTHERMAL IMAGING

Both optical and thermal structures can be shown. One should, however, think about the information one is interested in. If all one wants to see is optical structure, it is easier to use optical inspection techniques. If the problem is too difficult for optics (e.g. subsurface structure inspection in a black or shiny sample), then the new methods can be useful. One should also think about how to distinguish between optical and thermal structures, and it will be demonstrated that in this situation the difference between phase angle imaging and magnitude imaging is significant.

3.1 Optoacoustic Imaging and Microscopy

In optoacoustic imaging a thermal wave is generated locally by absorption of modulated focused optical radiation. But detection is an integral process because the thermal reaction of the whole sample is observed either by the optoacoustic cell or the piezoceramic material to which the sample is glued.

To keep things simple, first examples are presented where the sample surface is homogeneous within the optical penetration depth. A sample of this kind is shown in Fig. 2 [16]: Two thin spots of insulating material were glued to metal and then covered by a graphite layer. Optical inspection would not indicate that different material is hidden under the surface. However, as local heat transport is different, one can look far beyond the optical penetration depth into the material with the optoacoustic effect. This sample is a model for delamination of layers (e.g. paint).

The advantage as compared to optical imaging is greater the more opaque the sample is and the better heat is transported. Therefore metals and semiconductors should be of interest.

Metal subsurface inspection is demonstrated in Fig. 3. While Fig. 2 has been obtained with a CO_2 laser and microphone detection of signal magnitude, here an Argon ion laser was used for thermal wave generation. The

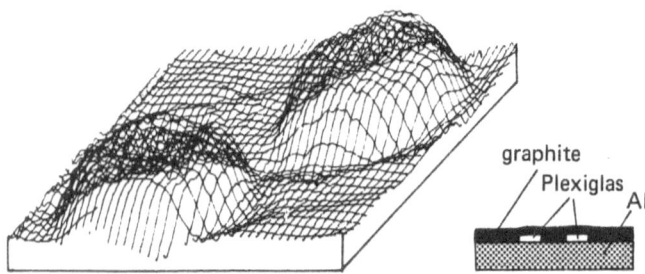

Figure 2: Delamination model

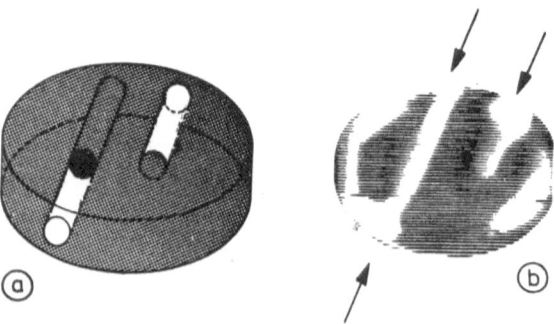

Figure 3: Subsurface hole analysis in aluminum. Hole diameter 1.5 mm, sample thickness 4.5 mm, modulation frequency 41 Hz

aluminum sample of 4.5 mm thickness was glued to piezoceramic material [19].
Two holes were drilled into the metal, one ended in the middle of the
sample, the other one was tightly refilled over half of its length (Fig.
3a) with aluminum. With X-rays one would find two half holes but the opto-
acoustic phase angle image shows the original hole even though material
was added afterwards (Fig. 3b) [20]. The image also shows a signal surface
distortion near the sample edge, a geometry effect reported earlier [19,21].
It is obvious that internal boundaries are well detected by optoacoustic
inspection. Similar examples have been reported by Thomas et al [22] and
Luukkala who imaged regions of plastic deformation in metal [23].

One advantage in optoacoustic spectroscopy is the possibility of
depth profiling: The layer of sample contributing to the signal magnitude
is about the thermal diffusion length which is inverse to the square root
of the modulation frequency [5,10]. Depth profiling is illustrated in Fig. 4
for two subsurface holes in aluminum [19,24]. For the signal magnitude the
hole ending at 0.4 mm under the sample surface is out of depth range at
the higher frequency, so the difference of the 18 Hz image and the 180 Hz
image would only reveal the geometry of a certain layer.

The phase angle image of the sample shows both holes still at the
higher modulation frequency [19]. This is because depth range is by about
a factor of 2 larger with the phase angle [25]. Another advantage of phase
angle images is described in the following.

The samples presented until now were only provided with thermal
structures, the surface was either black or shiny. But in reality a sample
under investigation may have additional optical structure which one is not
primarily interested in. In this situation one would like to know whether
the information on thermal structure is affected. If the metal surface

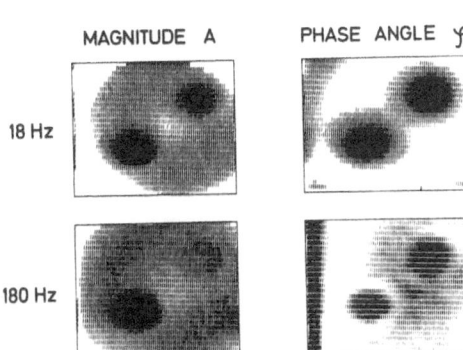

Figure 4:

Depth profiling in aluminum |19|

reflection is reduced locally by black paint, absorbed modulated power
and hence the magnitude A of the generated thermal wave increase. The
phase φ , however, does not change (if the layer of paint is thermally
thin) since it is a propagation time effect. Therefore the phase angle
image shows only thermal structures even in the presence of optical structures, while the magnitude image shows both [18]. To demonstrate this difference the sample of Fig. 4 was afterwards provided with two dots of black
paint. These dots are shown in Fig. 5a which was obtained with the optical
monitor. With the optoacoustic magnitude image (Fig. 5b) one cannot decide
which structure is optical and which is thermal, but the phase angle ig-
nores the black paint at the surface and reveals only thermal subsurface
structure (Fig. 5c).

Another example is shown in Fig. 6 where the sample consists of two
soldered different metals partially hidden under black paint. With the
phase angle the thermal boundary is well seen under the paint.

Besides detection of thermal subsurface structure another point of
interest is resolution. The range of the thermal wave used for imaging

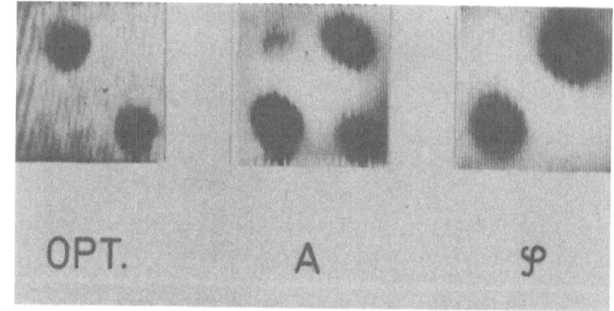

Figure 5: Optical structure on aluminum (left) is ignored in phase angle
           imaging (right).

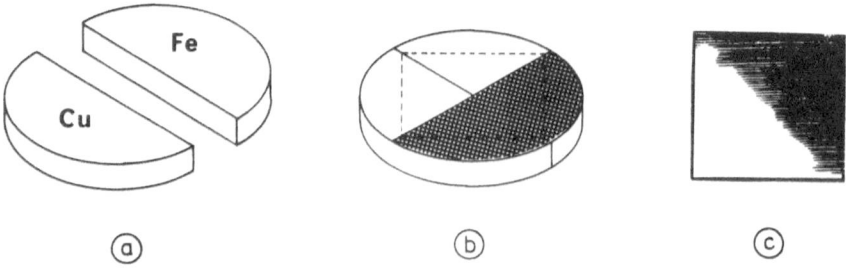

Figure 6: Near-surface thermal discontinuity

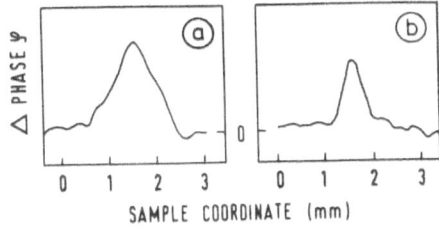

Figure 7:

Frequency dependence of resolution:
36 Hz (a), 90 Hz (b) |20|

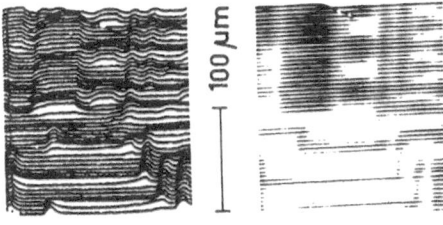

Figure 8:

Optoacoustic microscopy of an
integrated circuit

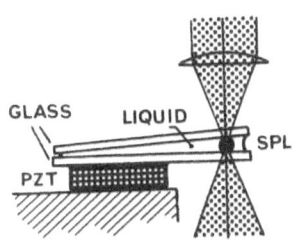

Figure 9: Principle of liquid coupling

decreases with increasing modulation frequency. If the laser beam is
scanned across the thermal discontinuity in Fig. 6, the change of phase
angle depends on modulation frequency as is illustrated by the results in
Fig. 7. Extrapolation to resolution required for microscopy indicates that
high frequencies are needed for thermal wave microscopy [14]. It could be
proved by resolution experiments that phase angle microscopy of an integrated circuit really shows thermal structures [18]. An example for opto-
acoustic phase angle microscopy is presented in Fig. 8 [26]. Total resolution is the convolution of the optical spot size and the generated thermal
wave. With blue light of an Ar-ion laser, larger aperture optics, and a
modulation frequency near 10^6 Hz a resolution of about 2 µm can be achieved.

Biological substances are also of interest for non-optical microscopy. The required frequency is beyond the range of microphones but these
samples cannot be glued to piezoceramic material. Microscopy of a hair
could be performed by using a liquid coupling which is illustrated in Fig.

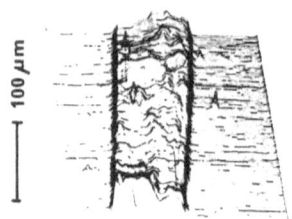

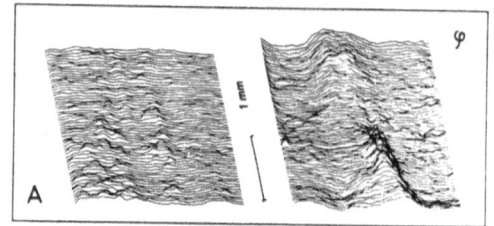

Figure 10: Optoacoustic
magnitude image of a hair,
length of 100 µm is indicated.

Figure 11: Phase angle image (right) and
magnitude image (left) of a green
leaf at 167 KHz |26|.

9 [26]. The signal magnitude image obtained at 291 KHz is shown in Fig. 10.
Magnitude and phase angle image differed when a green leaf was inspected at
167 KHz (Fig. 11) [27]. However, optical resolution had to be reduced here to
about 20 µm to avoid dangerous power densities on the sensitive material,
though the thermal contact with water provides some protection.

An obvious advantage of optoacoustic imaging as compared to optical
inspection is the fact that depth range depends on thermal and not on optical properties thereby allowing for nondestructive probing of subsurface
structures in opaque samples.

However, there are two features which are a drawback as compared to
optics: Primarily, optical inspection depends on a local response of the
sample and not on an integrated response distribution. Secondly, optical
inspection does not require physical contact with the sample. The next
chapter presents a method which combines to some extent the advantages of
optical and optoacoustic imaging.

3.2 Photothermal Imaging

Two methods have been reported for remote detection of modulated sample temperature. The "mirage effect" uses the deflection of a sensing beam
of light due to the warm layer of gas adjacent to the sample surface [8],
while photothermal detection is based on the fact that temperature modulation causes modulation of thermal infrared emission from the spot which is
illuminated [7]. Both techniques have been used recently by D. Fournier and
A.C. Boccara, P.E. Nordal and S.O. Kanstad, and M.V. Luukkala for front
surface spatial mapping and imaging [26]. The signal can be calculated from
the theoretical models of optoacoustic detection.

The situation is different if the rear surface is observed, as is
considered in the following.

Depth profiling - the fact that one can look into a depth which is about the thermal diffusion length or twice that value (depending on whether the signal magnitude A or its phase φ is used [25]) - is an important feature in optoacoustic imaging [19]. On the other hand, depth profiling in this case means that there is a depth behind which sample regions cannot be detected. This is because one observes an integral value to which thermal waves coming back from deep in the sample do not contribute. If the signal is plotted as a function of sample thickness, one finds that the complex vector converges in the Gaussian plane into an off-center point which means that the signal becomes independent of sample thickness (see dotted curves for 20 Hz in Fig. 12). Recent measurements by M.V. Luukkala [28] confirmed earlier calculations [25].

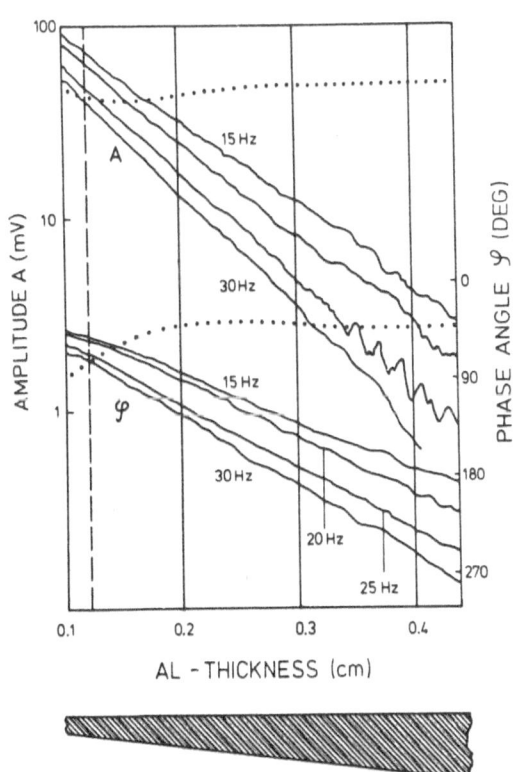

Figure 12: Magnitude A and phase angle φ of photothermal signal behind a wedged aluminum sample at different modulation frequencies [31]. Dotted curves show calculated values for optoacoustic signal at 20 Hz [25], dashed vertical line indicates thermal diffusion length at 20 Hz.

In the transmission arrangement, however, one observes only the thermal wave that has been propagated through the whole sample. Then in the complex plane the temperature amplitude converges into the origin, and the experiment confirms the well-known exponential decrease for the magnitude and the linear decrease for the phase of the thermal wave. Thermal wave propagation as a dynamic effect has been investigated first by A.J. Ångstrom [29], rear-surface illumination experiments with an optoacoustic cell were performed by M.J. Adams and G.F. Kirkbright [30].

Transmission measurements based on infrared detection of locally generated thermal waves in a wedged piece of aluminum indicated that sample thickness which can be inspected is not limited by thermal diffusion length but by noise (see Fig. 12) [31]. An example of scanned photothermal transmission imaging is shown in Fig. 13 where an aluminum sample of 3 mm thickness was provided with 1 mm diameter subsurface holes (dashed) and black

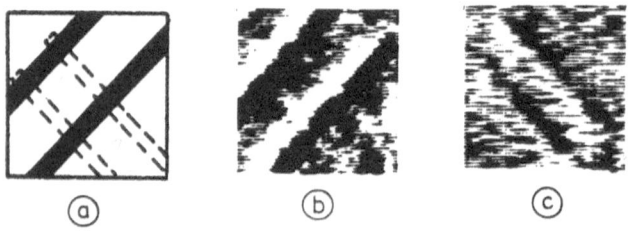

Figure 13: Magnitude (b) and phase angle (c) in photothermal transmission imaging of an aluminum sample (a) provided with 1 mm ϕ subsurface holes (dashed) and black lines on surface |17|.

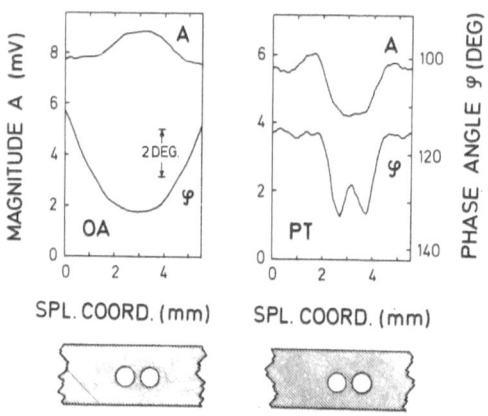

Figure 14: Two methods of thermal wave material analysis at 20.3 Hz. The same sample is used for optoacoustic microphone detection (left) and photothermal infrared rear surface inspection (right).

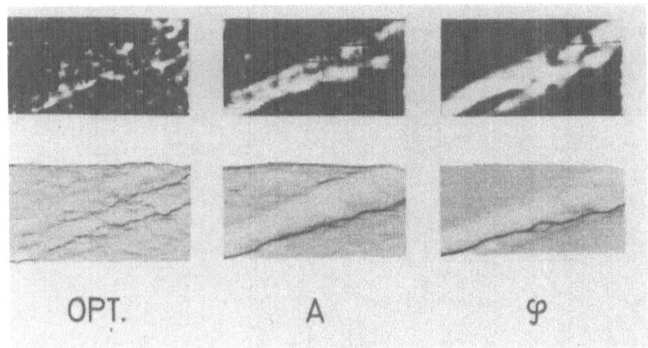

Figure 15: Welded stainless steel in photothermal rear surface imaging
(magnitude A and phase angle φ) as compared to simultaneously
recorded optical image (left).

lines across the surface. The magnitude image shows mainly optical surface
structure which is ignored in the signal phase image. Resolution in subsurface structure probing is better than with conventional optoacoustic detection (Fig. 14) since infrared inspection allows for local resolution of
the detected thermal wave whereas with optoacoustic methods the integral
of the thermal wave determines resolution. A realistic application for
nondestructive remote material inspection is demonstrated in Fig. 15: two
pieces of stainless steel were welded together. The seam in the rectangular
field of view (5 x 11 mm^2) is shown as a bright diagonal line. The dark
structures in the bright band indicate thermal discontinuities.

Another field of applications could be inspection of integrated circuits [17].

4. CONCLUSION

Both optoacoustic and photothermal imaging are methods of nondestructive material inspection which provide information on thermal structures.
Depth range is limited in optoacoustic and photothermal front surface imaging. This limit, however, depends on modulation frequency thus allowing
for depth profiling. In photothermal transmission probing, depth profiling
is not possible via frequency variation. But thickness of inspected material can be considerably larger and resolution is better at the same modulation frequency. Phase angle imaging is generally advantageous because it
allows to distinguish between optical and thermal structures.

Fields of applications could be inspection of metals and semiconductors, analysis of layers and biological samples. But the method is still in its infancy, and it is well possible we now have an answer to questions we do not yet know.

ACKNOWLEDGEMENT

The author acknowledges helpful discussions with Profs. B. Bullemer, P.C. Claspy, S. Perkowitz, and with Dr. A. Rosencwaig, Dr. P.E. Nordal, Dipl.-Ing. B. Lieder, and Dipl.-Ing. A. Ograbek. The author is also grateful to H. Bergauer, W. Funke, J. Pielmeier, and P. Wieczorek for skillful sample preparation and last not least to B. Baier for all her careful and patient work on the manuscript.

REFERENCES

|1| A.G. Bell, Am. J. Sci. 20, 305 (1880)

|2| A. Hordvik and H. Schlossberg, Appl. Opt. 16, 101 (1977)

|3| M.M. Farrow, R.K. Burnham, M. Auzanneau, S.L. Olsen, N. Purdie, and E.M. Eyring, Appl. Opt. 17, 1093 (1978)

|4| G. Busse and S. Perkowitz, Int. J. of Infrared and Submm. Waves 1, 139 (1980)

|5| Yoh-Han Pao, "Optoacoustic Spectroscopy and Detection", Academic Press (1977)

|6| A. Rosencwaig, "Photoacoustics and Photoacoustic Spectroscopy", John Wiley & Sons (1980)

|7| P.E. Nordal and S.O. Kanstad, Physica Scripta 20, 659 (1979)

|8| A.C. Boccara, D. Fournier, and J. Badoz, Appl. Phys. Lett. 36, 130 (1980)

|9| J.G. Parker, Appl. Opt. 12, 2974 (1973)

|10| A. Rosencwaig and A. Gersho, J. Appl. Phys. 47, 64 (1976)

|11| F.A. McDonald and G.C. Wetsel Jr., J. Appl. Phys. 49, 2313 (1978)

|12| Y.H. Wong, R.L. Thomas, and G.F. Hawkins, Appl. Phys. Lett. 32, 538 (1978)

|13| G. Busse in "Topical Meeting on Photoacoustic Spectroscopy", Optical Soc. of America, Ames/Iowa, August 1979

|14| A. Rosencwaig, Am. Lab. 11, 39 (1979)

|15| W.R. Harshbarger and M.B. Robin, Acc. Chem. Res. $\underline{6}$, 329 (1973)

|16| G. Busse and A. Ograbek, J. Appl. Phys. $\underline{51}$, 3576 (1980)

|17| G. Busse, Opt. Comm. $\underline{36}$, 441 (1981)

|18| A. Rosencwaig and G. Busse, Appl. Phys. Lett. $\underline{36}$, 725 (1980)

|19| G. Busse and A. Rosencwaig, Appl. Phys. Lett. $\underline{36}$, 815 (1980)

|20| G. Busse, IEEE Ultrasonics Symposium Proceedings, p. 622 (1980)

|21| R.S. Quimby and W.M. Yen, Appl. Phys. Lett. $\underline{35}$, 43 (1979)

|22| Y.H. Wong, R.L. Thomas, and J.J. Pouch, Appl. Phys. Lett. $\underline{35}$, 368 (1979)

|23| M. Luukkala and S.G. Askerov, Electronics Lett. $\underline{16}$, 84 (1980)

|24| G. Busse, Optics and Laser Techn. $\underline{12}$, 149 (1980)

|25| G. Busse, Appl. Phys. Lett. $\underline{35}$, 759 (1979)

|26| E.A. Ash (ed.), "Scanned Image Microscopy", Academic Press (1980)

|27| B. Lieder, Diplomarbeit Hochschule der Bundeswehr München (1980)

|28| A. Lehto, J. Jaarinen, T. Tiusanen, M. Jokinen, and M. Luukkala (to be published)

|29| M.A.J. Ångstrom, Phil. Mag. $\underline{25}$, 180 (1863)

|30| M.J. Adams and G.F. Kirkbright, Analyst $\underline{102}$, 281 (1977)

|31| G. Busse, Infrared Physics $\underline{20}$, 419 (1980)

7. Magnetic Resonance

EPR OF $CuSO_4 \cdot 5H_2O$ AND Fe(III)-TPP DETECTED BY THE
PHOTOACOUSTIC EFFECT

U.Netzelmann, E.v.Goldammer, J.Pelzl, and H.Lerchner
Abteilung für Physik und Astronomie
Ruhr-Universität, D-4630 Bochum 1, FRG

ABSTRACT

EPR spectra from polycrystalline samples of $CuSO_4 \cdot 5H_2O$ and Fe(III)-tetraphenylporphyrin have been studied by means of the photoacoustic effect. In Fe(III)-TPP three absorptions occur in the photoacoustic EPR spectrum which are characterized by their different dependency on the modulation frequency.

INTRODUCTION

The photoacoustic effect (PAE) has become an efficient tool for studies of processes where a generation of heat is involved. Besides its application in optical spectroscopy and ac-calorimetry /1,2,3/, PAE already has been used in an investigation on ferromagnetic resonances in metallic compounds /4,5/. Some EPR studies which make use of the PAE are in progress and are reported by several research groups /6,7/. The calorimetric detection is a well established method in paramagnetic resonance spectroscopy and originally it was proposed by Gorter in 1936. In the last decade a great number of experimental and theoretical work was published concerning thermal measurements of the heat released during the paramagnetic resonance process /8,9/. The essential advantage of

the PA detected EPR spectroscopy is the minor shift and distortion of the baseline in comparison to conventional EPR
experiments, especially for broad absorption lines where
high microwave power is required. Although the sensitivity
of PA detected EPR absorption is considerably smaller if
compared to a normal EPR experiment, the signal to noise
ratio is enhanced at very low temperatures caused by the decrease of the heat capacity and by the temperature dependence
of the transducer gas response which varies roughly with
$T^{-1/4}$ /10/.

Here, we report on an investigation of PA detected paramagnetic resonance on polycrystalline samples of $CuSO_4 \cdot 5H_2O$ and
Fe(III)-tetraphenylporphyrin at room temperature. The first
of the two compounds already has been studied by conventional
EPR and therefore it seems to be well suited for a comparison
of the two experimental methods, whereas the second compound
is of some interest with respect to its biological importance.

RESULTS AND DISCUSSION

All measurements were performed on a modified EPR spectrometer which allows to modulate the klystron frequency. In Fig.1
a block diagram of the spectrometer is shown together with
the PA detection cell used in the present investigation.
From its acoustic properties the latter represents a so-called
extended Helmholtz resonator. Theoretical and experimental
studies of the frequency response of such an acoustic resonator have been reported recently /10,11/. The PA detection
technique was tested on polycrystalline $CuSO_4 \cdot 5H_2O$ where EPR
absorption is well known. All recordings were taken at a
fixed modulation frequency ν_M varying the external magnetic
field strength. Figure 2 gives the result from a simultaneous
measurement of the amplitude A and the phase angle φ of a PA
detected EPR spectrum of $CuSO_4 \cdot 5H_2O$. Besides the expected
increase of the amplitude in the resonance case, there is a
net change $\Delta\varphi$ of the phase angle φ. The variation of the
amplitude depends on the modulation frequency. The relation
between A and $\Delta\varphi$ (for a definition - see Fig.3) at the resonance magnetic field strength B = 0.32 T as a function of
the modulation frequency ν_M is represented in Figure 3. The

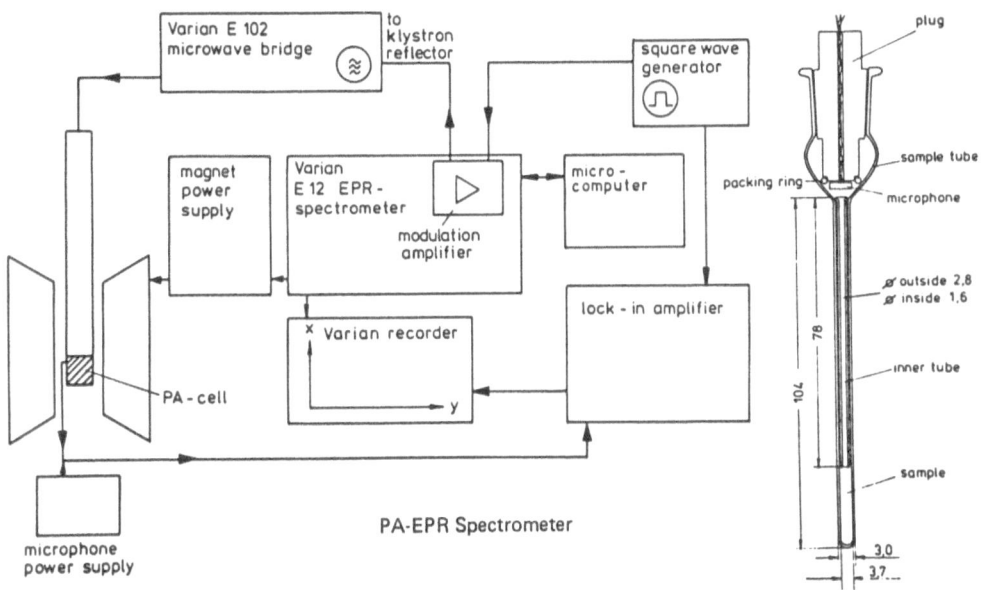

Fig.1 : Block diagram of the EPR spectrometer modified for
        photoacoustic detection. The sample cell containing
        the microphone is represented schematically at the
        right.

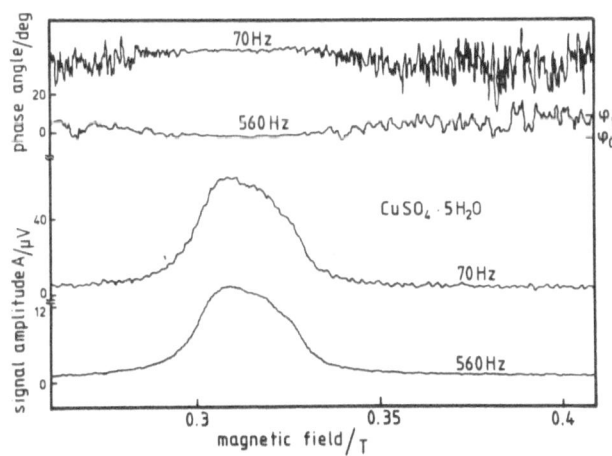

Fig.2 : EPR spectra of polycrystalline CuSO₄·5H₂O at room
        temperature detected by PA-EPR spectroscopy. The am-
        plitude and the phase angle are measured at vary-
        ing modulation frequencies. The microwave power was
        100 mW.

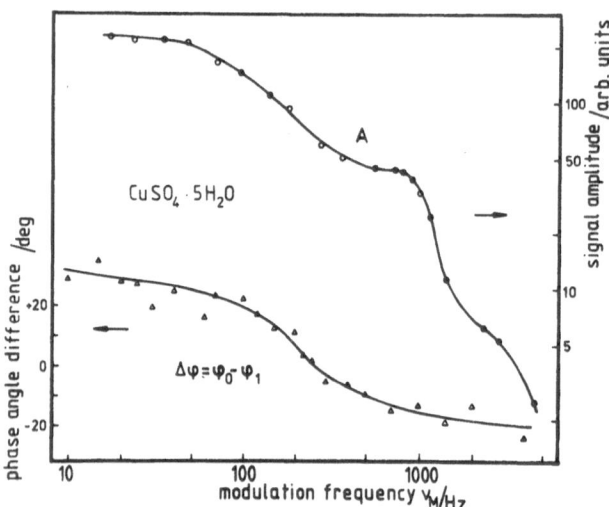

Fig.3 : Logarithmic plot of the amplitude A and the phase
        angle φ at varying modulation frequencies ν_M of
        $CuSO_4 \cdot 5H_2O$ at the resonance magnetic field B_{res}.

frequency dependence of A and φ results from the acoustic
transfer function based on the physical properties of the
acoustic resonator and from the variation of the photoacoustic effect with the periodicity of the radiation impact.
The existence of multiple resonances is a characteristic feature of an extended Helmholtz resonator. This is demonstrated
in Fig.3 by the distinct changes of the amplitude A with ν_M
at $\nu_M \gtrsim 0.5$ kHz. The deviations from the ν_M^{-1} -dependence of
A at $\nu_M \lesssim 50$ Hz can be rationalized by the microphone fre-
quency-response function with its lower limit at $\nu_M \approx 50$ Hz.
In the context of the Rosencwaig-Gersho theory /1/ we are
concerned with the case of an "optically" transparent solid
with a large thermal thickness at high modulation frequencies,
which becomes moderate at about 100 Hz, supposed the crys-
tallites are not in intermediate thermal contact with each
other. The resulting frequency variations of the phase angle
and the amplitude are of the right order of magnitude. A quantitative analysis of the results will be available if the
acoustic transfer function of the cell arrangement has been
determined by an independent measurement which is in progress.

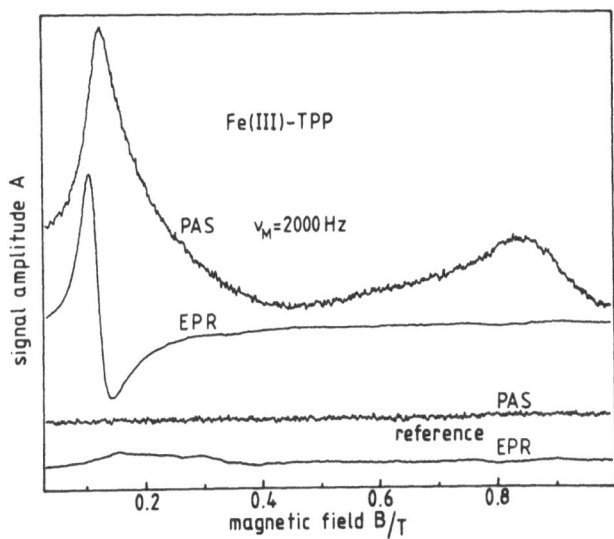

Fig.4 : EPR spectrum of Fe(III)-tetraphenylporphyrin at room
temperature recorded by conventional EPR (lower trace)
and by PA-EPR (upper trace). The reference spectrum
was taken from a MgO sample. The PAS and EPR spectra
were obtained by data accumulation (50 and 200 scans,
respectively).

While CuSO$_4$·5H$_2$O has been used in order to characterize the
experimental set up, the potential advantage of PA detected
EPR may be demonstrated by some preliminary studies on Fe(III)
tetraphenylporphyrin (Fe(III)-TPP) a synthetic derivate of the
porphyrins, which are of fundamental importance for a series
of biological processes. The lower part of Figure 4 shows
a conventional EPR spectrum of Fe(III)-TPP in comparison with
the recording obtained from the empty cavity. The upper part
of this figure gives the PA detected EPR spectrum of the same
sample together with a corresponding reference measurement on
a sample of MgO. All measurements are taken from powdered
samples at room temperature. As far as the background problem
is concerned, the improvement resulting from the PA detection
becomes evident, especially in the case of the weak resonance
line at B≈0.9 T. While the strong absorption at B = 0.135 T
corresponds to the well known EPR signal of high-spin ferri-

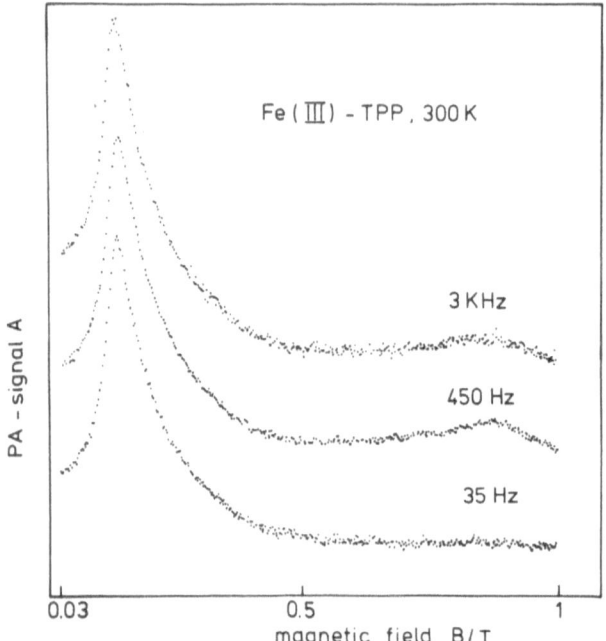

Fig.5 : PA-EPR spectra of Fe(III)-tetraphenylporphyrin at
        various modulation frequencies. All spectra are taken
        at room temperature. The microwave power was 150 mW.

porphyrins with $g_\perp \approx 6$, the weak absorption at B = 0.878 T
is caused by a transition with $g \approx 0.9$. The signal with $g_\parallel \approx 2$
which is known from EPR studies at lower temperatures is
masked by the broad resonance line at B = 0.135 T. The origin
of the high field absorption (B = 0.878 T) is not yet clear
and is a subject of current studies. An interesting feature
is revealed by the spectra measured as a function of the modulation frequency, as it is shown in Figure 5. The relative
intensity of the weak line at B = 0.878 T increases with increasing modulation frequencies ν_M, which possibly is caused
by paramagnetic centers located in the surface area of the
powder grains whereas the line at B = 0.135 T results ex-
clusively from bulk absorption.

CONCLUSION

Preliminary results obtained on $CuSO_4 \cdot 5H_2O$ and Fe(III)-TPP demonstrate the applicability and utility of photoacoustic detected EPR spectroscopy. The low sensitivity of a PA detection at room temperature can be overcome by use of data accumulation and/or by decreasing the temperature. The advantages common to calorimetric and PA detection are the reduction of base line problems, the minor restriction of the microwave frequency range, and the possibility of studying saturation phenomena. One special feature of PA-EPR may be its depth sensitivity. A possible example of such an application are the PA detected EPR spectra of some porphyrins which exhibit a different frequency dependence at high and low resonance magnetic field absorptions.

ACKNOWLEDGEMENT

This work was supported by the Deutsche Forschungsgemeinschaft

REFERENCES

1. A.Rosencwaig,"Photoacoustics and Photoacoustic Spectros-copy", Chem.Analysis Vol.57, eds.P.J.Elving, J.S.Wine-fordner, and I.M.Kolthoff, J.Wiley, N.Y.1980.

2. "Digest of the Topical Meeting on Photoacoustic Spectros-copy", Optical Society of America, Washington, D.C.1979.

3. Contributions of the present conference.

4. A.Cleves Numes, A.M.M.Monteiro, K.Skeff Neto, "Detection of ferromagnetic resonance by photoacoustic effect", Appl.Phys.Lett.35, 656 (1979).

5. C.Evora, R.Landers, and H.Vargas, "Photoacoustic detection of ferromagnetic resonance in films", Appl.Phys.Lett. 36, 864 (1980).

6. R.L.Melcher, "Thermoacoustic detection of electron para-magnetic resonance", Appl.Phys.Lett.37, 895 (1980).

7. H.Coufal, "Acoustic detection of Electron Spin Resonance", this conference.

8. W.S.Moore, "Thermal detection of paramagnetic resonance", Pure and Appl.Chem.40, 211 (1974).

9. J.Schmidt and I.Salomon, "High Sensitivity Magnetic Reso-nance by Bolometer Detection", J.Appl.Phys.37, 3719 (1966).

10. K.Klein, J.Pelzl, and N.Fütterer, "Photoacoustic cell for low temperature PAS", this conference.

11. O.Nordhaus and J.Pelzl, "Frequency dependence of resonant photoacoustic cells: The extended Helmholtz resonator", Appl.Phys., in press.

ACOUSTIC DETECTION OF MAGNETIC RESONANCE

H. Coufal

Physikdepartment E13, Technische Universität München, D-8046-
Garching, West Germany

ABSTRACT

The detection of magnetic resonance with photoacoustic techniques
is desribed and compared with conventional detection methods.
Applications of the acoustical detection method to other modulation spectroscopic techniques are discussed.

INTRODUCTION

Acoustic detection has proved to be a powerful tool in the
study of excitation and de-excitation processes. Particularly in
the visible and infrared spectral ranges, opto- /1/ or photoacoustic /2/ techniques are well established by now. Recently,
the spectral range was extended down into the microwave region
/3,4/; in this type of experiment, part of the amplitude modulated
(AM) incoming radiation is absorbed and causes acoustical waves
that can be detected.

In modulation spectroscopy /5/, however, one uses continuous
wave (CW) irradiation with constant amplitude: by applying a
suitable periodic perturbation to the physical property under
study - temperature, stress or strain, or external fields are
widely used for this purpose - the amplitude of the scattered
radiation is modulated. In very few experiments is a frequency
modulation of the incoming radiation converted by a dispersive
feature of the sample into an AM of the scattered radiation.

Modulation spectroscopy has been found to be a potent
technique in detecting sharp structures superimposed on a broad

background; it boasts high spectral resolution and high sensitivity. Photoacoustics, however, seems to be notable for its
rather low resolution and sensitivity; yet it is only sensitive
to the heat created within the sample and has unique depth profiling capabilities /6/. A combination of both techniques, therefore, should be of considerable interest. Modulation spectra recorded with one of these acoustic detection schemes should reflect
excitation and de-excitation processes, well understood from
classical modulation spectroscopy, and those processes due to the
detection of the absorbed energy as described, for example, by
the Rosencwaig-Gersho-Theory /7/ of photoacoustics. The acoustical detection of magnetic resonance (ADMR) was selected to test
these considerations.

EXPERIMENT

A simple device, shown in Fig.1, was designed to detect
acoustically magnetic resonance phenomena, using standard spectrometers and accessories. The sample is contained in a standard
quartz sample tube. A detector-housing, containing a conventional electret microphone with integrated preamplifier, is
coupled vacuum-tight to the sample tube. Inside, another quartz
tube serves as an acoustic waveguide. This connects the sample
volume with the volume in front of the microphone diaphragm, thus
forming an Helmholtz resonator. Its resonance frequency was determined to be 1.2 kHz at room temperature, using chopped light
to irradiate a carbon black absorber. Using an Air Products
LTD-3-110 Helitran to cool the sample tube to 20K shifted the
resonance to 0.3 kHz; this shows, in addition, that this cell is
very convenient for optical PA studies at variable temperature
and provides an excellent method for calibrating the frequency
response of the system.

The ADMR spectra shown below were taken with a standard
Varian V4502 x-band spectrometer at a power level of 1mW. Using
an AM of the microwave power, the FMR-Spectrum of 0.1 mg of YIG
powder was recorded (Fig.2). Replacing the sample with the same
amount of DPPH-powder, switching to CW microwave power and a
field-modulation with 20 Hz frequency and 1mT amplitude, the first

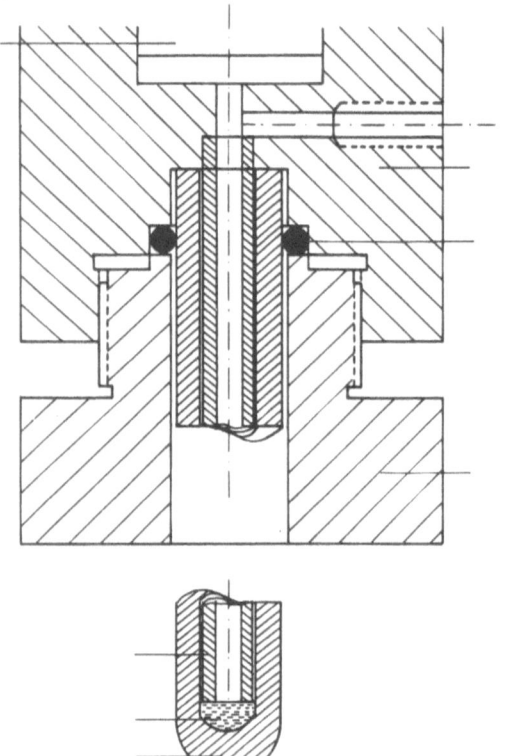

Figure 1:

Cell for the acoustical
detection of
magnetic resonance.

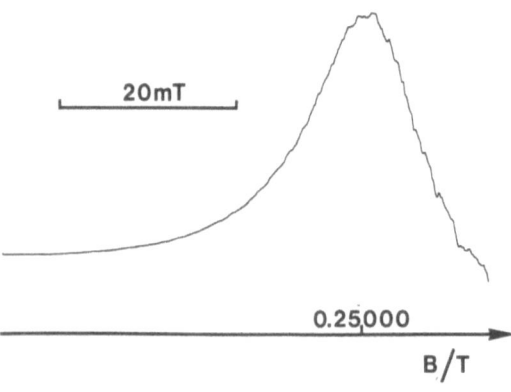

Figure 2:   Acoustically detected FMR-spectrum of
                                 YIG powder with pulsed excitation.

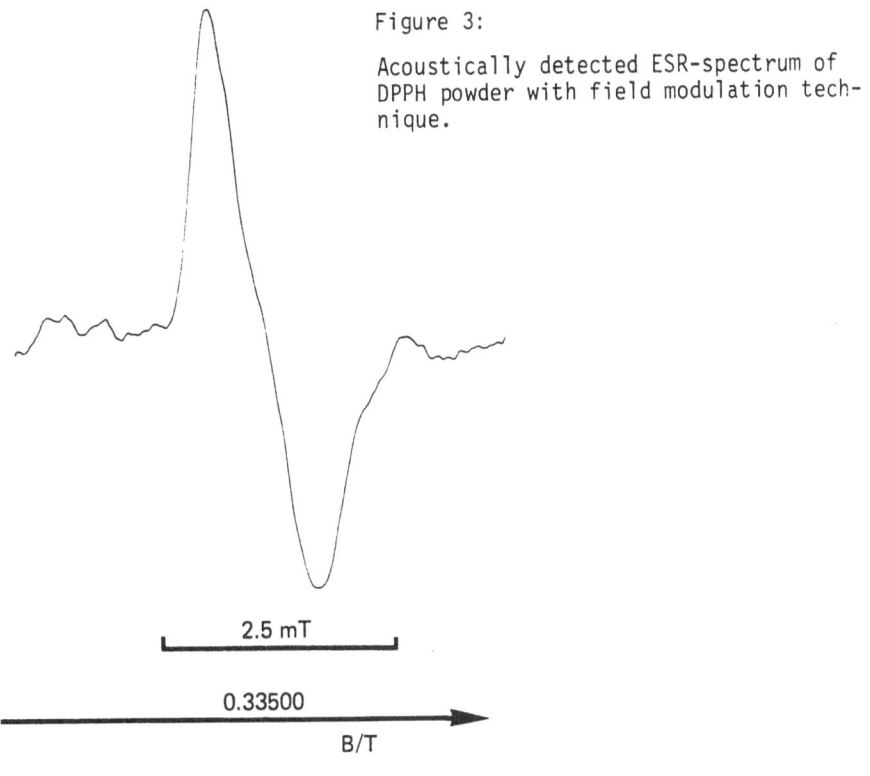

Figure 3:

Acoustically detected ESR-spectrum of
DPPH powder with field modulation technique.

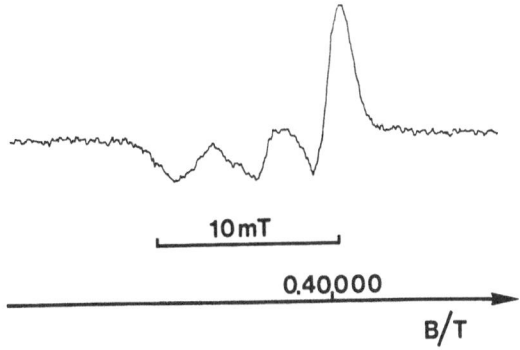

Figure 4:

Acoustically detected
ESR-spectrum of a
ruby single crystal
at 20 K.

derivative spectrum of the ESR absorption could be observed (Fig.
3). A similar spectrum was obtained from a 3x3x20 mm KCl single
crystal containing pitch. A ruby crystal with the same dimensions
showed, at room temperature, a poor signal to noise ratio S/N,
but lowering the sample temperature to 20 K increased the S/N
sufficiently to record the ADMR spectrum shown in Fig.4.

At a constant modulation amplitude of 1mT for all samples, the modulation frequency f_m was varied from 10 Hz to 274 Hz in steps of 12 Hz. An analysis of these data shows that the modulation frequency dependences within this limited frequency range can be described by a power law $f_m^{-\alpha}$. The exponent α for DPPH was found to be 1.05 $\pm$ 0.05, whereas for both single crystals, α = 1.5 $\pm$ 0.05 was determined. In the case of FMR in YIG, α = 1.3 + 0.05/- 0.10 was obtained. These exponents are for the ESR experiments in agreement with the Rosencwaig-Gersho-Theory, assuming optically transparent, thermally thin - DPPH powder - and thermally thick samples - the single crystals. And certainly, the ESR samples had a skin depth much larger than the sample size; whereas the thermal diffusion lengths at the modulation frequencies employed were larger than the grain size of the DPPH powder although smaller than the dimensions of the single crystals. In the FMR-spectra, this distinction was not possible; but the experimental findings are in good agreement with results by Evora et al /4/. Experiments by B. Melcher on the frequency dependence of the ADMR signal of DPPH at higher modulation frequencies with an α = 1.5 are consistent with the assumption that, in his case, the diffusion length became shorter than the grain size.

To emphasise the acoustical aspect of ADMR, the air in the acoustical system was replaced by helium. The observed 20% increase of the signal is in agreement with the theory /7/. Mixing the DPPH with 100 mg of pure SiO_2 powder did not affect the conventionally detected ESR, whereas the ADMR-signal was attenuated by two orders of magnitude due to the dramatic thermal change; also $\alpha \approx$ 1.5 was found.

On the other hand, the g-values derived from ADMR and conventionally detected magnetic resonance are in excellent agreement as are also the power and modulation amplitude dependence /8/. Minor differences can be attributed to the different detection techniques: conventional ESR and FMR detect a change in the Q of the microwave cavity with a suitable microwave bridge that needs proper balancing. Spurious absorptions in the cavity wall or an unbalanced bridge will cause background problems. In ADMR, however, only the heat caused by microwave absorption and subsequent spin-lattice relaxation within the sample are detected.

CONCLUSION

It should be evident by now that the methods and concepts of modulation spectroscopy can be combined with acoustical detection and its theories. The field modulation technique is, of course, only suitable for sharp lines, but it is nicely complemented by the RF amplitude modulation technique for wide lines /3, 4, 10, 11/.

By no means is acoustical detection a substitute for conventional ESR, for its sensitivity at room temperatures and 100 Hz modulation frequency is several orders of magnitude lower. But even here, it might provide a convenient tool for testing the proper tuning of spectrometers.

If, however, one takes advantage of the photoacoustic features of the signal as described for example by the Rosencwaig-Gersho-Theory /7/, ADMR might prove powerful in special applications. With the photoacoustic signal showing approximately $T^{-1/4}$ temperature and an $f_m^{-\alpha}$ modulation frequency dependence, ADMR will be as sensitive as conventional ESR when low temperature and low modulation frequencies are required; especially at low modulation frequencies is it a substitute for other thermal detection techniques /12/.

The application of ADMR to NMR is straightforward. Using powder samples, the sensitivities of conventional NMR and AD-NMR (the large solid gas interface enhancing the acoustically detected signal by orders of magnitude) are in fact comparable. Preliminary data on the AD-NMR of ^1H in $CaSO_4$ x $2H_2O$ support this claim. Using the techniques developed for time-domain photoacoustic spectroscopy /13/, spin-lattice relaxation times can be easily determined. Piezoelectric detection of the acoustic signal /14/ in connection with pulsed excitation should give direct access to very short relaxation times.

Taking advantage of the depth profiling capabilities /6/ of the acoustic detection scheme, surface and bulk resonances can be easily distinguished.

The author would like to thank Professor Dr. E. Lüscher, in whose laboratory most of the experiments were carried out, for his stimulating interest and support.

REFERENCES

/1/   Yoh-Han Pao, ed., 'Opto-acoustic Spectroscopy and Detection', Academic Press, NY (1977)

/2/   A. Rosencwaig, 'Photoacoustics and Photoacoustic Spectroscopy', Chem.Analysis 57, eds. P.J. Elving, J.S. Winefordner and I.M. Kolthoff, Wiley, NY (1980)

/3/   O.A.C. Nunes, A.M.M. Monteiro, K.S. Neto, Appl.Phys.Lett. 35, 656 (1979)

/4/   C. Evora, R. Landers and H. Vargas, Appl.Phys.Lett. 36, 864 (1980)

/5/   M. Cardona, 'Modulation Spectroscopy', Solid State Physics, Suppl. 11, Seitz and Turnbull, eds., Academic Press, NY (1969)

/6/   M.A. Afromowitz, P.S. Zeh, S. Yee, J.Appl.Phys. 48, 209 (1977)

/7/   A. Rosencwaig and A. Gersho, J.Appl.Phys. 47, 64 (1976)

/8/   B. Melcher, Appl.Phys.Lett. 37, 895 (1980)

/9/   H. Coufal

/10/   U. Netzelmann, E.v. Goldammer, J. Pelzl and H. Lerchner; this Volume

/11/   H. Vargas

/12/   W.S. Moore, Pure and Appl. Chem. 40, 211 (1974)

/13/   A. Mandelis and B.S.H. Royce, J.Opt.Soc.Am. 70, 474 (1980)

/14/   W. Jackson and N.M. Amer, J.Appl.Phys. 51, 3343 (1980)

APPLICATIONS OF THE PHOTOACOUSTIC EFFECT TO
FERROMAGNETIC RESONANCE AND ELECTRON
PARAMAGNETIC RESONANCE

H. Vargas*

Department of Physics,
University of Nottingham,
Nottingham.
NG7 2RD
U.K.

(*On leave from University of Campinas, SP, Brasil)

ABSTRACT

This article describes the rapid development in the application
of the photoacoustic method of detecting ferromagnetic and paramagnetic solids. Recent measurements of ferromagnetic resonance
(FMR) in thin films of Ni, Fe and $Fe_{80}B_{20}$ and of electron paramagnetic resonance (EPR) in polycrystalline samples of DPPH,
$CuSO_4 5H_2O$ and Fe-Tetraphenylporphyrin are discussed and details
are given of the various experimental arrangements used. The
photoacoustic technique is compared critically with more
conventional methods (thermal and electronic) of detecting EPR
and FMR.

1. INTRODUCTION

The last few years have seen a remarkable resurgence of
interest in the photoacoustic effect (PAE) as an alternative
method that permits absorption spectroscopy to be used on a
wide variety of optical absorption phenomena which involve the
generation of heat[1,2,3]. With the recent development of coherent
monochromatic optical radiation sources (lasers), modern microphones and electronics and properly designed acoustic chambers,
the application areas for the photoacoustic effect have widened[4].
The method is now used in many aspects of spectroscopic and non-
spectroscopic analysis of materials[4,5], and is tending to
diversify in its applications to match specific needs. This is
very apparent when one considers the wide range of conferences
and journals in which the uses of the PAE are reported.

The correct interpretation of this effect on solid samples has been given in the one-dimensional theory of Rosencwaig and Gersho[6], which we shall refer to as RG, while a number of extensions to cover different, more complex, situations have been advanced[7,8,9,10,11]. The primary source of the acoustic signal arises from the periodic heat flow from the sample to the surrounding gas as the solid is cyclically heated by the absorption of intensity-modulated or chopped light. The periodic flow of this heat into the gas produces pressure fluctuations which are detected as an acoustic signal by a sensitive microphone transducer attached to the cell. The PA method using the gas-microphone as a detector, although indirect, is a sensitive calorimetric technique for measuring how much of the radiant energy absorbed by the sample is actually converted into heat[1].

This has suggested the possibility of using the photo-acoustic effect as a calorimetric method to investigate, in the microwave region of the electromagnetic spectrum, any phenomena involving a rise in temperature during a physical process, electron paramagnetic resonance is an example of such a phenomenon. The heat released during the electron paramagnetic resonance (EPR) process can be detected using bolometric techniques as originally proposed by Gorter in 1936. If a sample containing paramagnetic impurities is placed in an appropriate magnetic field and irradiated by an RF field, provided that the magnetic resonance condition is satisfied, the sample will be heated and, as a result, will exhibit an increase in temperature. The kinetics of the heating of the sample during the variation of the external field leads to a temperature variation of the sample which follows the usual electron paramagnetic resonance (EPR) absorption line in a manner similar to the absorption detected by Zavoisky's conventional electronic detection method[12,13]. The resonance condition is then the same for both methods:

$$h\nu = g\beta B, \tag{1}$$

where h is Planck's constant, ν is the frequency of the electro-magnetic wave, g is the appropriate g-factor, β is the electronic Bohr magneton, and B is the external magnetic field.

To measure the temperature rise, and hence to detect the thermal EPR spectrum, a sensitive thermometer is attached to the sample in the thermal detection method. This technique is given the acronym TD-EPR.

In the photoacoustic method of detection, by chopping the microwave power absorbed by a paramagnetic sample inside the acoustically sealed cell containing a non-absorbing gas, the increase in temperature of the sample during the resonance causes pressure fluctuations in the surrounding gas which are measured by a microphone.

The first experiment on solid samples showing the possibility of using the photoacoustic effect to detect resonances in the microwave region of the electromagnetic spectrum was performed at the Department of Physics of the University of Campinas in 1978. The authors reported[14] the use of the photoacoustic effect to study the well-known phenomenon of ferromagnetic resonance (FMR) in Ni and Fe. Also detected was the EPR of the well-studied O_3^- (ozonide) free radical created by irradiation[15]. In their experiment, the sample contained in a Helmholtz cell coupled to an electret microphone was positioned in the shorted end of the microwave cavity of an ordinary X-band EPR spectrometer. The microwave power was amplitude modulated by applying square pulses to the Klystron reflector which provided 20 mW at 9 GHz. The EPR spectrometer was operated in the conventional manner. On sweeping the magnetic field, an increase in the photoacoustic signal was found corresponding to, in the case of Ni and Fe, the absorptive part of the permeability of these metals.

The magnetic field at which resonance occurred was similar to that already observed using microwave reflection measurements.

For the paramagnetic centre O_3^-, only a weak signal was observed although the g-value and linewidth observed agreed well with those obtained by the conventional EPR.

The idea of using the photoacoustic effect in the microwave spectral range is not new. In gaseous systems, photoacoustic signals have been detected in the millimetre, sub-millimetre and in the centimetre wavelength regions[16]. Recently[17], with a photoacoustic cell in a conventional EPR spectrometer, the absorption and relaxation of the microwave energy between Zeeman magnetic sub-levels of molecular oxygen was observed.

To date the photoacoustic method in the microwave region on
many solid samples have been used to detect FMR and EPR between
room temperature and liquid helium temperature[18-26].

Although most of the projects discussed in the literature
(published or sometimes in private communications) are still
in progress and not fully interpreted, some important features
of the PA method are worth noting. The PA method of detecting
EPR shares many of the advantages of TD-EPR over the conventional
or electronic method. These can be summarized as follows:
detection of broad resonant lines; improved sensitivity at low
temperatures where the specific heat of the crystal lattice falls
rapidly as T^3; simplicity of the experimental arrangement. It
is likely therefore that the future use of the photoacoustic method
of detection of EPR will be used to exploit these advantages.

The purpose of this review is two-fold. First it is to
introduce PAE to the non-specialist with no previous knowledge
of the subject, whether for general information or for providing
a basis for future work in the area. Secondly, it is hoped that
this review will be of service to those actively engaged in the
field.

Section 2 of this review will describe briefly the
experimental arrangement, with particular emphasis on the
adaptations that are necessary to ordinary EPR spectrometers
in order to make the measurements, and on the construction and
adaptation of cells in the spectrometer. In the following
sections we will discuss the results obtained on ferromagnetic
(section 3) and paramagnetic (section 4) substances and compare
the sensitivity of the photoacoustic method with the more
conventional techniques for observing the same physical parameters.

2.   EXPERIMENTAL TECHNIQUES

In the following sections (2.1 and 2.2) the spectrometer used
for photoacoustic detection at X-band of electron paramagnetic
resonance (EPR) and ferromagnetic resonance (FMR) will be
described. It consists of a Klystron or Gunn oscillator as a
source of the microwaves, a waveguide with frequency and power
measurement facilities, and the photoacoustic cell, which is the
heart of the instrument. A magnetic field can be provided by any

suitable magnet. In addition, one requires a pulse modulator to intensity modulate the microwave source.

In section 2.3 the design of the room temperature cell is discussed. In section 2.4, the design for the low temperature cell is described.

2.1: The PA Spectrometer

The microwave power is intensity modulated or chopped at a modulation frequency of say, ν_M, by applying square pulses to the Klystron reflector (ν_M should not be confused with the microwave frequency ν). The sample to be studied is placed in an acoustically sealed cell containing a gas and a sensitive microphone and is positioned in the shorted end of the waveguide between the poles of an electromagnet. Absorption of chopped microwaves by the sample causes its temperature to increase. Because the microwave is chopped, the temperature rise is periodic at the chopping frequency and it is this periodic temperature rise which, in turn, causes a modulation of pressure in the closed cell. The pressure modulation is an acoustic signal at the chopping frequency and it is this which is detected by the microphone. Hence, by monitoring the microphone signal during the variation of the external magnetic field, and plotting it as a function of the wavelength of the incident microwave radiation, an acoustic resonance of the sample is observed. The analog signal from the microphone is then amplified, phase detected by the lock-in amplifier and recorded as a function of the magnetic field strength. With this simple experimental arrangement the signal recorded by using amplitude modulated microwaves gives the absorption of the acoustic resonance.

The detected absorption is a direct measure of the heat generated in the sample by the incident radiation. Hence, no magnetic field modulation is required for the detection of the acoustic resonance, although this can be easily done.

2.2: The Combined PA-EPR Spectrometer

Figure 1 shows a complete spectrometer that has been recently developed[20,27]. The system involves a slight

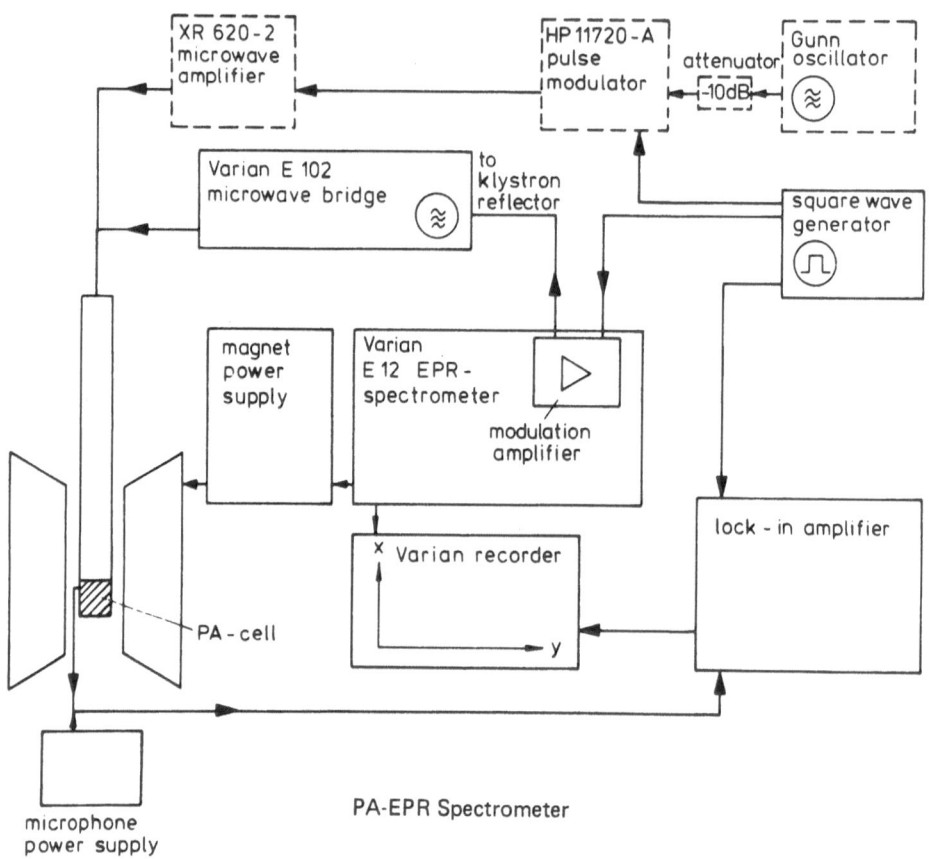

FIGURE 1   Block-diagram of the combined PA-EPR spectrometer. The
              dashed lines are modifications described in the text.

modification of the Klystron modulation of a standard Varian
EPR spectrometer, (see Figure 1, solid lines). The square
wave modulation signal is fed into the input of the cavity
driver amplifier which is normally used for Klystron triangle
wave modulation in the tuned mode of the microwave bridge.
As the square wave voltage is AC coupled to the Klystron
reflector, the mechanical frequency tuning of the Klystron
has to be adjusted in a manner so that the microwave frequency
is switched between the full-resonance and off-resonance states
of the cavity. The efficiency of modulation can be monitored
on the EPR oscilloscope. A square wave generator provides both
the Klystron modulation and lock-in amplifier reference signal.

The DC output of the lock-in amplifier is fed to the external
input of the spectrometer recorder. This arrangement allows
simultaneous recording of the PA signal and the conventional
EPR signal. The dashed line in Figure 1 is another modification
in the modulation of the Klystron[20,27], using also a standard
EPR spectrometer. Whereas in the version described above the
maximum microwave source power was P = 100 mW, up to 2 W can be
provided by the arrangement indicated by the dashed line in
Figure 1. The microwave source was a Gunn oscillator which is
tunable from 7 GHz to 12.5 GHz, having a power output of 20 mW.
A pulse modulator (Hewlett Packard Type 11720A) was used to
chop the microwaves ($r_{on/off}$ > 80 dB), which was then amplified
by a travelling wavetube amplifier (Kaltec XR6202). In order
to protect the power amplifier and also to improve the dynamic
range of the modulator, a 10 dB attenuator was inserted between
the oscillator and the modulator. Reflection from the guide and
the power amplifier could be reduced with the help of a slit
guide tuner.

2.3: The Room Temperature Cell Design

The photoacoustic cell is the heart of the PA-spectrometer
and in most designs an electret or capacitor microphone is used
as the pressure transducer. The design of the cell involves a
complex optimization procedure in order to achieve the necessary
high signal-to-noise ratio for photoacoustic studies in the
microwave region. As in the case involving optical excitation,
some precautions are necessary[1]. Since the signal in the photoacoustic cell used for solid samples varies inversely with the
gas volume, one should attempt to minimize the gas volume.
Furthermore, the distance between the sample and the cell
window should always be greater than the thermal diffusion
length of the gas, since it is the boundary layer of gas that
acts as an acoustic piston generating the signal in the cell.
At the present time there have been few cell designs for studies
in the microwave region. All the cell designs have used
inexpensive electret microphones (Knowles Electronics, BT-1959,
BT-1751) with internal FET pre-amplifier. All these microphones
have a maximum sensitivity of 10 mV at a 1.3 V polarization
with a flat response between approximately 100 Hz and 10 kHz.

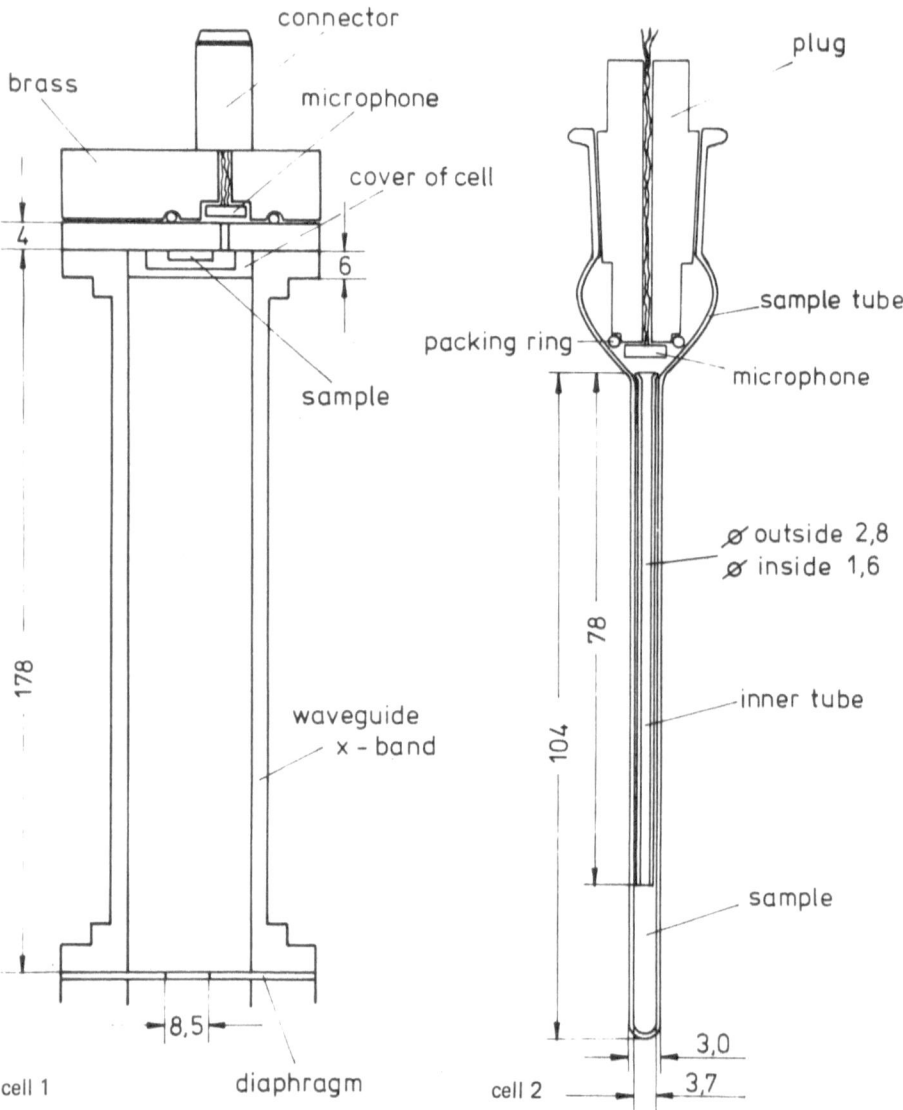

FIGURE 2   Construction details of photoacoustic cells. Cell 1 is
            machined in the microwave short. Cell 2 fits into the
            X-band cavity replacing the sample tube.

At lower frequencies, the microphones show a drop in the

frequency response.

     Figure 2 (cell 1) depicts a cell proposed by Evora[19].

The acoustic cell (a Helmholtz resonator) was made of lucite,

a low-loss material in the microwave region, and has dimensions

$2 \times 6 \times 17 \text{ mm}^3$. The microphone chamber (dimensions
$2 \times 5 \times 8 \text{ mm}^3$) was machined in the back of the microwave short.
The pressure fluctuation in the sample chamber was transmitted
to the microphone chamber via a small hole (1 mm in diameter
and 4 mm long). The specific frequency range was determined by
the volume of the microphone compartment, the channel area and
the length. The basic Helmholtz resonance frequency ν_R is given
by:

$$\nu_R = \frac{C}{2\sqrt{\pi}} D \left(\frac{V_1 + V_2}{V_1 V_2 (L + 1.7D)} \right)^{\frac{1}{2}}$$

where C is the velocity of sound in air, V_1 and V_2 are the volume
of the sample and microphone chambers respectively, L and D are
the length and diameter of the duct connecting both chambers.
The resonance frequency was around 6000 Hz. With this cell,
it is possible to measure directly the dependence of the PA
signal up to 1500 Hz with no correction due to the cell resonant
response around 6000 Hz.

Another very simple cell is shown in Figure 2 (Cell 2).
It was designed[20] to fit into the X-band cavity replacing the
sample tube. Acoustically, the cell construction corresponds
to that of an extended Helmholtz resonator[28]. All parts except
the plug are made out of quartz. The plug supports the microphone (Knowles Electronics BT-1959). With this quartz-tube cell
simultaneous recording of the photoacoustic signal and the
conventional EPR signal can be achieved.

The cell developed by McCann[22] consists of a
copper cylinder with a narrow hole in the centre. By
soldering the copper cylinder to the end of the waveguide, a
high Q value is obtained. The cell, with a volume of 0.05 cm^3,
is machined in the bottom of the cylinder and is sealed with a
quartz window using vacuum grease. The microphone used was a
Knowles BT-1751. Because of its small entrance port, the
incorporation in the cell is rather simple. Furthermore, the
gas volume in the cell is small. Due to the high Q of this
design, only small microwave power is necessary.

2.4: Low Temperature Cell Design

The construction of cells suitable at low temperatures is
still being explored. The cells are constructed in the form of
a Helmholtz resonator. With some modifications, cell 2 of
Figure 2 allows the variations of the temperature between 77 K
and 300 K. The whole system is cooled by direct immersion in
liquid nitrogen. The main problem with this configuration is
the noise from the boiling nitrogen. This cell was recently
modified to allow variation of temperature from 90 K to 320 K[24].

The same arrangement was independently developed by A. Vasson
and A.-M. Vasson[25]. It consists of a variable temperature
conventional EPR system with a continuous gas flow cryostat.
The microphone is kept at room temperature. The connecting
tube between the sample cell is of stainless steel, having an
inner diameter of 1.0 mm and length of 250 mm. For gas (e.g.
He) substitution, the cell is equipped with a small inlet and
a valve to equilibrate the gas pressure when measurements are
performed at different temperatures. The body of the sample
chamber is made from brass and the microphone is a Knowles
BT-1751.

3. APPLICATIONS TO FERROMAGNETIC RESONANCE

Ferromagnetic resonance is the analogue of electron paramagnetic resonance (EPR) and nuclear magnetic resonance (NMR).
As is well known, in this method the frequency and the magnetic
field at which resonance occurs depend on the geometry of the
sample[29]. Several workers[30,31] using microwave reflection
measurements have already studied the FMR of Ni and Fe. Both
the real and imaginary parts of the high frequency permeability
of flat specimens of Ni and Fe have been investigated[32] at
X- and Q-band with a constant field B_o parallel to the plane
of the specimen.

In this section, the detection of FMR of Ni, Fe and $Fe_{80}B_{20}$
using PAE will be described. The observed data will be compared
qualitatively and quantitatively with theoretical models for
the generation of the photoacoustic signals as developed by
Rosencwaig and Gersho[6].

3.1: Qualitative Description of the Signals

The photoacoustic cell used in these experiments is shown in Figure 2 (cell 1). The flat sample investigated was placed inside the cell containing air at room temperature. The geometry was such that both the applied DC magnetic field and the intensity-modulated microwave source were mutually perpendicular.

Figure 3 shows the signals for a Ni thin film as a function of external magnetic field[14,19]. At 0.15 T the PA signal peaks due to FMR. The spectrum observed is similar to that in the same metal using microwave resonance reflection methods[32]. Similar spectra are observed for a series of samples.

Figure 4 shows the spectrum for the Fe specimen. The resonance occurs at 0.1 T at 12.1 GHz. Again, the characteristics of the resonance measured using the PA effect are identical to those observed by conventional microwave reflection methods. For a flat specimen, as in the case of Fe, measured with a magnetic

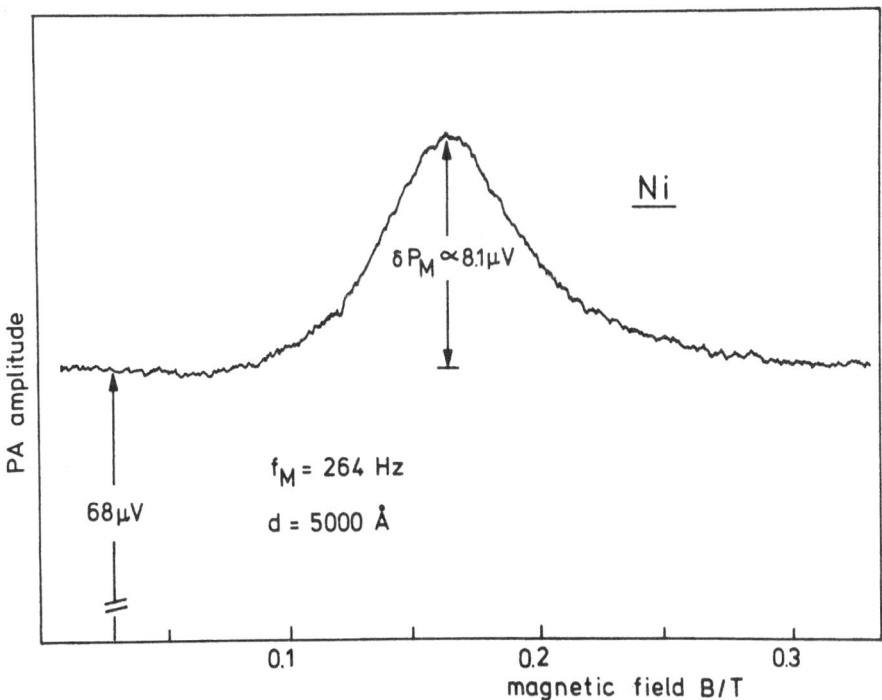

FIGURE 3   Photoacoustic detection of FMR in Ni. The increase of the PA signal δP_M and the value at B = 0, δP_B are described in the text.

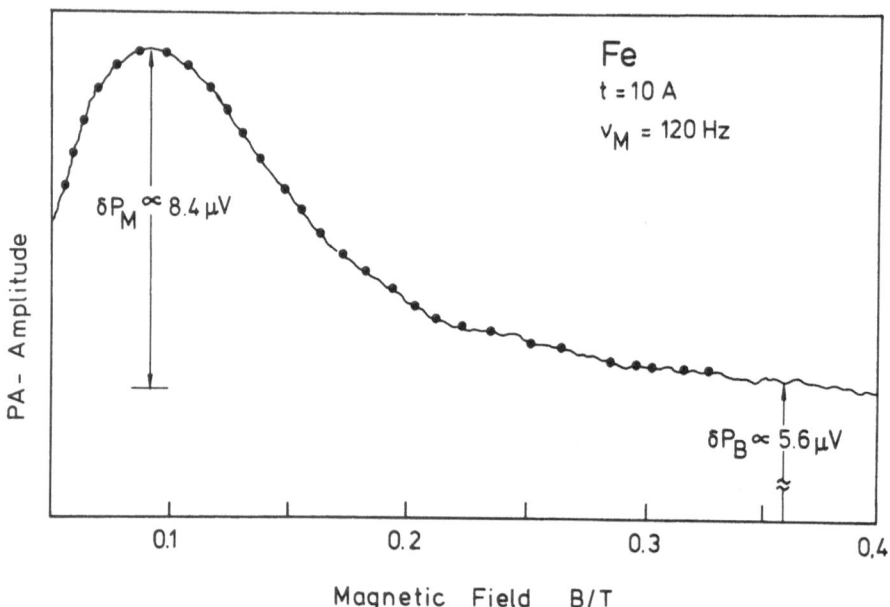

FIGURE 4   Photoacoustic detection of FMR in Fe. The points in
              the curve are field values which satisfy resonance
              conditions described in the text.

field B_o parallel to the plane of the sample, the resonance
condition for both methods is given by:

$$h\nu = g\beta \, B_{o \; eff} \qquad (2)$$

The field at which resonance occurs satisfies the condition of
resonance when $B_{o \; eff}$ is given by:

$$B_{o \; eff} = [(B_o + 4\pi M) B_o]^{\frac{1}{2}}. \qquad (3)$$

This is shown in Figure 4. The spectrum observed shows a well-
defined maximum when the value of $B_{o \; eff}$ in the resonance condition
(2) is given by equation (3).

    The absorption agrees very well with the spectra observed by
Griffiths[30] when measuring the variation of the high frequency
losses with applied magnetic field for Fe at wavelengths of 1 to
3 cm. One striking feature of all the observed spectra is the
small amplitude of the resonance peak relative to the $B_o = 0$
amplitude. From microwave reflection measurements it is known

that the absorption at resonance increases by a factor of about 10 over the zero field value for both Ni and Fe films.

Another important feature of the PA signal is its dependence on modulation frequency ν_M. The amplitude of the photoacoustic signal increases with decreasing modulation frequency. This $1/\omega$ frequency dependence ($\omega = 2\pi\nu_M$) has been thoroughly tested by a number of experimental studies and agrees with the one-dimensional theory under certain conditions. This theory[6] leads to the following conclusions. For an optically transparent, absorbing or opaque solid sample the intensity of the PA signal should vary with chopping frequency, as $\omega^{-3/2}$ in the so-called thermally thick regime (where the thermal diffusion length of the sample μ_S is less than the sample thickness t, and less than the optical penetration length α^{-1} (α is the optical absorption coefficient). However, the signal varies as ω^{-1} in the thermally thin regime where one of the two previously given conditions is not fulfilled[6].

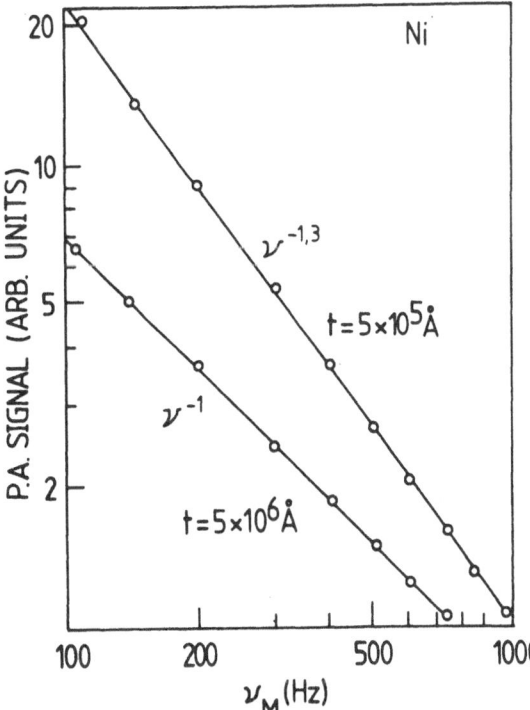

FIGURE 5   Photoacoustic signal intensity as a function of modulation frequency for Ni thin films.

Figure 5 shows the photoacoustic signal as a function of
the modulation frequency for Ni. The signal decays as $\nu_{M1}^{-1.3}$.
In the following, all the qualitative aspects observed will be
compared with the predictions of the one-dimensional theory.

3.2: Relation of PA Signal Observed to FMR

We shall discuss now the photoacoustic signal dependence
on both magnetic field and the chopping frequency. These
features can be understood considering first, the classical
problem of a sample being periodically heated by a modulated
microwave beam which can be obtained from the expression:

$$P = -\text{div } \vec{S} \qquad (4)$$

where $\vec{S}$ is the Poynting vector of the microwave in the metallic
sample. Using the expression for the electric and magnetic
fields inside of the metallic sample[33] one obtains:

$$\vec{S} = \hat{z} \frac{c^2 |B_o|^2}{(8\Pi)^2 \sigma} (2k_I) e^{-2k_I x} \qquad (5)$$

where k_I is the imaginary part of the propagating vector inside
of the metallic sample:

$$k_I = \text{Im} (\frac{1}{c} \sqrt{4\Pi\sigma\omega\mu}) \qquad (6)$$

(σ is the conductivity of Ni, ω is the frequency of the microwave,
μ is the complex transverse permeability of Ni: $\mu = \mu_1 + j\mu_2$).
The modulated heating power per unit volume can then be
written as:

$$P(x,t) = \left(\frac{cI_o}{8\Pi\sigma} \right) \beta^2 e^{-\beta x} [1 + \cos(2\pi\nu_M t)] \qquad (7).$$

where $I_o = \frac{c|B_o|^2}{8\Pi\sigma}$ (B_o is the intensity of the microwave magnetic
field on the surface) and:

$$\beta = 2k_j = \frac{2}{c} \sqrt{2\Pi\sigma\omega\mu_R} \qquad (8)$$

$$\mu_R = (\mu_1^2 + \mu_2^2) + \mu_2 \qquad (9)$$

The theoretical treatment of the PA effect from solids was
developed by Rosencwaig and Gersho (referred to as the RG theory).

Although the RG[6] theory uses a simple one-dimensional piston
model for the expansion of the gas near the solid surface, it
correctly predicts PA signals for a number of experimental
conditions. Rosencwaig and Gersho considered the problem of a
sample being periodically heated by a chopped light beam.
According to these authors the heating power per unit volume,
P, as a function of the penetration x, is given by:

$$P(x,t) = \frac{1}{2} \beta I_o e^{-\beta x} [1 + \cos(2\pi\nu_M t)] \qquad (10)$$

where β is the optical absorption coefficient and I_o is the
average incident power.

Using the following prescription:

$$I_o \beta \rightarrow \frac{c I_o}{8\Pi\sigma} \beta^2 \qquad (11)$$

we can translate all the RG's results for our problem.
Most important is eq. (21) of RG paper for the variable Q,
which is directly related to the pressure fluctuations in the
acoustic cell:

$$\delta P = Q \exp [j(wt - \pi/4)]$$

Since the samples are pure metallic crystals the thermal
conductivity is much bigger than the respective quantities for
the backing (glass or brass) with which the samples are in
contact, and for the gas. In this case the expression for Q
reduces to:

$$Q = \frac{j P_o}{2\sqrt{2} T_o} \frac{1}{a_g l_g} \frac{c I_o}{8\Pi\sigma} \frac{\beta}{k_s} [\coth(\sigma_s t) - \frac{\exp(-\beta t)}{\sin(\sigma_s t)}] \qquad (12)$$

where P_o is the ambient pressure in the sample chamber, T_o is
the ambient temperature, l_g is the dimension of the gas chamber
parallel to the direction of the microwave beam propagation,
$a_g = (\Pi\nu_M/\alpha_g)^{1/2}$, α_j is the thermal diffusion constant of the
gas, $\sigma_s = (1 + j) (\Pi\nu_M/\alpha_s)$, α_s is the thermal diffusion constant
of the sample, t is the sample thickness and k_s is the sample
thermal conductivity.

Qualitatively, the dependence of the PA signal at resonance
can be explained by expression (12). The increase in the PA

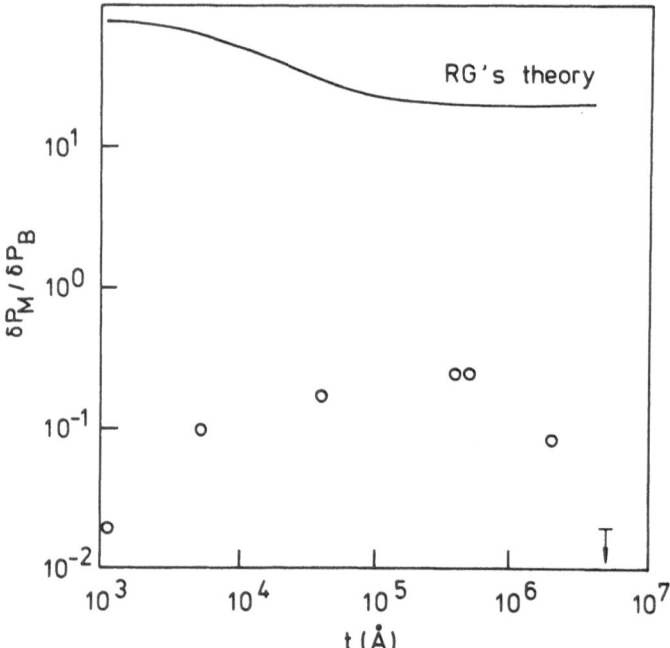

FIGURE 6   Variation of the ratio $\delta P_M/\delta P_B$ versus film thickness
              t of Ni films.

signal corresponds to the resonant absorptive behaviour of β
through its dependence on the effective permeability μ_R given
by equation (9). Assuming that the transport properties
(electrical and thermal conductivity) do not vary with the
magnetic field, using expression (12) it is possible to
determine the relation $\delta P_M/\delta P_B$ for each Ni film thickness.

The δP_M corresponds to the increase of the PA signal at
resonance and δP_B is the value of the photoacoustic signal
at $B_o = 0$ (see Figure 3). The theoretical values for $\delta P_M/\delta P_B$
versus t are shown in Figure 6.

Equation 12 based on the RG model cannot explain
quantitatively the experimental observations. One of the
reasons for this disagreement is the small amplitude of the
resonance peak relative to the amplitude at $B_o = 0$. It is
unclear whether the background arises from the microwave
absorption at the cell walls, or from the surface of the
films (each of which was prepared by different methods)
leading to over-absorption inside the sample. Since the

microwave penetration is only about 10^{-4} to 10^{-5} cm, any
strains or contamination of the sample can give a spurious
contribution to the PA signal. Also it is worth noting that
the gas microphone works well when the solid sample has a
large surface-to-volume ratio as in the case of powdered
samples. With the microwave power used in the experiments
(20 mW) the signals observed are small ($\sim \mu V$). Hence spurious
reflections from the cell walls and surface contamination of the
films can cause an appreciable background level.

3.3: The Frequency Dependence of the PA Signal

As is shown in Figure 5, the amplitude of the PA signal for Ni films varies with the chopping frequency as $\nu^{-1.3}$
when the film thickness is 5×10^5 Å (50 μm) and as ν^{-1} when the
thickness is 5×10^6 Å (500 μm). For Fe films a ν^{-1} variation is
found for both 10^7 Å (1000 μm) and 10^5 Å (10 μm) thick films. To
interpret these results one needs to recall the three important
parameters, sample thickness t, optical (microwave) absorption
length α^{-1} (α is the absorption coefficient) and thermal
diffusion length μ_s, which is given by :

$$\mu_s = (2k/\omega \rho C)^{\frac{1}{2}}$$

where k is the thermal conductivity, ω is the chopping frequency,
ρ is the density and C is the specific heat. μ_s is typically
270 μm at 100 Hz. The value of α^{-1} is around 1 μm,
the typical value for microwave penetration in metals. In
opaque metals ($\alpha^{-1} \ll t$) which are thermally thin ($\mu_s \gg t$); the
signal varies as ν^{-1} according to the RG theory. In thermally
thick opaque solids ($t > \mu_s > \alpha^{-1}$) the signal should also vary
as α^{-1}. These theoretical predictions are shown in Figure 5.
For the 50 μm sample, which shows the $\nu^{-1.3}$ dependence,
there is disagreement with the theoretical predictions. The
500 μm sample on the other hand supports the predictions of the
theory. For the Fe samples the ν^{-1} dependence predicted by the
theory is observed. The $\nu^{-3/2}$ dependence is observed only if
$\mu_s < t$ and if $\mu_s < \alpha^{-1}$. The anomalous behaviour of Ni indicates
that great care should be taken in the preparation of samples.
The spectra of Ni prepared by different methods give different
results. When the absorption is small, as it is in metals

(absorption lengths of 1 μm to 0.1 μm), the surface contamination
or careless forms of preparation can lead to spurious results.
Even using the conventional method of microwave reflection, a
slight strain on the surface of the material can lower the
amplitude of the ferromagnetic resonance by appreciable values.

The data also show that the signal is larger for thermally
thin than for thermally thick solids. So within the experimental
uncertainty, the ratio of signal strengths of different samples
(different reflectivities and methods of preparation) support the
prediction of RG theory.

3.4: Conclusions and Perspectives

We have discussed the detection of ferromagnetic resonance
by the photoacoustic effect. The data presented, although
preliminary, lead to the following conclusions. The method has
special significance for measuring the absorptive part of the
permeability of ferromagnetic metals. The resonant magnetic
field, at which the PA occurs, the linewidth and g-values are
in excellent accord with those reported using more conventional
FMR techniques. Quantitatively the dependence of the PA signal
on both the magnetic field and the chopping frequency agrees only
approximately with the standard theories for the generation of the
photoacoustic signal.

With sufficient microwave power, the method is capable of
high sensitivity. Because of its inherent simplicity, the
method could provide a powerful complementary tool to the well-
established methods for studying ferromagnetic materials.

4. APPLICATIONS TO ELECTRON PARAMAGNETIC RESONANCE

In the previous section we discussed the detection of FMR
using the PA effect. The experiments have proved that the use of
microphones to monitor indirectly the increase of temperature during
the absorption gives a reasonable agreement and reproducibility
with the measured values obtained by more conventional methods
for detection FMR in metals.

This section discusses the applicability of the PA method to
the detection of the electron paramagnetic resonance (EPR). In
particular a comparison will be made between the sensitivity of
both methods.

4.1: Qualitative Aspects of EPR-Detected Signals

The PA-EPR detected signal is directly proportional to the microwave power, whereas a conventionally detected signal is proportional to the square root of the power. Figure 7 shows the dependence of the PA signal[22] as a function of the microwave power for the sample (<< 1 mg) of DPPH at a chopper frequency of 100 Hz. The increase in sensitivity is independent of linewidth because the magnetic field is swept linearly with no modulation.

Figure 8 shows a simultaneous recording[20] of the PA signal and of the conventionally detected EPR signal at room temperature, for a sample of $CuSO_4 5H_2O$ in powder form. There is no doubt as to the superiority of the conventional EPR method with respect to sensitivity and signal-to-noise ratio. With regard to the limiting values of the microwave power used in both experiments, the PA detection is of the order of 10^3 to 10^4 less sensitive. In the simultaneous experiments shown in the Figure 8 decreasing the

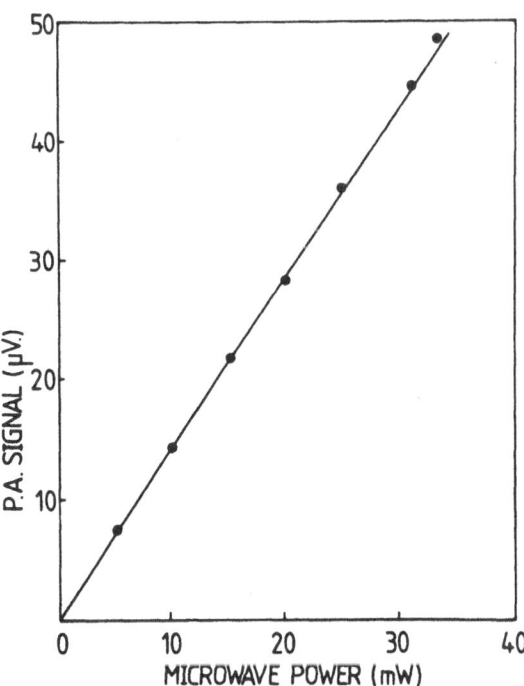

FIGURE 7   Dependence of the signal as a function of the microwave power in DPPH at 100 Hz.

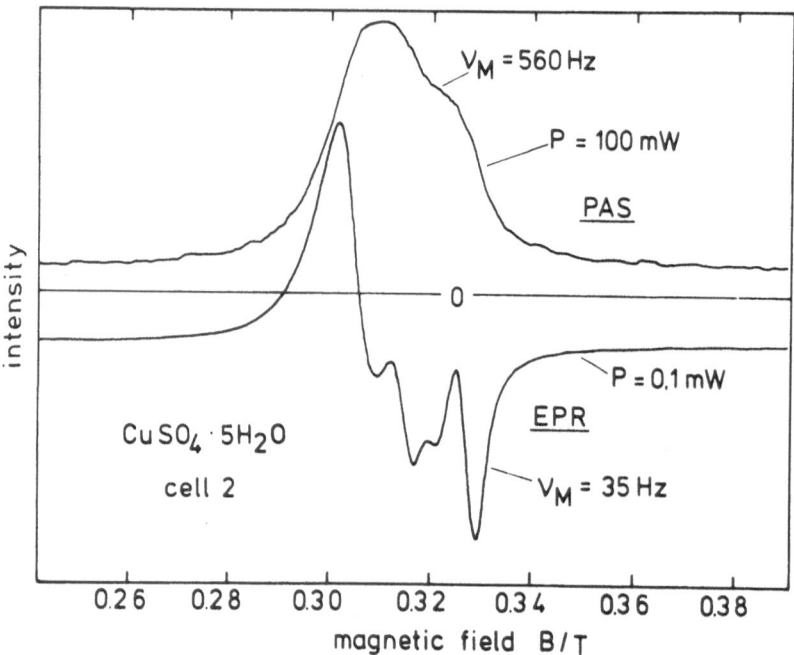

FIGURE 8   EPR spectra of CuSO$_4$5H$_2$O polycrystalline sample at
                300 °K detected by PA method and by conventional EPR
                simultaneously.

frequency of modulation and increasing the incident power, the
sensitivity of PA can gain a factor of 10^2. However, it is still
less sensitive than the conventional EPR method at room temperature,
particularly for samples such as CuSO$_4$5H$_2$O and DPPH, which have
long relaxation times, T_1, and narrow lines[34,35]. This unfavourable
condition of PA-EPR as compared with conventional EPR alters
considerably in the case of broad resonance lines, and at low
temperatures. An example of this is the FE(III)-
tetraphenylporphyrin) recently reported[20,24]. In this metallo-
organic complex the detection by conventional EPR suffers from
severe limitations. The EPR spectra consist of broad lines and
the shape is asymmetric, making assignment of hidden, or weak,
lines extremely difficult.

These measurements also show the feasibility of PA-EPR technique
for the characterisation of surface and sub-surface structures.
The depth of these structures can be controlled by monitoring the

frequency of modulation ν_M, and thus increasing or decreasing the thermal diffusion length in the samples, defined as:

$$\mu_s = \left(\frac{2k}{\omega \rho C} \right)^{\frac{1}{2}}$$

where $\omega = 2\pi\nu_M$, ρ is the mass density, C is the specific heat (per unit mass) and k is the thermal conductivity[1]. With decreasing temperature the intensity of the signal becomes considerably enhanced due to two effects[1,36]: (a) the heat capacity of the solid decreases rapidly ($\sim T^3$) at temperatures below the Debye temperature θ, (b) the pressure response of the surrounding gas medium varies as T^{-2}. Figure 9 shows the ratio of the PA-EPR detected signal to the conventionally detected EPR signal[25] as a function of temperature for a single crystal of $CuSO_4 5H_2O$. Descending from room temperature to liquid helium temperature an enhancement of the order of 10^3 was observed. Although the

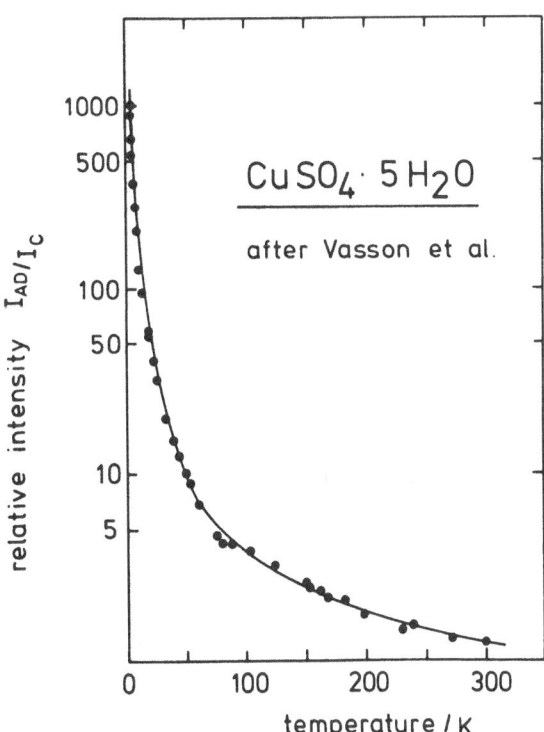

FIGURE 9   Temperature dependence of relative intensities of PA-EPR spectra and conventional EPR of Cu^{2+} in $CuSO_4 5H_2O$ single crystal.

Cu^{2+} is not a particularly favourable example, especially because
of its long relaxation time T_1, the results clearly illustrate
the potential of the PA-EPR method at low temperature[25].

4.2: Qualitative Comparison Between the PA-Method and the Conventional EPR Technique

The PA results show some close similarities with the TD-EPR
method[37,38,39]. As in the case of thermally detected EPR (TD-EPR)
the photoacoustic method is very sensitive at low temperature and
suitable for studying resonance lines with broad and asymmetrical
lines. Taking into account at this qualitative level the close
analogy between the methods of detection, some comparisons of
the PA method with conventional EPR can be made. These comparisons
are made in terms of the theory of Schmidt and Solomon which was
developed for TD-EPR.

As shown by these authors[13], the power absorbed at resonance
by a system of N_s spins, S, with (2S + 1) equally spaced energy
levels, is given by:

$$\pi \; = \; \frac{H_1^2}{H_1^2 + h^2} \left(\frac{h^2 \nu^2}{kT} \right) \left(\frac{1}{T_1} \right) \frac{S(S + 1)}{3} N_s \qquad (13)$$

where H_1 is the microwave magnetic field, $h = (\gamma^2 T_1 T_2)^{-\frac{1}{2}}$ is the
field required to saturate the magnetization to one half its
equilibrium value, ν is the microwave frequency, T the absolute
temperature, and T_1 the spin lattice relaxation time. The
process described above causes the sample to heat up. The change
in temperature of the sample is monitored using, for example, a
carbon resistance thermometer.

In the case of PA detection expression (13) can be used
when the power absorption π is chopped at the frequency ν_M.
Hence, when expression (13) is multiplied by $(1 + \exp(i\nu_M t))/2$,
we have a close analogy between the PA method and TD-EPR.
Inspection of equation (13) indicates that the absorbed power π
is larger for a line with short spin-lattice relaxation time T_1.
For such resonance (with a broad and asymmetrical line) that
saturates at high microwave power, the photoacoustic detection
gives results that conventional EPR fails to produce because
high imput microwave power can be used without the associated

problem of microwave bridge balancing which occurs in normal
EPR. It can also be seen that like TD-EPR, π increases as ν^2,
which makes the PA method more sensitive as the microwave
frequency is increased.

Since the PA method detects only the heat caused by microwave absorption there is practically no limitation in frequency.
This is not true for normal EPR because as the frequency is
increased, the conventional microwave diodes tend to become
rapidly more noisy and less efficient. Hence improvement in
sensitivity over the X-band can be obtained for spins with
large g-values or for high magnetic field (small g-values).
The photoacoustic detection of Fe(III)-TPP discussed in the
last section is a good example of such an improvement in
sensitivity.

Finally it is worth noting that when the PA method is
compared with the TD-EPR, it shows some advantages. The first
one is that the PA method shows no deformation of the base
line due to magnetoresistance effects. Secondly, the room
temperature microphone used in the PA method is inherently
insensitive to the effects of temperature, magnetic field and
microwave absorption that can change the output signal of a
cooled bolometer.

4.3: The Amplitude and Phase Angle of the Signal

The physical process operative in a PA experiment may be
enumerated as:

1.   excitation of the sample using a pulsed microwave source.

2.   generation of heat in the sample via thermal radiationless
     processes.

3.   excitation of sound in the detector gas.

4.   sound detection by a microphone, and

5.   microphone signal amplification and recording.

The key feature of the PA effect is that the application of
exciting radiation which is sine-wave modulated in intensity
will result in pressure variations which are also sine-wave
modulated and of the same frequency. However, the time at

which the absorbed microwave power reaches its maximum does
not coincide with that at which the pressure amplitude is a
maximum. This is due to the fact that the physical processes
intervening between sample absorption (excitation) and sound
detection require a finite length of time. Consequently, a
phase difference between the incoming wave train and the out-
going acoustic wave train will be observed. There are, then,
two important observables in a PA experiment, the amplitude of
pressure variations (sound intensity) and phase lag (or phase
angle ϕ).

In those cases, where the spin-lattice relaxation rate is much
faster than the modulation frequency, the PA-detected phase angle
shift observed is a result only of the heat diffusion in the solid.
Again the frequency dependence of the thermal diffusion length
shows the possibility to determine from a phase analysis whether
concentration gradients of the paramagnetic centres arise from
the surface or from the bulk of the samples. We note also
the possibility of determining saturation parameters and relaxation times by measuring the phase shift as a function of
modulation frequency. The photoacoustic phase shift method
should be applicable to the study of samples which have very
 long relaxation times, where low modulation frequency are
necessary. In this case, inhomogeneities in the magnetic field
would not be important as in the case for narrow line width
studies by conventional EPR.

     In Figure 10, the dip height of the phase angle and the
amplitude of the signal from DPPH is plotted against the
chopping frequency. The amplitude decreases more like the
increase power of the frequency than the expected $\nu^{-3/2}$
dependence, in the high frequency optically transparent limit[6].

     The same measurements on DPPH were recently reported by
Melcher[23] showing the expected $\nu_M^{-3/2}$ dependence in the high
frequency limit. The difference between the data shown in
Figure 10 and Melcher's results indicates that a quantitative
interpretation has to be made with great care because the
experiments are concerned with powder samples in a cell arrangement which exhibits acoustic resonances.

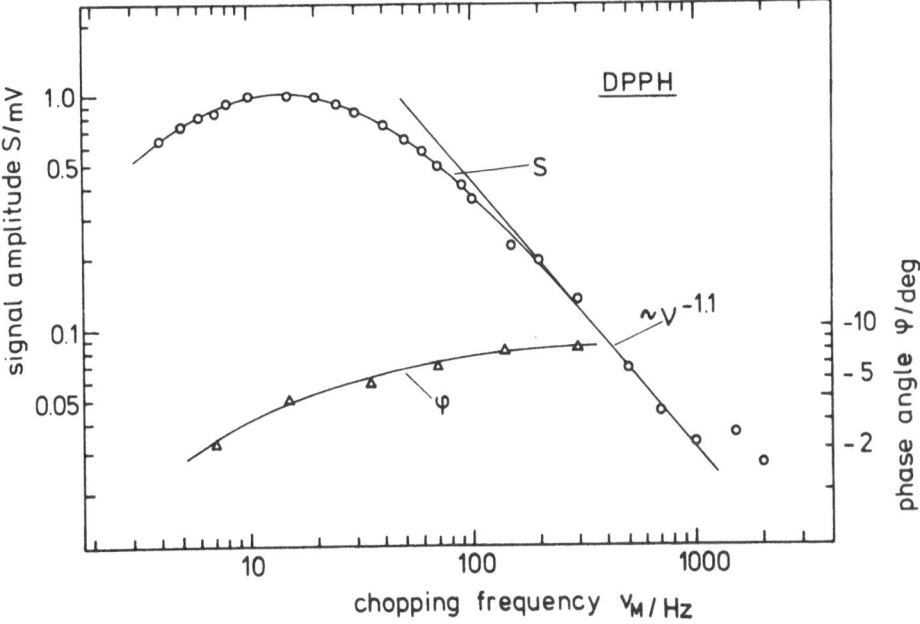

FIGURE 10   Log-log plot of the frequency dependence of the
            amplitude and phase angle ϕ.

4.4: Conclusions

We have discussed the detection of electron paramagnetic
resonance (EPR) using the photoacoustic method. The results
presented on samples of DPPH, $CuSO_4 5H_2O$ and Fe(III)-tetraphenyl-
porphyrin at room and low temperature lead to the following
conclusions:

1.   Photoacoustic detection of paramagnetic resonance is a
     powerful new technique complementing thermal detection
     (TD-EPR) using bolometric techniques.

2.   At room temperature for resonance lines with long spin-
     lattice relaxation times its sensitivity is several orders
     of magnitude lower than that for conventional EPR. For
     broad and asymmetrical resonance lines (short relaxation
     time T_1), the photoacoustic detection shows clear advantages
     over the conventional method.

3.   At low temperature in favourable cases, the sensitivities
     of the PA method and the conventional are comparable,
     especially at liquid helium temperatures.

4. It is possible to determine by analysing the magnitude
    or the phase of the PA-EPR detected signal whether the
    signal arises from the surface or from the bulk of the
    material. This depth profiling capability is a feature
    unique to the photoacoustic technique.

ACKNOWLEDGEMENTS

The author would like to thank Dr. L. Eaves (Nottingham
University) and Professor J. Pelzl (Ruhr-Universität,FRG) for
encouragement and assistance throughout the course of this
work. Many thanks are due to many members of the Physics
Department for the interest they have shown in this work,
among them, Professors E. R. Andrew, K. W. H. Stevens and
Dr. W. S. Moore. He would also like to thank Drs. A. Vasson
and A.-M. Vasson, and U. Netzelmann for permission to
reproduce diagrams from their work. Finally, the author is
grateful to the FAPESP, Sao Paulo and to the Royal Society
for support during this work.

REFERENCES

1. A. Rosencwaig, 'Photoacoustic spectroscopy', Adv. Electron.
    Electron Phys., $\underline{46}$, 207, (1978).

2. C. C. Ghizoni, M. A. A. Siqueira, H. Vargas and L. C. M. Miranda
    'On the use of the photoacoustic cell for investigating the
    electron-phonon interaction in semiconductors', Appl. Phys.
    Lett., $\underline{32}$, 554, (1978).

3. R. Florian, J. Pelzl, M. Rosenberg, H. Vargas and R. Wernhardt,
    'Photoacoustic detection of phase transitions', Phys. Stat.
    Sol. (a), $\underline{48}$, K35, (1978).

4. Digest of the Topical meeting on photoacoustic spectroscopy,
    Ed. Optical Society of America, Washington D.C., (1979).

5. This Conference.

6. A. Rosencwaig and A. Gersho, 'Theory of the photoacoustic
    effect with solids', J. App. Phys., $\underline{47}$, 64, (1976).

7. F. A. McDonald and G. C. Wetsel, Jr., 'Generalized theory of
    the photoacoustic effect', J. Appl. Phys., $\underline{49}$, 2313, (1978).

8. Herbert S. Bennet and Richard A. Forman, 'Frequency dependence
    of photoacoustic spectroscopy: Surface and bulk absorption
    coefficients', J. Appl. Phys., $\underline{48}$, 1432, (1977).

9. C. L. Cesar, H. Vargas, J. A. Meyer and L. C. M. Miranda, 'Photoacoustic effect in solids', Phys. Rev. Lett., 42 1570, (1979).

10. L. C. Aamodt, J. C. Murphy and J. G. Parker, 'Size considerations in the design of cells for photoacoustic spectroscopy', J. Appl. Phys., 48, 927, (1977).

11. C. A. S. Lima, L. C. M. Miranda and R. Santos, 'Sample-gas thermal contact resistance and photoacoustic signal generation', J. Appl. Phys., 52, 137, (1981).

12. L. Bojko, W. Nowy, J. Wiechula and B. Sujak, 'Exploitation of the heating of the paramagnetic substances as a method of investigation of the electron paramagnetic resonance (EPR)', Acta Physica Poloniac, 4, 533, (1967).

13. J. Schmidt and I. Solomon, 'High-sensitivity magnetic resonance by bolometer detection', J. Appl. Phys., 37, 3719, (1966).

14. H. Vargas, C. Evora and L. C. M. Miranda, 'Photoacoustic effect in the microwave region', Meeting of Physical Society of Brasil, Fortaleza (CE), (1978).

15. N. F. Leite, E. C. Silva, H. Vargas and C. Rettori, 'ESR of O_3^- trapped in γ-irradiated $NaClO_3$', Solid State Comm., 28, 961, (1978).

16. S. P. Belov, A. V. Burenin, L. I. Gershtein, V. V. Korolikhin and A. F. Krupnov, 'High-sensitivity millimetre and sub-millimetre wide-range microwave spectroscopy of gases', Opt. Spektrosk, 35, 295, (1973).

17. Gerald Diebold and David L. McFadden, 'Observation of the optoacoustic effect in the microwave region', Appl. Phys. Lett., 29, 447, (1976).

18. O. A. Cleves Nunes, A. M. M. Monteiro, K. Skeffueto, 'Detection of ferromagnetic resonance by photoacoustic effect', Appl. Phys. Lett., 35, 656, (1979).

19. C. Evora, R. Landers and H. Vargas, 'Photoacoustic detection of ferromagnetic resonance in films', Appl. Phys. Lett., 36, 864, (1980).

20. U. Netzelmann, 'Photoacoustic detection of the EPR in solids', Thesis to be submitted for the degree of Master, Ruhr Universitat Bochum, FRG.

21. (Private Communication).

22. J. McCann, 'Thermal effects in magnetic resonance', Thesis to be submitted to the University of Nottingham for the degree of Doctor of Philosophy.

23. R. L. Melcher, 'Thermoacoustic detection of electron paramagnetic resonance', (Private Communication).

24. E. V. Goldammer, U. Netzelmann, J. Pelzl and H. Vargas,
     'Detection of EPR in metallo-organic complexes using the
     photoacoustic effect', to be submitted.

25. A. Vasson and A.-M. Vasson, 'Acoustic detection of EPR
     between room temperature and liquid helium temperature', J.
     Phys. C: Solid St. Phys., in press.

26. H. Coufal, 'Acoustic detection of electron spin resonance',
     This Conference.

27. U. Netzelmann, J.Pelzl, E. V. Goldammer and H. Lerchner,
     'Photoacoustic detection of the EPR in solids', This Conference.

28. O. Nordhaus and J. Pelzl, 'Frequency dependence of resonant
     photoacoustic cells: The extended Helmholtz resonator, App.
     Phys., in press.

29. Charles Kittel, 'On the theory of ferromagnetic resonance
     absorption', Phys. Rev., 73, 155, (1948).

30. J. H. E. Griffiths, 'Anomalous high-frequency resistance
     of ferromagnetic metals', Nature, 158, 670, (1946).

31. K. J. Standley and K. H. Reich, 'Ferromagnetic resonance in
     nickel and in some of its alloys', Proc. Phys. Soc. B-68,
     713, (1955).

32. N. Bloembergen, 'On the ferromagnetic resonance in nickel
     and supermalloy', Phys. Rev., 78, 572, (1950).

33. J. D. Jackson, 'Classical electrodynamics', Wiley, N. York,
     1962,

34. F. K. Kneubuhl, 'Line shapes of electron paramagnetic
     resonance signals produced by powders, glasses and viscous
     liquids', J. Chem. Phys., 33, 1074, (1960).

35. J. P. Goldsborough, M. Mandel and G. E. Pake, 'Influence
     of exchange interactions on paramagnetic relaxation times',
     Phys. Rev. Lett., 4, 13, (1960).

36. K. Klein, J. Pelzl and H. Futterer, 'Photoacoustic cell for
     low temperature PAS', This Conference.

37. T. M. Al. Sharbati, 'Thermal detection of electron para-
     magnetic resonance', Ph.D. Thesis, University of Nottingham
     (1973).

38. I. A. Clark, 'EPR of iron group transition metal ions by
     thermal detection', Ph.D. Thesis, University of Nottingham,
     (1975).

39. J. R. Fletcher, J. M. Grimshaw, A. P. Knowles and W. S. Moore,
     'An EPR study of $3d^4$ ions in Al_2O_3 by thermal detection', J.
     Phys. C: Solid St. Phys., 13, 6391, (1980).

8. Instrumentation

INSTRUMENTATION FOR PHOTOACOUSTIC SPECTROSCOPY AND CALORIMETRY OF LIQUIDS
AND SOLIDS

P.S. Bechthold

Institut für Festkörperforschung
Kernforschungsanlage Jülich GmbH
Postfach 1913, D-5170 Jülich
W.Germany

ABSTRACT

This article deals with some specific aspects of photoacoustic instrumentation. Various incoherent and coherent light sources are discussed
which allow to cover the whole UV-VIS-IR spectral range. Advances of IR
photoacoustic spectroscopy are briefly outlined. The principles and per-
formances of gas-microphone and thermometric detectors for liquid and
solid state photoacoustic measurements are described with emphasis on
devices which can be operated at variable temperatures. It is reviewed
how nonradiative relaxation processes can be studied in the 1.3 K to
>1000 K temperature range by photoacoustic spectroscopy and calorimetry.
The response of Helmholtz resonant cells is shortly discussed. Electronics
and data aquisition are briefly considered.

1. INTRODUCTION

A photoacoustic spectrometer is composed of three main parts:
1. the modulated or pulsed light source including all optical components
   like monochromators, mirrors, etc.
2. the detection system, e.g. a photoacoustic cell equipped with a microphone, and a suitable power monitor which permits signal normalization
3. the electronics used for data aquisition and processing.

We shall discuss these three parts in the given order. It will be
pointed out how the whole UV-VIS-IR spectral range can be covered and
what possibilities exist to study nonradiative relaxation processes in
the 1.3 K to >1000 K temperature range.

2. LIGHT SOURCES

Strong light sources are required in photoacoustic experiments because acoustic detectors like microphones and piezoelectric receivers
impose fundamental detector noise limitations. The choice of the light
source for the spectral region to be studied depends on the power emitted,
spectral radiance ($W sr^{-1}cm^{-2}nm^{-1}$), stability, lifetime and last but not
least, costs. Two types of light sources have principally to be distinguished: 1. The incoherent emitters of continuum radiation to be discussed in
section 2.1 and second the coherent laser sources, which will be treated
in section 2.2. The latter are of course superior in spectral radiance ,
but the incoherent sources cover a broader spectral range than any specific laser and in addition are much cheaper.

2.1 INCOHERENT SOURCES

The incoherent sources suitable for photoacoustic spectroscopy mainly fall into two major categories which are the arc discharges and the
incandescent sources /1,2/. The most popular incoherent source in photoacoustic experiments on liquid and solid materials is the high pressure
xenon arc discharge which is an efficient emitter in the 200 to 2700 nm
spectral range. Three photoacoustic spectrometers using this light source
are presently commercially available /3/.

In a xenon lamp the arc is struck between two electrodes of tungsten
or thoriated tungsten-for the higher power arcs-and is contained within
a fused quartz envelope. For improved performance in the ultraviolet it
can also be enveloped by suprasil quartz. In this case its radiation
will produce toxic ozone in the surrounding air and a ventilation system
will be required. The xenon gas filling of the cold lamp is at a pressure
of a few atmospheres. During operation this pressure raises up to 70 atm.
Therefore Xe-lamps are subject to possible explosion. They must be operated in a protective housing. Even when cold the lamps must not be handled
without protective face shield and clothing.

The arc gap is usually a few mm wide. Arc temperatures are in the
range 8000 to 10000 K. Forced air cooling or water cooling is necessary
to prevent the lamp from overheating. The brightest arcs are the small
ones operating with a very short gap, but they also show the greatest non-
uniformity across the source as well as a function of angle. In addition

they show the strongest intensity fluctuations due to arc wandering
on the electrodes. So they are not necessarily the most suitable for
photoacoustic applications. Commercial Xe-bulbs range in size from 75 W
to 30 kW. The lamps used in photoacoustic measurements are usually in the
300 W to 1000 W power range. The spectral radiance of a 1 kW xenon arc
lamp at 550 nm is about 1 W $cm^{-2}sr^{-1}nm^{-1}$. This value closely corresponds
to that of the sun at sea level.

In a photoacoustic spectrometer the Xe-lamp is mechanically or electronically modulated and focused on the entrance slit of a grating monochromator having as high a throughput as possible. The monochromator
should therefore exhibit a high dispersion of typically 0.25 mm/nm in the
UV-VIS region to 0.06 mm/nm in the near infrared and a low relative aperture (typical aperture f/4 with a focal length of 0.25 m). The accessory
optics should be achromatic and without absorption over the whole spectral
range from the ultraviolet to the near infrared. It therefore preferentially contains reflective components.

Incandescent light sources occasionally used for visible and near
infrared photoacoustic spectroscopy are the tungsten- and tungsten-halo-
gen lamps. Their color temperatures are about 2600 and 3200 K, respectively. At wavelengths below 3 μm their spectra are similar to that of a
black body source of the same temperature. The increasing absorption of
their glass or quartz envelopes hinders their application at longer
wavelengths.

In tungsten halogen lamps some halogen gas is added to the lamps
atmosphere. It reacts with the evaporated tungsten and forms a volatile
tungsten halogenide which is stable at lower temperatures but decomposes
on the hot filament surface. Thereby the filament is regenerated which
provides a longer life of the lamp and allows the lamp to be operated at
higher temperatures. This yields a stronger output and extends it in the
blue and ultraviolet range. In addition with the quartz envelope the size
of the lamp can be strongly reduced.

In Fig. 1 the spectral radiance of a 1000 W Xe arc lamp is compared
to that of a 1000 W quartz tungsten halogen lamp which is more stable and
the spectrum of which is less structured. The photoacoustic signals obtained with the tungsten halogen lamp are therefore easier to normalize.
The 1000 W mercury lamp (Fig. 1) although brighter in the UV spectral
range is not so valuable for spectroscopic applications, because the

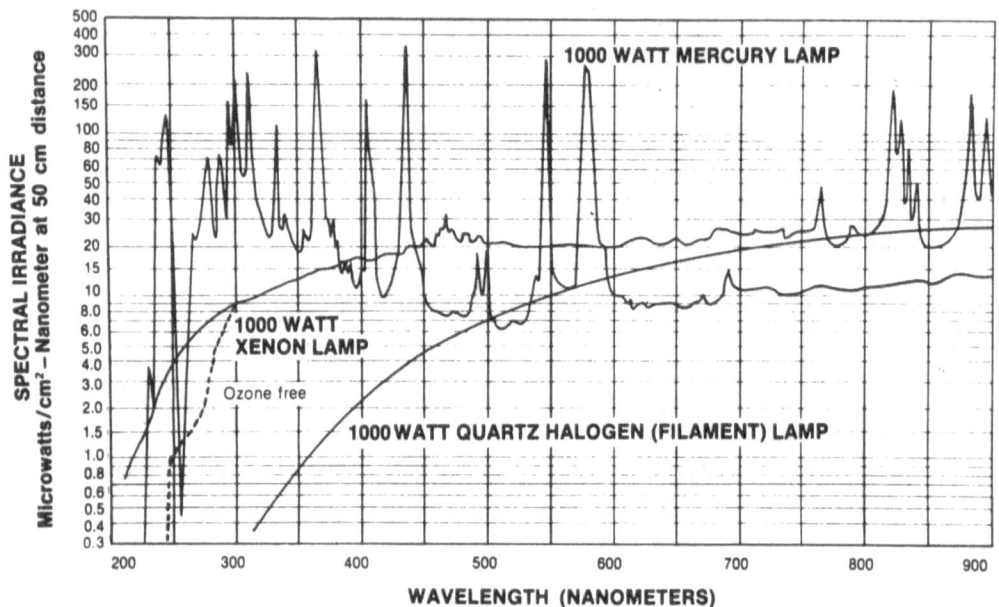

Figure 1: Output spectra of three UV-VIS lamps taken with a resolution of 2 nm. The units on the vertical axis refer to the bare lamps and can be increased with a collecting rear reflector and proper focusing of the light (by permission of ORIEL Corporation).

very structured output is unsuitable for a proper source compensation. It is, however, applicable in calorimetric studies of materials (especially organic substances) which are transparent in the VIS-NIR region but strongly absorb in the UV. The argon arc source has a less structured output and is also superior to the Xe arc in the UV spectral range. Therefore, it should prove particularly useful for UV-applications.

The spectra of solids and liquids in the UV-VIS and near IR range are mainly broad and not very specific. In contrast, in the mid-IR region they show relatively sharp and specific structures which can be used to identify molecules, molecular groups, and surface adsorbed species. Therefore, intermediate infrared photoacoustic spectroscopy should be particularly important for analytical studies, investigations of conformational changes in molecules, as well as studies of surface layers and surface reactions (catalysis, corrosion).

The photoacoustic technique allows to avoid the dispersion induced distortions which often occur in conventional infrared absorption spectra when powdered samples are pressed into KBr or other
pellets /4/. When the refractive index of the matrix material matches
that of the sample the reduction of light scattering is linked to an
enhanced transmission of the composite material (Christiansen effect).
This effect frequently occurs in regions of anomalous dispersion which
accompany the regions of resonant absorption. It therefore distorts the
measured shape of the absorption bands and simulates additional structures.

Despite of the promising features of infrared photoacoustic spectroscopy, the majority of all studies up to now are confined to the UV-VIS
spectral range. The reason is the lack of suitably bright broad band IR
light sources. Low current carbon arcs /5/ can in principle be used, but
they tend to be very unstable, and have a limited lifetime.

Two attempts have recently been reported to overcome the source problem. Low and Parodi /6,7/ used a specially designed incandescent carbon
rod source, as previously described by Boyd et al /8/. The rod of 0.8 cm
diameter and 12.7 cm length operates at temperatures up to 2700 K in a
purified argon atmosphere at a pressure of 1.7 atm. It is contained in a
water cooled housing which is supplied with a KBr optical window. In the
middle the rod has a reduced V-shaped cross section at a length of 3 cm.
This reduced cross section increases the resistance and therefore the
temperature of this section of the rod. The slot of this section is used
as the light source. Its emissivity is close to unity so that the radiation curve closely resembles that of a black body radiator.

Using this light source, a mechanical chopper, a slightly modified
commercial infrared monochromator and a photoacoustic cell equipped with
a condenser microphone and a KBr optical window, Low and Parodi were
able to take IR photoacoustic spectra in the 4000-880 cm^{-1} range.

To observe surface attached species they used a special cell in which
the sample could be transferred by means of an externally applied magnet
from the acoustic chamber to a furnace where reactions could be carried
out at high temperature /7/. They were able to observe surface species
on silica and alumina powders which were present in submonolayer coverages.

The other approach to overcome the light source problem in the infrared range is to use Fourier-transform infrared photoacoustic spectroscopy
/4,9/. This technique which has also been used in the visible part of

the spectrum /10/ employs a Michelson interferometer instead of a dispersive monochromator and allows to collect data from all spectral frequencies simultaneously throughout the measurement. During the measurement one of the mirrors of the interferometer is moved which results in
an interferogram recorded as a photoacoustic signal. This interferogram
is then numerically Fourier-transformed to a conventional spectrum. The
spectral throughput of the Michelson interferometer is up to 50 times
higher than the throughput of the grating monochromator. It therefore
strongly reduces the data collection time.

Conventional IR light sources like globars and Nernst glowers may
be used with the Fourier-transform technique. Sufficient results in the
4800 to 800 cm^{-1} spectral range with a resolution of 8 cm^{-1} have been
reported /9/.

Obviously, the combination of a stronger light source with the
Fourier-transform technique should lead to even more improved results.
In this context two other potentially useful light sources should be
mentioned. They are the incandescent tungsten slit source equipped with
an IR-transparent window as developed by Taylor et al./11/ and the
synchrotron /12/. The synchrotron provides a bright light source which is
tunable from the far infrared to the soft X-ray region. The radiance of
"DORIS" in Hamburg exceeds that of a 6000 K black body radiator in its
whole spectral range /12/. In the infrared DORIS is up to two orders of
magnitude brighter than the 6000 K black body radiation. It is,therefore,
suited especially for far infrared Fourier-transform PAS. In addition
synchrotron radiation covers the widest tuning range in the VUV and near
X-ray region. Further applications should arise from this feature.

2.2 COHERENT SOURCES (LASERS)

The outstanding features of laser light are its extreme degrees of
temporal and spatial coherence. While the first gives rise to the lasers
high spectral purity, the latter controls the optical phase across the
front of the output beam and leads to the very small beam divergence of
typically 0.2-6 mrad. This narrow beam width allows the light to be focused
to an essentially diffraction limited spot, so that enormeous energy densities of quasimonochromatic coherent light can be achieved. The spectral
radiance of lasers exceeds that of the incoherent sources by many orders
of magnitude. These properties recommend the laser as a most valuable
tool for many kinds of spectroscopic applications, especially when high

spectral resolution or high power levels are required. In addition, lasers allow nonlinear optical effects to be observed like two photon absorption and four wave optical mixing, which can not be obtained with the incoherent sources. In this chapter we shall mainly be interested in tunable lasers /13/, although fixed frequency lasers can be used in many photoacoustic experiments, like measurements at low concentrations /14/, photoacoustic microscopy and imaging /15/ and calorimetric investigations /16,17/. Some fixed frequency lasers and their harmonic radiation are also needed for pumping the tunable sources, namely the ruby laser (λ=694 nm, pulsed), the Nd:YAG laser (λ=1064 nm, pulsed and cw), the argon and krypton ion lasers (several lines throughout the UV-VIS range, Kr: also IR), the N_2 laser (λ=337 nm, pulsed) and various excimer lasers (pulsed, several broad UV-lines depending on the gas filling).

It is possible to cover most of the UV- to IR-spectral range by using several tunable lasers (Fig. 2). However, a specific tunable laser can only be used in a limited wavelength range without changing of the mirrors or the laser active medium, e.g. the dye in a dye-laser. In some cases the mirrors of the pump laser have to be changed, too.

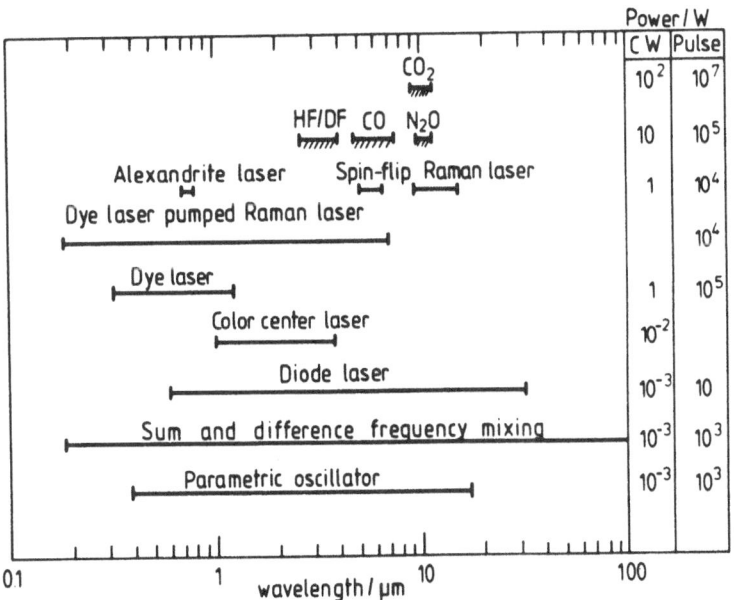

Figure 2: Overall tuning ranges of some important continuously tunable (——) and step tunable ($\mathit{7\!7\!7}$) lasers and nonlinear optical sources. Typical values of cw and pulsed peak output powers are listed in the columns on the right side of the diagram.

Therefore spectroscopic investigations in a larger spectral range are always time consuming and of substantial experimental complexity.

In dye lasers the fluorescent radiation of excited dye molecules is used to obtain laser emission. Dye lasers can continuously be tuned over the 340-1240 nm spectral range with a spectral width of 1 Å or less. Dyes have to be exchanged to span the full spectral region, each dye lasing over ∿50 to 100 nm. Power conversion efficiences up to 50 % have been obtained with laser dyes. Dye lasers are commercially available as cw and pulsed lasers. Cw lasers are pumped by argon or krypton ion lasers. Pulsed lasers are pumped by Nd:YAG- or ruby lasers and their harmonics, N_2 lasers, excimer lasers or flash lamps. If pumped by flash lamps the tuning range reduces to 420-760 nm due to photochemical decomposition of some laser dyes or rapid triplet population during the pump pulse and subsequent triplet-triplet absorption. The use of an argon ion laser as the pump laser also limits the IR tuning range of the dye laser to about 700 nm because the output of the argon laser does not significantly overlap with the absorption spectra of the IR laser dyes.

The tunability of coherent sources can be extended towards shorter UV wavelengths by nonlinear frequency conversion in noncentrosymmetric crystals. Optical second harmonic generation occurs in such a crystal when an intense fundamental laser beam causes distortions of the induced polarization at twice the optical input frequency /19/. Using various crystals the tuning range of coherent optical radiation has been extended to 217 nm by frequency doubling of dye lasers /20/.
Even shorter ultraviolet wavelengths can be generated, when two laser beams of different frequencies are mixed in such a crystal to generate the sum frequency. This way the tuning range has been extended to 185 nm with available nonlinear crystals /21/. The conversion efficiencies in such mixing processes strongly depend on the power levels of the fundamental laser beams and the properties of the nonlinear crystals. The generated UV radiation therefore covers the μW to MW power range. High power lasers can be frequency doubled with 30%-50% conversion efficiencies.
VUV generation in crystals is limited by absorption. In this case higher order optical mixing in metal vapors and gases can be used but with much less efficiency /22/. The shortest wavelength generated up to now is 38 nm ≅ 32.6 eV /23/.

A variety of tunable coherent sources has been developed to cover the infrared spectral range. A recent important development are the color-center lasers /24/. Their range of frequency tuning is the near infrared from 800 nm to 3.3 μm. The laser active media consist of alkali halide crystals which contain a high concentration of color centers. Several crystals are needed to cover the whole spectral range. The specific crystal is mounted on a cold finger which is cooled to liquid nitrogen temperature. Suitable pump lasers are dye lasers, argon and krypton ion lasers, ruby lasers and Nd:YAG lasers, also the emission lines at 1.32 and 1.34 μm.

Tunable near IR laser emission has also been achieved with transition-metal doped crystals. Using a Nd:YAG laser as a pump laser source continuous tunability has been demonstrated from 1.61 to 1.74 μm for Ni:MgF$_2$ and from 1.63 to 2.08 μm for Co:MgF$_2$ /25/. A recent development is the very bright alexandrite (BeAl$_2$O$_4$:Cr^{3+})-laser /26/ which is tunable between 700 and 815 nm. Average power outputs up to 35 W have been achieved with this solid state laser.

The recombination radiation of semiconductor diodes can be used for laser action and can be tuned from 0.6-3 2 μm with various semiconductor diodes of changing composition /27,28/.

The diodes are manufactured from single crystalline alloy semiconductors like GaAs, Al$_x$Ga$_{1-x}$As, In$_x$Ga$_{1-x}$As, Pb$_{1-x}$Sn$_x$Se, Pb$_x$Cd$_{1-x}$S, PbS$_{1-x}$Se$_x$, and Pb$_{1-x}$Sn$_x$Te. The desired emission range is determined by the alloy composition, x, which is precisely controlled during crystal growth. Population inversion can be achieved by electron injection into the conduction band or by optical pumping. The laser beam of diode lasers diverges sharply because it emerges from the small front surface of the diode. The diodes are operated at cryogenic temperatures which are provided by closed cycle refrigerators. Since the bandgap of the semiconductor depends on the temperature, the laser frequency can coarsly be tuned by varying the temperature. The fine tuning is utilized by variation of the bias current which produces small temperature changes. Tuning can also be accomplished by a magnetic field or hydrostatic pressure. Semiconductor lasers can be operated in cw as well as pulsed mode. The typical output power of semiconductor lasers is considerably small ranging from 100 μW to 1 mW, for cw operation. Powers greater than 1 mW are available in selected wavelength regions only. Pulsed lasers give peak powers to some ten watts.

Another important near infrared coherent light source is the
parametric oscillator /29,30/. Parametric generation also originates in a second order nonlinear optical effect. When an incoming
laser photon is irradiated on a nonlinear crystal like $LiNbO_3$ it can
break up into a "signal" and an "idler" photon spontaneously, $\omega_p = \omega_s + \omega_i$
(energy conservation). In general a broad spectrum of frequencies is generated in this parametric fluorescence process. It can be interpreted
as a mixing of the pump photons with the zero point fluctuations of the
electromagnetic field at the signal and idler frequencies. In principle,
the parametric fluorescence light is radiated into a wide solid angle,but
preferential directions are demanded by momentum matching conditions
$(k_p = k_s + k_i)$. For parametric oscillation the nonlinear crystal is placed
inside an optical resonator that provides resonances at either ω_s or ω_i or
both. The laser pump beam at ω_p is properly focused into the crystal. When
momentum matching is achieved on the resonator axis and the gain of the
nonlinear parametric generation is sufficient to overcome the resonator
losses at the signal and idler frequencies, the device will oscillate
in a manner similar to a laser. Varying the refractive indices of the nonlinear crystal, e.g. by heating it, shifts the momentum matching condition on the resonator axis to other frequencies and thereby tunes the
output of the parametric oscillator. It seems possible to design parametric oscillators which are tunable over a very wide wavelength range
/30/. Two commercial systems exist, only one of which is currently available /31/. Both use $LiNbO_3$ single crystals as the nonlinear material. The
older system uses a frequency doubled Nd:YAG laser as the pump source and
covers a tuning range from 650 nm to 3.6 μm. The other system uses a flash-
lamp excited dye laser as pump source and covers the spectral range from
730 nm to 2.6 μm. Its average power exceeds 1 mW in the whole spectral
range.

Besides parametric oscillation and sum frequency mixing also difference frequency generation /32/ of two laser beams can be realized in
asymmetric nonlinear crystals. With one tunable and one fixed frequency
laser one can get tunable outputs at longer wavelengths.This effect can
be used to fill the spectral gaps between the various infrared lasers.
Wavelengths up to the 3 mm range have already been generated by using this
technique, but often the power levels are too small for photoacoustic experiments. This is even worse with the four wave parametric mixing process
which originates in a third order nonlinear process.

The Spin-flip Raman Laser and the Dye Laser pumped Raman Laser are both based on a third order nonlinear optical effect, the stimulated Raman scattering. In a spin flip Raman laser /28,33/ a single crystal, e.g. n-type InSb is cooled down to liquid helium temperature and kept in an external magnetic field. In this field the energy levels of the conduction electrons split into two sublevels according to the spin orientation in the field. The separation between the sublevels is proportional to the magnetic field strength. When the crystal is irradiated by an intense CO or CO_2 laser, stimulated Raman scattering accompanied by a spin flip of an electron may occur. Stokes as well as anti-Stokes components can be observed. The frequency of the scattered radiation can be tuned by sweeping the magnetic field strength. With a pulsed CO_2 laser as the pump source tuning ranges of 9 to 14.6 μm are provided. Using a CO laser for pumping the tuning range extends from 5-6 μm. In the latter range pulsed as well as cw operation is possible. Cw powers of 1 W and in pulsed operation, peak powers of 100 W to 1 kW are commercially available /34/.

Stimulated Raman scattering in high pressure H_2 gas (5-20 bar) has successfully been used to shift the output of very high peak power dye lasers and frequency doubled dye lasers to cover the spectral ranges from 185 to 880 nm and from 700 nm to 7 μm without a gap /35,36/. These light sources require good spatial mode quality of the dye laser provided by a frequency doubled Nd:YAG pump laser. At conversion efficiencies above 1 % peak powers of several kW of Stokes and anti-Stokes radiation have been obtained.

The 'free electron laser', which in principle could be continuously tuned between 100 nm and 1 mm, is presently under development. In this laser electrons are forced to oscillate at the frequency of a periodic magnetic field. So far it has been operated at a single wavelength of 3.417 μm with an average power of 0.36 W and a peak power of 7 kW /38/.

In addition to the above described continuously tunable sources of laser radiation, a number of molecular gas laser systems exist which are step tunable in the 2.8 to 12 μm region /39/. The CO_2 laser operates on lines spaced by ~ 2 cm^{-1} in the 9 to 11 μm range. N_2O and CS_2 lasers operate in the 9.45 to 11.2 μm (spacing ~ 1 cm^{-1}) and 11 to 11.5 μm (spacing ~ 0.2 cm^{-1}) regions, respectively. The CO laser operates between 4.7 and 7.5 μm on lines which are separated by ~ 4 cm^{-1}.

HF and DF, HCl and DCl, and HBr and DBr lasers can be operated in the 2.8-3.4 μm, 3.8-4.2 μm, and 4.3-4.6 μm regions, respectively.

In Fig. 2 the tuning ranges and approximate power levels of some important tunable sources of laser radiation are summarized. It is demonstrated that the near UV-FIR spectral range can be covered entirely by various coherent sources.

3. PHOTOACOUSTIC OR OPTOACOUSTIC DETECTION

In a photoacoustic or optoacoustic experiment the sample is illuminated by the periodically modulated or pulsed light source. At least part of the light energy absorbed by the sample is converted into heat. Various techniques have been developed to measure,directly or indirectly, the periodical or transient heat generation. These include photoacoustic detection by means of a gas-microphone cell, piezoelectric detection /40/, thermooptical deflection ("mirage effect" and thermal lensing) /41/, photothermal radiometry /42/ and thermometric methods. Here we shall concentrate on gas-microphone cells, particularly those which can be operated at variable temperatures.

In addition,we briefly review the thermometric methods which,although relatively insensitive at room temperature,appear to be very promising at temperatures below 4.2 K.

3.1 GAS-MICROPHONE PHOTOACOUSTIC CELLS

The gas-microphone cell is the most commonly used photoacoustic detector. Usually the sample and a nonabsorbing gas are enclosed in a gas tight chamber which has an optical window and also contains an electret or condenser microphone. The light induced heat flow from the sample to the gas produces pressure changes in the gas which are sensed by the microphone. Temperature variations of 10^{-5} to 10^{-6} degrees at the sample-gas boundary are readily detected.

The inner gas volume of the cell should be kept small, because the photoacoustic signal grows as the inverse active gas volume of the cell. The reduction of the cell volume, however, reaches a lower limit when thermal and thermoviscous loss to the cell window and walls become dominant /43-46/.

Therefore, the sample to cell window and wall to wall distances inside the cell should always exceed the thermal diffusion length of the gas μ_g. This quantity is defined by $\mu_g = \sqrt{\frac{\alpha_g}{\pi f}}$ where α_g is the thermal diffusivity of the gas and f is the modulation frequency of the light source. The lowest modulation frequency to be used, therefore, determines the reasonable cell dimensions. Two cell designs have principally to be distinguished. In the "nonresonant cell" the microphone is located in close proximity to the sample /43/. To reduce the spurious signal it is protected from direct illumination. Nonresonant cells are operated at frequencies which are far below the acoustic resonance frequencies of their cell cavity. The frequency response of these cells most closely follows the predictions of the Rosencwaig-Gersho theoretical model /43,47/. The second design is the "Helmholtz resonant cell" /43/. In this cell the microphone is separated from the sample chamber by a thin long tube. Such cells show acoustic resonances at frequencies which are well within the frequency range used in photoacoustic experiments (fig. 4). The resonances can be interpreted in terms of the Helmholtz resonator model, as outlined in the appendix.

Numerous resonant and nonresonant gas-microphone photoacoustic cells for liquid and solid state studies are described in the literature /7,16, 17,43,44,48-58/. Almost all of them are confined to room temperature operation /7,43,44,48-53/.

Since the photoacoustic technique measures the internal heating of the sample it can be used for spectroscopic as well as calorimetric applications. Material inhomogeneities /15/ and phase transitions/16,17,55,59/ can be studied on a local scale. When used at low temperatures a better spectroscopic resolution is accompanied by an increase of fluorescent life times and a decrease of chemical reaction rates. The latter is especially important when intermediates of photochemical cycles are to be studied. Photochemical intermediates can preferentially be populated and trapped   at low temperatures /60,61/. In addition the photoacoustic signal increases when temperature decreases /16,56-58,60/. Therefore, low temperature operation appears to be particularly attractive. Nonresonant as well as resonant cells have been used. Pichon et al. /16/ used a nonresonant cell which was equipped with a specially designed electret microphone and a MOSFET preamplifier. It was operated in a free running gas-flow cryostat at temperatures down to 5 K (see contribution of Fournier and Boccara /41/ for further details). The first low temperature experiment was performed by Murphy and Aamodt /54/. They used a Helmholtz resonant design and cooled their sample cell by direct immersion in liquid nitrogen. However the direct contact of their sample cell and sound transmitting duct to the boiling coolant and the signal attenuation in the long sound transmitting duct caused severe acoustic noise problems.

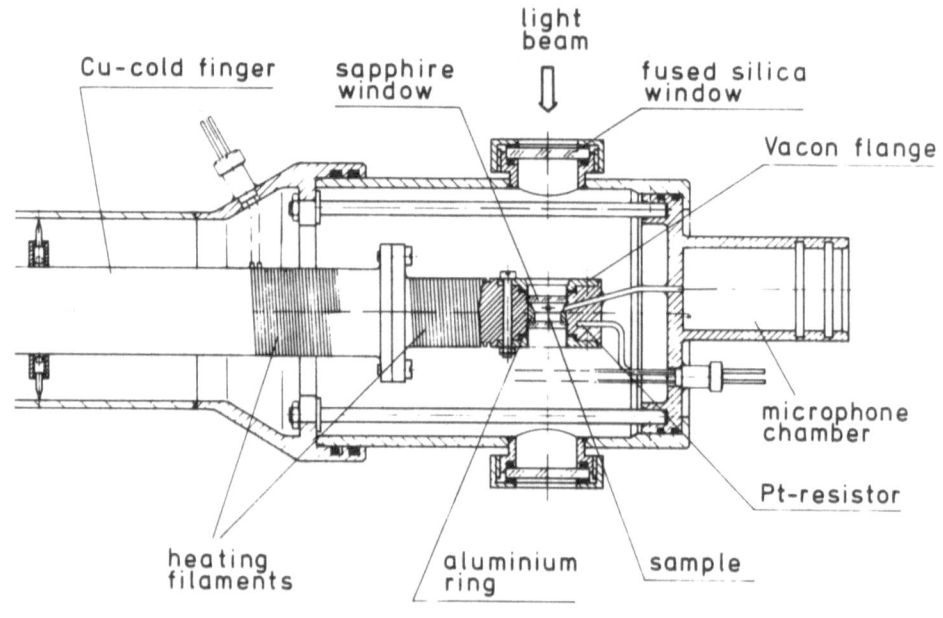

Figure 3: Setup of a Helmholtz resonant low temperature photoacoustic cell suitable for temperatures between 90 K and 320 K /54/.

Fig. 3 shows a resonant cell /57/ which largely reduces these problems. The much shorter sound transmitting stainless steel tube does not contact the boiling coolant. Also the vacuum isolated sample chamber is not directly immersed in the liquid nitrogen but contacts it from a distance via a copper cold finger. The temperature of the sample can continuously be varied between 90 K and 320 K by two separate coaxial heating filaments and an external action control circuit. The microphone chamber is attached to the outer side of the vacuum chamber and is kept at room temperature during operation. To avoid convection of the gas the microphone chamber is slightly lifted with respect to the cooled sample chamber.

Using this cell we will now discuss some basic properties of the photoacoustic signal. Fig. 4a,b represent the frequency response of the photoacoustic amplitude and phase of a carbon black sample at 90 K and 295 K /57/. The absolute values of the amplitude refer to the root mean square of the microphone output. The phase is given in degrees and contains a constant instrumental contribution. The light source was a 2 mW

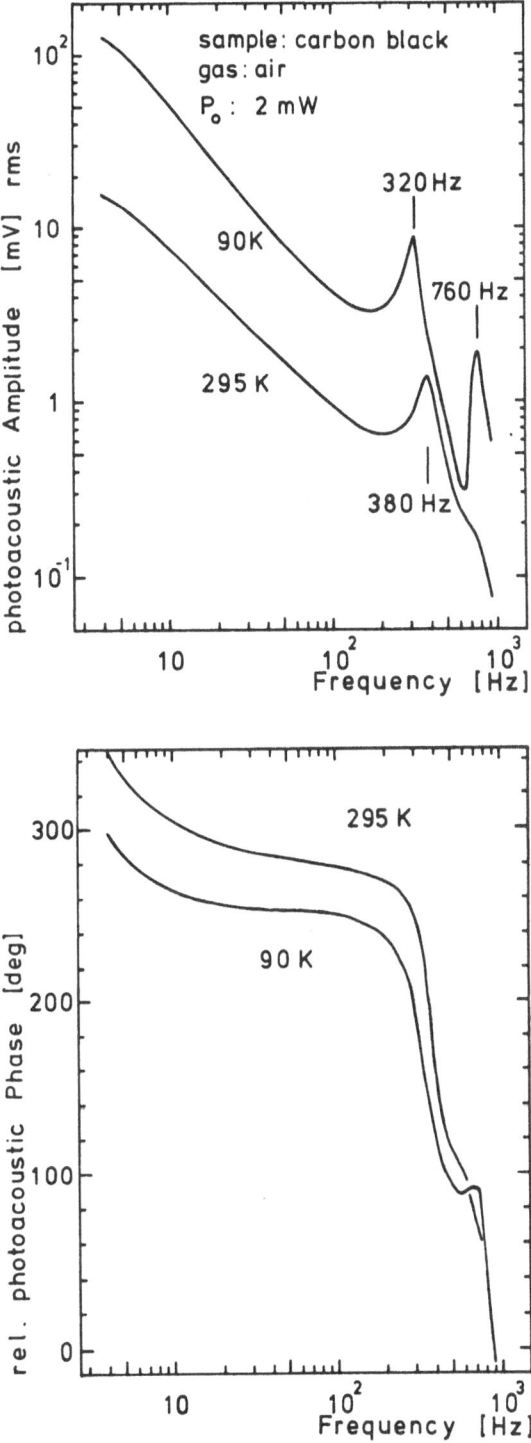

Figure 4a+b:Frequency dependence of the photoacoustic amplitude and phase
at 90 K and 295 K as obtained with a carbon black sample. The absolute values
of the amplitude refer to the root mean square of the microphone output.
The phase contains a constant instrumental contribution.

He-Ne laser. Air at atmospheric pressure was used as the sound transmitting gas.

In the figures the acoustic signal caused by the sample is convoluted
with the frequency response of the sample cell and the detection electronics. As expected the amplitude increases as temperature decreases. Two
distinct resonances are observed at 90 K. They are shifted to higher frequencies at higher temperatures due to the simultaneous increase of the
velocity of sound in air. The resonance enhancement of the amplitude is
linked to a dual step phase change as predicted by the extended Helmholtz resonator model /53/. According to the Rosencwaig-Gersho theory /47/

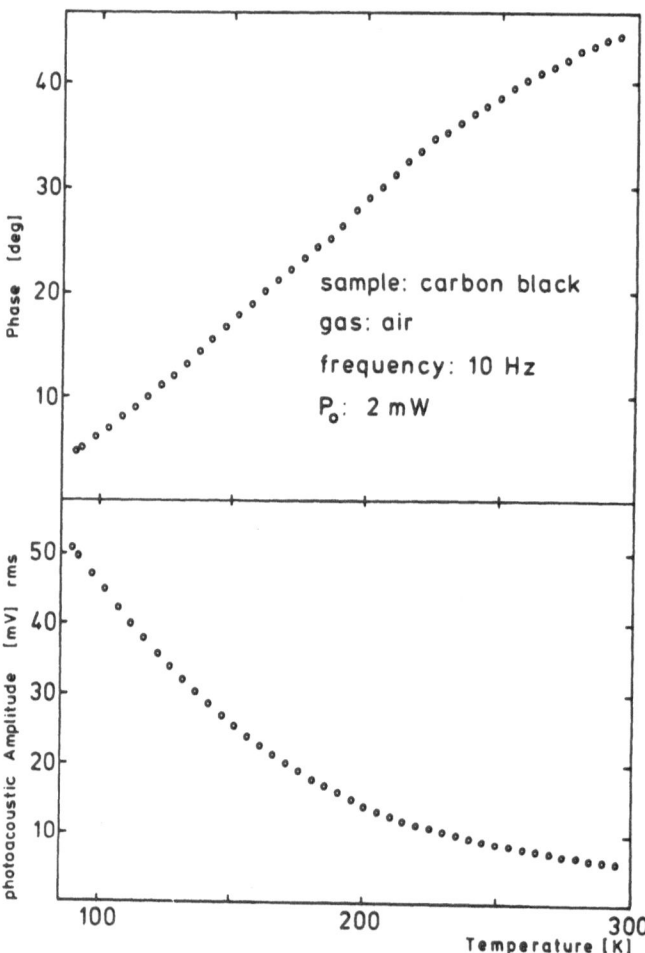

Figure 5: Temperature dependence of the photoacoustic signal at a modulation frequency of 10 Hz /60/.

the saturated amplitude of the carbon black signal should fall off as the inverse chopping frequency, and the phase at the sample surface should be constant $-\frac{\pi}{2}$ with respect to the modulated light source. This is approximately realized at intermediate frequencies despite of the constant instrumental phase shift. The slope of the amplitude decreases at low modulation frequencies. This is accompanied by a simultaneous increase of the photoacoustic phase and is mainly caused by a lateral heat flow via the gas to the cell window and walls /46,56/. A minor contribution, especially in the phase change results from the frequency response of the electronic system /56/. Fig. 5 shows the temperature-dependence of the photoacoustic signal at a modulation frequency of 10 Hz /60/. The increase of the photoacoustic amplitude with decreasing temperature is expected and has even been quantified in the case of the nonresonant cell /16/. In contrast, the observed phase changes are larger than expected. Theoretically the phase of a saturated signal is constant $(-\frac{\pi}{2})$ at the sample surface /47/. Therefore, the changes should be attributed to the temperature dependent parameters of the gas. However, an estimate in terms of the Helmholtz resonator model fails to explain the observed magnitude /56/.

Further investigations with a precisely characterized sample, an accurately defined cell geometry, and a precise knowledge of the temperature gradient in the sound transmitting duct are therefore necessary to clear up this point. So far the results of fig.4 and 5 can be used to characterize the performance of the variable temperature cell. The utility of the cell is demonstrated elsewhere /60,61/.

Let us now turn towards high temperature operation /17,55,56/. To protect the microphone from serious damage in this case it has to be cooled /55/ or has to be separated from the hot region in a Helmholtz resonant arrangement /17,56/. The latter approach seems to be the more appropriate because a larger temperature range can be studied and a smaller thermal gradient is generated inside the cell. Therefore, a better temperature homogeneity can be achieved at the sample. Fig. 6 shows a resonant high temperature cell which was designed for temperatures up to 1050 K /56/. Fig. 7 shows its sample chamber in greater detail. As the low temperature cell, it can be heated by two separate coaxial heating filaments. The cell is made from stainless steel with a fused quartz optical window. The sample chamber is supported by a ceramic isolation material and is placed in a stainless steel container which can be filled by an inert gas to prevent corrosion. The microphone chamber is fixed to the outer wall of this container. Again both chambers are connected by

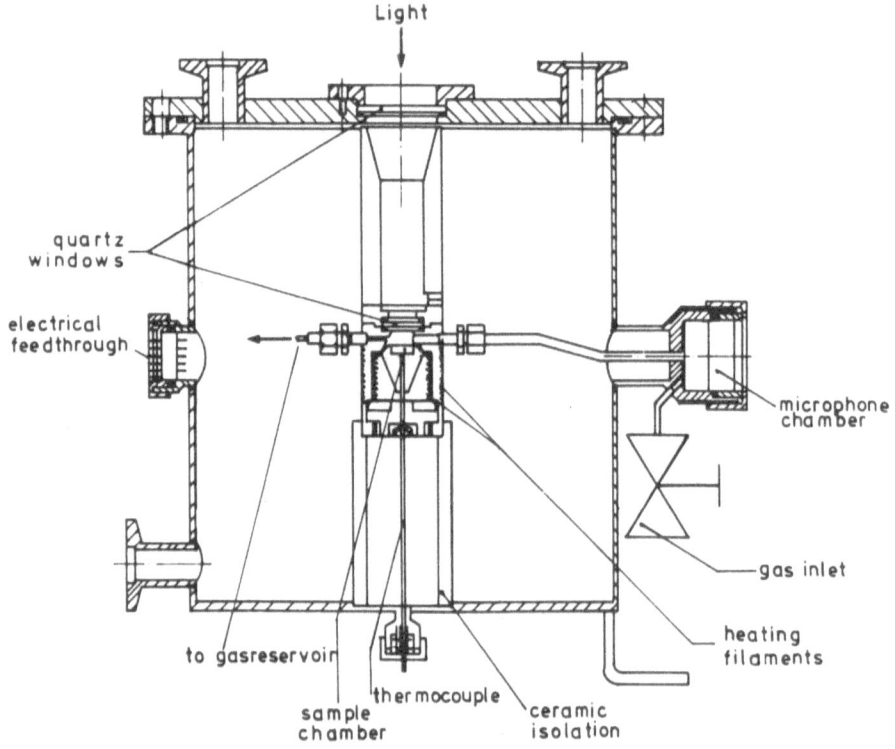

Figure 6: High temperature photoacoustic cell designed for temperatures up to 1050 K /56/.

a thin walled stainless steel tube. To avoid convection of the gas inside the cell and to protect the microphone from direct thermal irradiation, the sample chamber is slightly lifted with respect to the microphone chamber.

The most delicate part of the cell is its optical window. The sample chamber is therefore constructed in such a way that it can be opened without opening of the window. Its sample tray is metal to metal sealed against a polished conical fitting. The lock consists of a modified bajonet joint.

The cell shows a qualitatively similar frequency- and temperature-response as the low temperature cell /56/. It has already been used to study phase transitions in metal hydrogen interstitial alloys /17/. As an example fig. 8a,b show the photoacoustic amplitude and phase for a $TaH_{0.5}$ sample during heating and subsequent cooling. A first order phase tran-

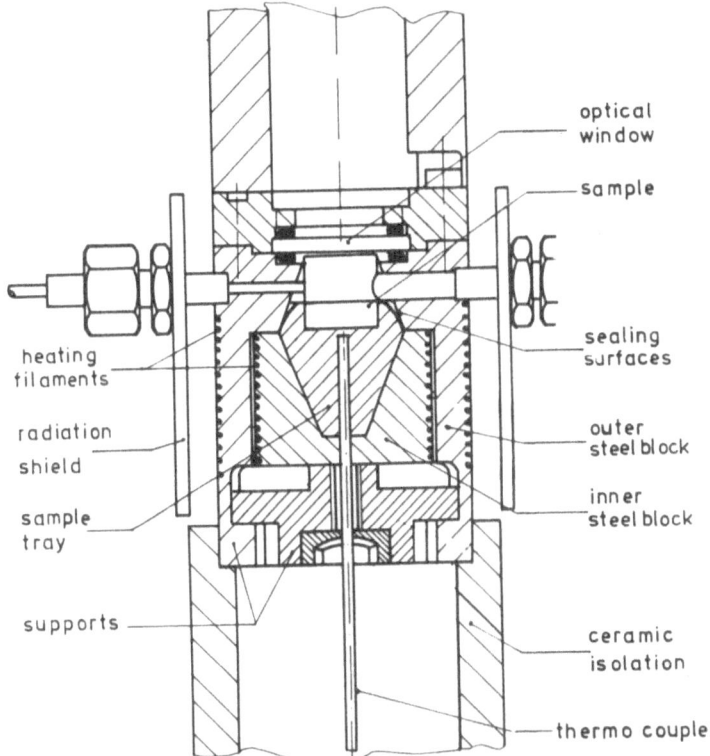

Figure 7: Sample chamber of the high temperature photoacoustic cell.

sition is found at 335 K. The observed behaviour of the photoacoustic signal is in qualitative agreement with the predictions of the Korpiun-Tilgner theoretical model /62/.

Besides in gas-microphone photoacoustic cells condenser microphones can be employed to measure directly the volume changes of the sample which follow the nonradiative energy release. In that case the membrane of the microphone mechanically contacts the surface of the sample (e.g. a liquid) to be studied /63-65/. Membrane displacements of the order of 1 Å are readily detected. Volume changes down to 10^{-10} cm^3 were already observed. Relaxation times of 100 μs can be resolved. The method was used to study quantum efficiencies and enthalpy changes during photochemical reactions. Instead of capacitor microphones piezoelectric transducers can also be applied in such experiments, especially, when pulsed light sources are used /14,40,64,66-68/. Absorption coefficients of liquids as small as

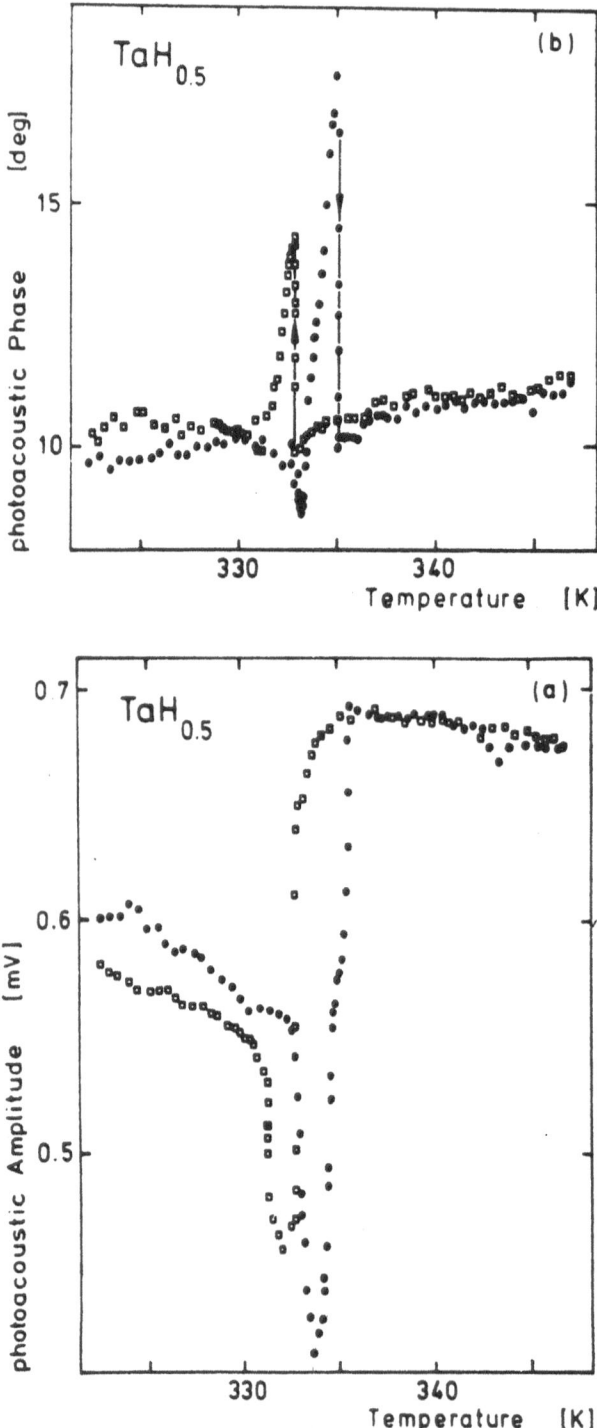

Figure 8: Photoacoustic detection of a first order phase transition in
TaH$_{0.5}$ recorded by stepwise increasing (●) and decreaseing (□) the sample
temperature /17/.

10^{-7} cm^{-1} can be measured by this technique /66,67/. It has already been applied at low temperatures to study the absorption of liquid methane /67/.

3.2 THERMOMETRIC METHODS

The most obvious approach to measure light induced heat is to use thermocouples /69-72/, thermopiles, thermistors /73/ or other resistance thermometers /74-77/, which are placed in close proximity to the sample. These devices are free of acoustic noise problems which should be advantageous in some applications. Although at temperatures close to room temperature they are less sensitive than gas microphone cells, they have proved particularly useful for absolute determinations of thermal diffusivities /70,72,74/. Temperature oscillations of 10^{-3} degrees can readily be detected /70,73/. Thermometric methods become advantageous at temperatures below 4.2 K, where germanium or carbon resistors are used /75-77/. The sensitivity of these resistance thermometers increases by several orders of magnitude on cooling down from room temperature to 1.5 K. At this temperature a temperature rise of 1 mK decreases their resistance by about 100 Ω /76/. In addition, the drastically reduced specific heat of the sample ($\propto T^3$) increases the magnitude of the temperature oscillations at the low temperatures and thereby also increases the resulting signal. A further decrease of temperature to 0.3 K will even more increase the sensitivity. Optical absorption coefficients as small as 10^{-7} cm^{-1} are expected to be within the reach of this detection technique /76,77/. Such low temperature measurements can also be performed using superconducting thin films as detectors /78,79,80/. If the film is operated at its transition temperature, small changes in the temperature of the film will result in large changes of its resistance. The film can directly be deposited on the sample or can be placed at a distance when operated in superfluid helium. In the latter case heat pulses can ballistically be transported via second sound from the sample to the detector over distances of several centimeters without diffusion or appreciable attenuation /78,79,81/. The dynamic range of the detector, which is extremely small due to the narrow temperature region of the superconducting transition, can be shifted by an externally applied magnetic field. In practical cases the magnetic field serves to tune the transition temperature of the film to the Dewar ambient /79,80/. Operation in a resonant cavity should provide an additional signal enhancement. Due to Smith and Laguna /78/ the method is capable of detecting tem-

perature amplitudes down to 3×10^{-9} degrees. The low response time of 10
to 100 ns of these thin film bolometers will allow for measurements of
nonradiative lifetimes.

4. DATA AQUISITION AND SOURCE COMPENSATION

The choice of the electronic subsystem used for processing of the
preamplified detector signal mainly depends on the temporal behaviour of
the selected light source whether periodically chopped or pulsed. With
a periodically modulated source a lock-in amplifier tuned to the modulation frequency is used. The simplest lock-ins have a single phase sensitive detector, which means, that the phase must be appropriately set
to produce a maximum signal output. Since the phase of the photoacoustic
signal varies with the absorption coefficient of the sample, only in phase
or quadrature components can be measured with such lock-ins when the
wavelength of the light is continuously scanned. Therefore, modern lock-
ins often contain two phase sensitive detectors operating in quadrature
with each other. They also contain a vector sum circuit which permits
the simultaneous readout of signal magnitude and phase. Ratio options are
also available which can be used for signal normalization (see below).
With pulsed light sources boxcar signal averagers or transient recorders
are prefentially used for signal processing.

The ideal light source for spectroscopic applications should exhibit a
constant radiance throughout its emission spectrum. However, real light
sources show characteristic spectral structures and emission lines. In
addition many light sources drift with time or flicker. Therefore signal
normalization will be required, real time source compensation being the
most advantageous.

An uncorrected digitized sample spectrum recorded with a single beam
photoacoustic spectrometer can be normalized by a point to point division
with the spectrum of a reference material, e.g. carbon black, which exhibits photoacoustic saturation in the entire spectral region of interest.
However, short time fluctuations of the light source can only be eliminated with double beam spectrometers. The most perfect source compensation
is obtained with two matched photoacoustic cells, one containing the sample
and the other containing the reference material. This technique can be
employed in all spectral regions but has the inherent disadvantage that

the available source power has to be divided between the two cells. Therefore, other power monitors like semiconductor diodes or pyroelectric detectors needing less intensity for detection are often applied. When a
pyroelectric detector is utilized in the UV-NIR-range, only about 10 % of
the source power is being used in the reference channel. Unfortunately,
the spectral response of a pyroelectric is not identical with that of a
photoacoustic cell. Therefore, for a proper source compensation in a second
step the spectrum of the sample, normalized with respect to the pyroelectric
detector, has to be divided by a previously recorded photoacoustic reference
spectrum which is also normalized to the pyroelectric detector. Fig. 9 com-
pairs the power spectrum of a xenon arc illumination system, as obtained
with a powdered carbon black sample in a photoacoustic cell, to that
measured with the pyroelectric detector. Fig. 9a corresponds to the convolution of the xenon arc output spectrum of fig.1 with the throughput of
the monochromator and the additional optics. The photoacoustic spectrum
normalized to the pyroelectric output is shown in fig. 9b. The step at
800 nm in curve b is associated with a grating change in the monochromator.
Under favourable conditions even this perturbation can completely be eliminated by the procedure outlined above. Photoacoustic spectrometers often contain microprocessors for data storage and additional
data processing.

5. CONCLUSION

Various light sources used to cover the UV to IR spectral range have
been described. The broad band incoherent sources coupled to a monochromator are the most suitable to rapidly scan a broad spectral range. In the
IR range there is still a need for stronger sources, although Fourier transform measurements result in considerable improvement. Fixed frequency
lasers are successfully applied in calorimetric investigations. Tunable
lasers should be utilized when high resolution is required, e.g. in the
measurements of excitonic energy levels, or special features in a limited
spectral range are to be studied, like weak absorbing overtone bands etc.
Time resolved measurements are possible when pulsed lasers are used together with piezoelectric detectors. The photoacoustic and thermometric
methods together with photo-thermal deflection techniques /41/ and photothermal radiometry /42/ allow to cover the temperature range from 1.3 K to
the highest attainable temperatures. Shot noise may however set an upper
limit at very high temperatures.

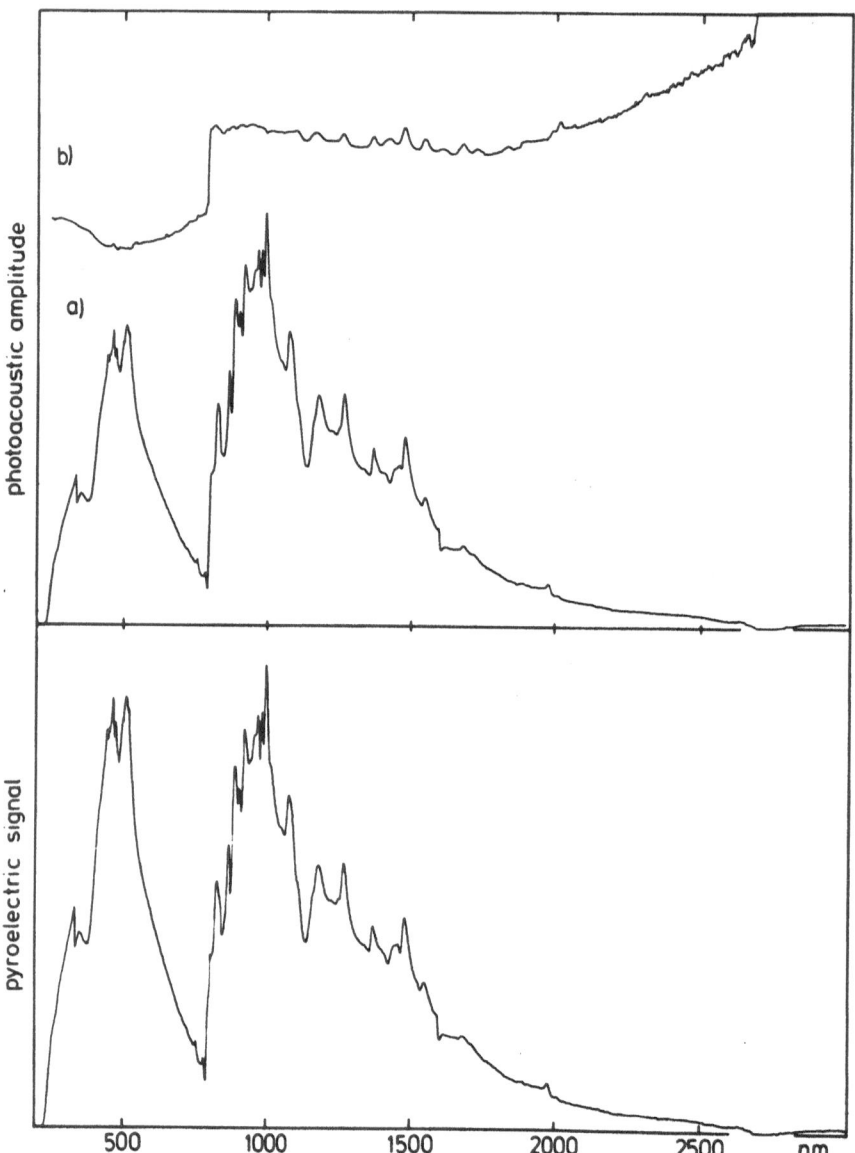

Figure 9: Power spectra of a xenon arc-monochromator illumination system as measured with a powdered carbon black sample in a photoacoustic cell and a pyroelectric detector. The photoacoustic spectrum normalized to the pyroelectric output is shown in curve b. The step at 800 nm is due to a grating change in the monochromator.

The sensitivity of photothermal radiometry increases as the third power of temperature, whereas the photoacoustic amplitude increases with decreasing temperature. Both methods are therefore complementary in temperature dependent measurements. When these methods are used together with microscopic and imaging techniques calorimetric and spectroscopic measurements can be performed on a local scale.

Nonradiative relaxation measurements are therefore capable of many new applications especially in biological materials, frozen solutions and other strongly light scattering materials which are difficult to study by any other means.

APPENDIX: THE HELMHOLTZ RESONATOR MODEL

A Helmholtz resonator in its simplest form consists of a volume V connected to a small cylindrical tube of length 1 and cross sectional area A (fig. 10). The gas column in the tube performs forced oscillations imposed by an external pressure modulation P(t). The gas in the tube is considered to move like a rigid piston without any compression. The restoring force exerted by the compressed gas is then given by:

$$F = K \frac{\Delta V}{V} \cdot A = K \frac{A \cdot x}{V} A$$

where x is the displacement of the piston from its equilibrium position. K is the bulk modulus of elasticity defined as the inverse adiabatic compressibility of the gas:

$$\frac{1}{K} = - \frac{1}{V} \left(\frac{\Delta V}{\Delta P}\right)_S$$

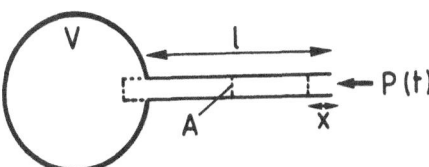

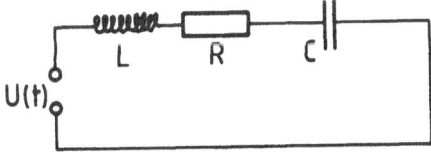

Figure 10:

Simple Helmholtz resonator and its analoguous electrical circuit.

It is related to the velocity of sound c and the density ρ of the gas by $K = \rho c^2$ /82/. The equation of motion of the gas piston is given by

$$P(t) \cdot A = \rho l A \ \ddot{X} + A^2 R_{ac} \ \dot{X} + \rho c^2 \ \frac{A^2 x}{V} \ .$$

The damping term $A^2 R_{ac} \ \dot{X}$ will be specified below. Replacing the volume displacement $A \cdot x$ by X leads to

$$P(t) = \frac{\rho l}{A} \ \ddot{X} + R_{ac} \ \dot{X} + \frac{\rho c^2}{V} \ X$$

We compare this equation with the equation of a series electrical LRC-circuit (fig. 10) to which a periodic electromotive force U(t) is applied:

$$U(t) = L \ \ddot{Q} + R \ \dot{Q} + \frac{1}{C} \ Q$$

Q is the electric charge.

It is obvious that an electro-acoustic analogy exists when we define an acoustic inertance and an acoustic capacitance by $L_{ac} = \frac{\rho l}{A}$ and $C_{ac} = \frac{V}{\rho c^2}$. The corresponding electrical and acoustical quantities are listed in table 1.

The value of the acoustic resistance R_{ac} can be obtained from Poisseuille's law which describes the non-turbulent steady flow of a gas

Table 1:

electroacoustic analogy

electric quantity	acoustic quantity
electromotive force U	acoustic pressure p
charge Q	volume displacement X
current $I = \frac{dQ}{dt}$	volume velocity $\frac{dX}{dt}$
Resistance R	acoustic resistance R_{ac}
Capacitance C	acoustic capacitance C_{ac}
Inductance L	Inertance L_{ac}

in a tube of uniform cross sectional area. In this case the volume velocity is given by /82/:

$$v = \frac{\Delta PA^2}{8\pi\eta l} \, .$$

Δp is the difference of pressure at the ends of the tube and η is the
coefficient of viscosity of the gas. The acoustic resistance is then

$$R_{ac} = \frac{\Delta p}{v} = \frac{8\pi\eta l}{A^2} \, .$$

To a Helmholtz resonant photoacoustic cell a second cavity is added
at the other end of the sound transmitting channel. The arrangement and
the corresponding electrical circuit are given in fig. 11 of Pelzl's contribution /83/ (see also fig. 2 of ref. /53/). His notation will be adopted
in the following. The voltages U_1 and U_2 correspond to the pressures p_1
and p_2 in the sample and microphone cells, respectively.

A variety of effects contribute to the frequency response of the
photoacoustic signal:
1. the pressure generating process in the sample chamber,
2. the losses due to three dimensional heat flow to the cell windows and
   walls,
3. the acoustic response of the Helmholtz resonant cell, and
4. the response of the detection electronics.

To evaluate the acoustic response of the resonator arrangement alone
and eliminate the influence of the other effects, Nordhaus and Pelzl define
an acoustic transfer function /53,83/. They normalize the acoustic signal
of the resonant cell with respect to a nonresonant cell, which has the
same dimensions as the sample chamber of the Helmholtz resonator. Its
electrical analogue is also shown in figs. 11 and 2, respectively, of
ref. /83/ and /53/. The voltage U_0 and current I_0 represent the pressure
oscillations and volume velocity of the generated photoacoustic signal.
Their relationship

$$U_0 = I_0 \, (i\omega C_1)^{-1}$$

is in accordance /84/ with the corresponding acoustic quantities of the
Rosencwaig-Gersho theory /47/. It is assumed that the addition of the
sound transmitting duct and the second chamber leaves the pressure generating process unchanged. To calculate the acoustic transfer function
we take advantage of the electroacoustic analogy. Applying the Kirch-

hoff's laws to the electrical circuit which corresponds to the Helmholtz resonant cell, we get:

$$I_0 = I_1 + I_2$$

$$U_{C_1} = I_1(i\omega C_1)^{-1} = I_2(i\omega L+R) + U_{C_2} = I_2(i\omega L+R+(i\omega C_2)^{-1})$$

The transfer function is thus

$$\frac{U_{C_2}}{U_0} = \frac{A_{C_2}}{A_0} \exp i(\varphi_0 - \varphi_{C_2}) \quad \text{with} \quad \frac{A_{C_2}}{A_0} = [\omega^2 C_2^2(R^2+((\omega C_r)^{-1}-\omega L)^2]^{-1/2}$$

and $\quad \varphi_{C_2} - \varphi_0 = \text{arc tan} \dfrac{R}{(\omega C_r)^{-1}-\omega L} \quad$ where $C_r = \dfrac{C_1 C_2}{C_1+C_2}$.

The resonance frequency is given by

$$\omega_R = (\omega_0^2 - 2\delta^2)^{1/2}$$ where ω_0 is the resonance frequency of the undamped Helmholtz resonator $\quad \omega_0 = (LC_r)^{-1/2} \cong c(\dfrac{A}{IV_r})^{1/2}$

with $C_r \cong \dfrac{V_r}{\rho^2 c}$.

$\delta = \dfrac{R}{2L} \cong \dfrac{4\pi\eta}{\rho A}$ is the damping constant.

When the circuit elements of the Helmholtz resonator are small in size compared to the wavelength of sound, the above model can reasonably explain the frequency response of the photoacoustic cell /53/. The model fails, however, to explain the multiple resonances which are observed when longer sound transmitting ducts are used. In this case the extended Helmholtz resonator model has to be used, which treats the interconnecting tube as an acoustic transmission line /53,83/.

With this model theoretical predictions are found in excellent agreement with experimental data.

ACKNOWLEDGEMENTS

The author would like to thank J.Pelzl and M.Campagna for many helpful comments and Mrs. A.Böhme for typing of the manuscript.

REFERENCES

1. M.W.Cann, 'Light sources in the 0.15-20 μm spectral range', Appl. Opt. $\underline{8}$, 1645 (1969)

2. J.A.Gelbwachs, 'Tunable radiation sources in the ultraviolet and visible spectral regions (0.1-1.0 μm)' in: Yoh-Han Pao (Ed.) 'Optoacoustic spectroscopy and detection', Academic Press, New York (1977)

3. Suppliers: (1) EG&G Princeton Applied Research, Princeton, N.J.
              (2) EDT Research, London
              (3) Gilford Instruments, Oberlin, Ohio

4. G.Laufer, J.T.Huneke, B.S.H.Royce, Y.C.Teng, 'Elimination of dispersion-induced distortion in infrared absorption spectra by use of photoacoustic spectroscopy', Appl. Phys. Lett. $\underline{37}$, 517 (1980)

5. C.S.Rupert, 'A water cooled carbon arc source for infrared spectroscopy', J. Opt. Soc. Am. $\underline{42}$, 684 (1952) and literature cited therein

6. M.J.D.Low, G.A.Parodi, 'Infrared photoacoustic spectra of surface species in the 4000-2000 cm^{-1} region using a broad band source', Spectros. Lett. $\underline{11}$, 581 (1978)
    - 'Infrared photoacoustic spectra of solids', ibid. $\underline{13}$, 151 (1980)
    - 'Carbon as reference for normalizing infrared photoacoustic spectra', ibid. $\underline{13}$, 663 (1980)
    - 'An infrared photoacoustic spectrometer', Infrared Phys. $\underline{20}$, 333 (1980)

7. - 'Infrared photoacoustic spectroscopy of solids and surface species' Appl. Spectrosc. $\underline{34}$, 76 (1980)
    - 'Infrared photoacoustic spectroscopy of surfaces', J. Mol. Struct. $\underline{61}$, 119 (1980)

8. W.J.Boyd, D.E.Jennings, W.E.Blass, N.M.Gailar, 'Carbon rod furnace infrared source', Rev. Sci. Instr. $\underline{45}$, 1286 (1974)

9. M.G.Rockley, 'Fourier-transformed infrared photoacoustic spectroscopy of polystyrene film', Chem. Phys. Lett. $\underline{68}$, 455 (1979)
    - 'Reasons for the distortion of the Fourier-transformed infrared photoacoustic spectroscopy of ammonium sulfate powder', ibid. $\underline{75}$, 370 (1980)
    - 'Fourier-transformed infrared photoacoustic spectroscopy of solids', Appl. Spectrosc. 34, 405 (1980)
    M.G.Rockley, J.P.Devlin, 'Photoacoustic infrared spectra (IR-PAS) of aged and fresh-cleaved coal surfaces', ibid. $\underline{34}$, 407 (1980)
    M.G.Rockley, D.M.Davis, H.H.Richardson, 'Fourier-transformed infrared photoacoustic spectroscopy of biological materials', Science $\underline{210}$, 918 (1980)
    - 'Quantitative analysis of a binary mixture by Fourier-transformed infrared photoacoustic spectroscopy', Appl. Spectrosc. $\underline{35}$, 185 (1981)
    - G.Busse, B.Bullemer, 'Optoacoustic Fourier-transform spectroscopy in the infrared', this conference

10. M.M.Farrow, R.K.Burnham, E.M.Eyring, 'Fourier-transform photoacoustic spectroscopy', Appl. Phys. Lett. 33, 735 (1978)
    - L.B.Lloyd, St.M.Riseman, R.K.Burnham, E.M.Eyring, 'Fourier-transform photoacoustic spectrometer', Rev. Sci. Instrum. 51, 1488 (1980)
    - D.Débarre, A.C.Boccara, D.Fournier,'Fourier transform spectroscopy for the visible and the near infrared', this conference

11. J.H.Taylor, C.S.Rupert, J.Strong, 'An incandescent tungsten source for infrared spectroscopy', J. Opt. Soc. Am. 41, 626 (1951)

12. C.Kunz (Ed.), 'Synchrotron Radiation, Techniques and Applications', Springer Verlag, Berlin-Heidelberg-New York (1979)

13. J.Kuhl, W.Schmidt, 'Tunable coherent light sources', Appl. Phys. 3, 251 (1974)
    - M.J.Colles, C.R.Pidgeon, 'Tunable lasers', Rep. Prog. Phys. 38, 329 (1975)

14. W.Lahmann, H.J.Ludewig,H.Welling, 'Opto-acoustic trace analysis in liquids with the frequency-modulated beam of an argon ion laser', Anal. Chem. 49, 549 (1977)
    - S.Oda, T.Sawada, 'Laser-induced photoacoustic detector for high-performance liquid chromatography', ibid. 53, 471 (1981)

15. Y.H.Wong, R.L.Thomas, G.F.Hawkins, 'Surface and subsurface structure of solids by laser photoacoustic spectroscopy', Appl. Phys. Lett. 32, 538 (1978)
    - H.K.Wickramasinghe, R.C.Bray, V.Jipson, C.F.Quate, J.R.Salcedo, 'Photoacoustics on a microscopic scale', Appl. Phys. Lett. 33, 923 (1978)
    - J.J.Pouch, R.L.Thomas, Y.H.Wong, J.Schuldies, J.Srinivasan, 'Scanning photoacoustic microscopy for nondestructive evaluation', J. Opt. Soc. Am. 70, 562 (1980)
    - R.L.Thomas, J.J.Pouch, Y.H.Wong, L.D.Favro, P.K.Kuo, 'Subsurface flow detection in metals by photoacoustic microscopy', J. Appl. Phys. 51, 1152 (1980)
    - P.K.Khandelwal, P.W.Heitman, A.J.Silversmith, T.D.Wakefield, 'Surface flow detection in structural ceramics by scanning photoacoustic spectroscopy', Appl. Phys. Lett. 37, 779 (1980)
    - M.Luukkala, S.G.Askerov, 'Detection of plastic Deformation in Metals with photoacoustic microscope', Electronics Lett. 16, 84 (1980)
    - A.Rosencwaig, G.Busse, 'High resolution photoacoustic thermal-wave microscopy', Appl. Phys. Lett. 36, 725 (1980)
    - G.Busse, 'Imaging with the optoacoustic effect', Optics and Laser Technology, June (1980), p. 149
    - G.Busse, 'Applications of optoacoustic and photothermal imaging and microscopy', 1980 Ultraconics symposium Proc. p. 622

16. C.Pichon, M.LeLiboux, D.Fournier, A.C.Boccara, 'Variable temperature photoacoustic effect: application to phase transition', Appl. Phys. Lett. 35, 435 (1979)

17. P.S.Bechthold, M.Campagna, T.Schober, 'Phase transitions in metal-hydrogen interstitial alloys by temperature dependent photoacoustic measurements', Solid State Comm. 36, 225 (1980)

18. F.P.Schäfer (Ed.), 'Dye Lasers', 2nd edition, Springer Verlag, Berlin-
     Heidelberg-New York (1977)
     - C.V.Shank, 'Physics of dye lasers', Rev. Mod. Phys. $\underline{47}$, 649 (1975)

19. N.Bloembergen, 'Nonlinear Optics', Benjamin, New York (1975)

20. H.J.Dewey, 'Second-harmonic generation in $KB_5O_8 \cdot 4H_2O$ from 217.1 to
     315.0 nm', IEEE J. Quantum Electron. $\underline{QE-12}$, 303 (1976)

21. R.E.Stickel, F.B.Dunning, 'Generation of tunable coherent vacuum uv
     radiation in KB5', Appl. Opt. $\underline{17}$, 981 (1978)

22. J.J.Wynne, P.P.Sorokin, 'Optical mixing in atomic vapors', in ref.32
     - J.Reintjes, 'Frequency mixing in the extreme ultraviolet', Appl.
     Opt. $\underline{19}$, 3889 (1980)

23. J.Reintjes, C.Y.She, R.C.Eckardt, N.E.Karangelen, R.A.Andrews,
     R.C.Elton, 'Seventh harmonic conversion of mode-locked laser pulses
     to 38 nm', Appl. Phys. Lett. $\underline{30}$, 480 (1977)

24. H.Welling, D.Fröhlich, 'Progress in tunable lasers' in: Festkörper-
     probleme XIX, H.J.Queisser (Ed.) Vieweg, Braunschweig (1979) p.403
     - V.A.Arkhangel'skaya, P.P.Feofilov, 'Tunable lasers utilizing color
     centers in ionic crystals (review)', Sov. J.Quantum Electron. $\underline{10}$,
     657 (1980)

25. P.F.Moulton, A.Mooradian, 'Broadly tunable cw-operation of Ni : MgF_2
     and Co : MgF_2 lasers', Appl. Phys. Lett. $\underline{35}$, 838 (1975)

26. J.C.Walling, O.G.Petersen, H.P.Jenssen, R.C.Morris, E.W.O'Dell,
     'Tunable alexandrite laser', IEEE J. Quantum Electron. $\underline{QE-16}$,
     1302 (1980)

27. I.Melngailis, A.Mooradian, 'Tunable semiconductor diode lasers and
     applications' in: 'Laser applications to optics and spectroscopy',
     S.F.Jacobs, M.Sargent, M.O.Scully, J.F.Scott (Ed.), Addison Wesley
     Comp., Reading Mass. (1975)

28. A.Mooradian, 'Tunable semiconductor lasers' in: 'Nonlinear Optics',
     Ph.G.Harper, B.S.Wherrett (Ed.), Academic Press, London-New York
     (1977)

29. S.E.Harris, 'Tunable optical parametric oscillators', Proc. IEEE $\underline{57}$,
     2096 (1969)

30. R.L.Byer, R.L.Herbst, 'Parametric oscillation and mixing' in ref.32
     - R.L.Byer, 'Parametric Oscillators and nonlinear materials', in:
     'Nonlinear Optics', Ph.G.Harper, B.S.Wherrett (Ed.), Academic Press
     London-New York (1977)

31. Chromatix, Mountain View, California

32. Y.R.Shen (Ed.), 'Nonlinear infrared generation', Springer Verlag,
     Berlin-Heidelberg-New York (1977)

33. C.K.N.Patel, E.D.Shaw, 'Tunable Stimulated Raman Scattering from
     mobile carriers in Semiconductors', Phys. Rev. B3, 1279 (1971)
     - C.K.N.Patel, 'High resolution spectroscopy and atmospheric and
       stratospheric detection of minor constituents with the use of
       tunable spin flip Raman lasers', Phil. Trans. R. Soc. London
       A293, 257 (1979) and literature cited therein
     - H.G.Häfele, 'Spin flip raman laser', Appl. Phys. 5, 97 (1974)

34. Edingburgh Instruments, Edingburgh, Scotland

35. V.Wilke, W.Schmidt, 'Tunable coherent radiation source covering
     a spectral range from 185-880 nm', Appl. Phys. 18, 177-181 (1979)

36. W.Hartig, W.Schmidt, 'A broadly tunable IR waveguide Raman laser
     pumped by a Dye laser', Appl. Phys. 8, 235 (1979)

37. H.Motz, 'Undulators and "free-electron lasers" ', Contemp. Phys.
     20, 547 (1979)

38. D.A.G.Deacon, L.R.Elias, J.M.J.Madey, G.J.Ramian, H.A.Schwettman,
     T.I.Smith, 'First operation of a free electron laser', Phys. Rev.
     Lett. 38, 892 (1977)

39. C.K.N.Patel, 'Laser Detection of Pollution', Science 202, 157 (1978)
     and literature cited therein

40. C.K.N.Patel, 'Optoacoustic effect in condensed media', this confe-
     rence
     - H.D.Breuer, 'Piezoelectric detection in photoacoustic spectroscopy:
       Theory and application', (this conference)
     - M.W.Sigrist, 'Generation of acoustic waves by pulsed lasers', this
       conference

41. D.Fournier, A.C.Boccara, 'Photoacoustic spectroscopies', this con-
     ference

42. P.E.Nordal, S.O.Kanstad, 'Photothermal radiometry', this confe-
     rence

43. A.Rosencwaig, 'Photoacoustic spectroscopy of solids', Rev. Sci.
     Instr. 48, 1133 (1977)

44. L.C.Aamodt, J.C.Murphy, J.G.Parker, 'Size considerations in the
     design of cells for photoacoustic spectroscopy', J. Appl. Phys.
     48, 927 (1977)

45. F.A.McDonald, G.C.Wetsel, 'Generalized theory of the photoacoustic
     effect', J. Appl. Phys. 49, 2313 (1978)

46. R.S.Quimby, W.M.Yen, 'Three dimensional heat flow effects in
     photoacoustic spectroscopy of solids', Appl. Phys. Lett. 35, 43
     (1979)
     -H.C.Chow, 'Theory of three dimensional photoacoustic effect with
      solids', J. Appl. Phys. 51, 4053 (1980)
     -F.A.McDonald, 'Three-dimensional heat flow in the photoacoustic
      effect', Appl. Phys. Lett. 36, 123 (1980)
     -F.A.McDonald, 'Three-dimensional heat flow in the photoacoustic
      effect - II: Cell-wall conduction', J. Appl. Phys. 52, 381 (1981)
     -R.S.Quimby, W.M.Yen, 'On the adequacy of one-dimensional treatments
      of the photoacoustic effect', J. Appl. Phys. 51, 1252 (1980)

47. A.Rosencwaig, A.Gersho, 'Theory of the photoacoustic effect with
     solids', J. Appl. Phys. 47, 64 (1976)

48. E.L.Kerr, 'The alphaphone - a method for measuring thin-film
     absorption at laser wavelengths', Appl. Opt. 12, 2520 (1973)
     -J.G.Parker, 'Optical absorption in glass: investigation using an
     acoustic technique', Appl. Opt. 12, 2974 (1973)
     -S.A.Schleusener, J.D.Lindberg, K.O.White, R.L.Johnson, 'Spectro-
     phone measurements of infrared laser energy absorption by atmos-
     pheric dust', Appl. Opt. 15, 2546 (1976)
     -M.J.Adams, A.A.King. G.F.Kirkbright, 'Analytical optoacoustic
     spectrometry, part I: instrument assembly and performance characte-
     ristics', The Analyst 101, 73 (1976)
     -M.J.Adams, B.C.Beadle, G.F.Kirkbright, 'Analytical optoacoustic
     spectrometry, part IV: a double beam optoacoustic spectrometer for
     use with solid and liquid samples in the ultraviolet,visible and
     near-infrared regions of the spectrum', The Analyst 102, 569 (1977)
     -J.F.McClelland, R.N.Kniseley, 'Photoacoustic spectroscopy with
     condensed samples', Appl. Opt. 15, 2658 (1976)
     -J.F.McClelland, R.N.Kniseley, 'Scattered light effects in photo-
     acoustic spectroscopy', Appl. Opt. 15, 2967 (1976)

     - R.C.Gray, V.A.Fishman, A.J.Bard, 'Simple cell for examination of
     solids and liquids by photoacoustic spectroscopy', Anal. Chem. 49,
     697 (1977)
     -D.M.Munroe, H.S.Reichard, 'Practical PAS of solids', Am. Lab. 9(2),
     119 (1977)
     -J.C.Murphy, L.C.Aamodt, 'The photothermophone, a device for absolute
     calibration of photoacoustic spectrometers', Appl. Phys. Lett. 31,
     728 (1977)
     -W.G.Ferrell, Y.Haven, 'High-performance cell for photoacoustic
     spectroscopy', J. Appl. Phys. 48, 3984 (1977)
     -P.E.Nordal, S.O.Kanstad, 'Infrared photoacoustic spectroscopy for
     studying surfaces and surface related effects', Int. J. Quantum Chem.
     XII, Suppl. 2, 115 (1977)

     -E.M.Monahan, A.W.Nolle, 'Quantitative study of a photoacoustic
     system for powdered samples', J. Appl. Phys. 48, 3519 (1977)
49. R.S.Quimby, P.M.Selzer, W.M.Yen, 'Photoacoustic cell design: reso-
     nant enhancement and background signals', Appl. Opt. 16, 2630 (1977)

50. D.Betteridge, H.E.Hallam, P.J.Meyler, 'Cell effects in optoacoustic
     spectrometry', Fresenius Z. Anal. Chem. 290, 353 (1978)
     -N.C.Fernelius, T.W.Haas, 'Resonant photoacoustic cells constructed
     from UHV hardware', Appl. Opt. 17, 3348 (1978)
     -D.Cahen, I.Lerner, A.Auerbach, 'A simple set-up for single and
     differential photoacoustic spectroscopy', Rev. Sci. Instrum. 49,
     1206 (1978)
     -S.O.Kanstad, P.E.Nordal, 'Open membrane spectrophone for photoacoustic
     spectroscopy', Opt. Comm. 26, 367 (1978)
     -P.E.Nordal, S.O.Kanstad, 'Photoacoustic reflection-absorption spec-
     troscopy (PARAS) of thin oxide films on aluminium', Opt. Comm. 24,
     95 (1978)
     -H.E.Eaton, J.D.Stuart, 'Microcomputer assisted, single beam, photo-
     acoustic spectrometer system for the study of solids', Anal. Chem.
     50, 587 (1978)
     -L.C.Aamodt, J.C.Murphy, 'Size considerations in the design of cells
     for photoacoustic spectroscopy II. Pulsed excitation response', J.
     Appl. Phys. 49, 3036 (1978)

-J.M.McDavid, K.L.Lee, S.S.Yee, M.A.Afromowitz, 'Photoacoustic deter-
mination of the optical absorptance of highly transparent solids',
J. Appl. Phys. 49, 6112 (1978)
-D.Durcharme, A.Tessier, R.M.Leblanc, 'Design and characteristics of a
cell for photoacoustic spectroscopy of condensed matter', Rev. Sci.
Instrum. 50, 1461 (1979)
-T.T.Wang, J.M.McDavid, S.S.Yee, 'Photoacoustic detection of localized
absorption regions', Appl. Opt. 18, 2354 (1979)
-Z.Yasa, N.M.Amer, H.Rosen, A.D.A.Hansen, T.Novakov, 'Photoacoustic
investigation of urban aerosol particles', Appl. Opt. 18, 2528 (1979)
-A.Fujishima, H.Masuda, K.Honda, 'Studies of electrode surface changes
in situ by photoacoustic spectroscopy', Chem. Lett., 1063 (1979)
-K.Kato, Y.Sugitani, 'Cell design for simultaneous measurement of
photoacoustic and fluorescence spectra', Bull. Chem. Soc. Japan 52,
3733 (1979)
-R.E.Blank, T.Wakefield, 'Double-beam photoacoustic spectrometer for
use in the ultraviolet, visible and near infrared spectral regions',
Anal. Chem. 51, 50 (1979)
-D.Cahen, H.Garty, 'Sample cells for photoacoustic measurements, Anal.
Chem. 51, 1865 (1979)
-N.C.Fernelius, 'Helmholtz resonance effect in photoacoustic cells',
Appl. Opt. 18, 1784 (1979)

51. R.W.Shaw, 'Helmholtz resonance cells for pulsed dye laser-excited
high resolution optoacoustic spectroscopy', Appl. Phys. Lett. 35,
253 (1979)

52. G.Busse, D.Herboeck, 'Differential Helmholtz resonator as an opto-
acoustic detector', Appl. Opt. 18, 3959 (1979)
-M.F.Cox, G.N.Coleman, T.W.McCreary, 'Double beam-in-time photoacoustic
spectrometer', Anal. Chem. 52, 1420 (1980)
-A.C.Tam, Y.H.Wong, 'Optimization of optoacoustic cell for depth profiling studies of semiconductor surfaces', Appl. Phys. Lett. 36,
471 (1980)
-P.Poulet, R.Unterreiner, J.Chambron, 'Spectroscopie photoacoustique
des liquides et des solides: Realisation d'un spectrometre spectres
photoacoustique de colorants et rendements quantiques', J. Biophys.
et Méd.Nucl. 4, 83 (1980)
-P.Helander, I.Lundström, D.McQueen, 'Light scattering effects in
photoacoustic spectroscopy', J. Appl. Phys. 51, 3841 (1980)
-V.A.Fishman, A.J.Bard, 'Open-ended photoacoustic spectroscopy cell
for thin-layer chromatography and other applications', Anal. Chem.
53, 102 (1981)
-U.Sander, H.-H.Strehblow, J.K.Dohrmann, 'In situ photoacoustic
spectroscopy of thin oxide layers on metal electrodes. Copper in
alkaline solution', J. Phys. Chem. 85, 447 (1981)

53. O.Nordhaus, J.Pelzl, 'Frequency dependence of resonant photoacoustic
cells: the extended Helmholtz resonator', Appl. Phys. (in press)

54. J.C.Murphy, L.C.Aamodt, 'Photoacoustic spectroscopy of luminescent
solids: Ruby', J. Appl. Phys. 48, 3502 (1977)

55. M.A.A.Siqueira, C.C.Ghizoni, J.I.Vargas, E.A.Menezes, H.Vargas,
L.C.M.Miranda, 'On the use of the photoacoustic effect for investigating phase-transitions in solids', J. Appl. Phys. 51, 1403 (1980)

56. P.S.Bechthold, 'Photoacoustic cell for the temeprature range
    300-1050 K', J. Photoacoustics (in press)

57. P.S.Bechthold, M.Campagna, J.Chatzipetros, 'Variable temperature
    photoacoustic spectroscopy. I. Instrumentation', Opt. Comm. 36,
    369 (1981)

58. H.Coufal, U.Möller, S.Schneider, 'Photoacoustic cells for measure-
    ments at various temperatures and pressures - design and charac-
    terization', (this conference)
    -K.Klein, J.Pelzl, H.Fütterer, 'Photoacoustic cell for low temperature
    PAS',this conference

59. R.Florian, J.Pelzl, M.Rosenberg, H.Vargas, R.Wernhardt, 'Photo-
    acoustic detection of phase transitions', phys. stat. sol. (a)48,
    K35 (1978)
    -P.Korpiun, J.Baumann, E.Lüscher, E.Papamokos, R.Tilgner, 'Photo-
    acoustic effect at first order phase transitions at increasing and
    decreasing temperature', phys. stat. sol. (a)58, K13 (1980)
    -W.Knoll, J.Baumann, P.Korpiun, U.Theilen, 'Phase separation in
    chlorophyll A containing dipalmitoyllecithin vesicles: a fluores-
    cence and photoacoustic study', Biochem. Biophys. Res. Comm. 96,
    968 (1980)

60. P.S.Bechthold, M.Campagna, 'Variable temperature photoacoustic
    spectroscopy. II. Temperature characteristic and applications',
    Opt. Comm. 36, 373 (1981)

61. P.S.Bechthold, K.-D.Kohl, W.Sperling, 'The photocycles of bacterio-
    rhodopsin in buffered aqueous suspension studied at low tempera-
    tures by means of photoacoustic spectroscopy', this conference

62. P.Korpiun, R.Tilgner, 'The photoacoustic effect at first-order phase
    transition', J. Appl. Phys. 51, 6115 (1980)
    -P.Korpiun, R.Tilgner, 'The photoacoustic effect at phase transitions
    in thermally thin and thick samples',phys.stat.sol.(a) 67, 201 (1981)
    -P.Korpiun, 'The PAE at phase transitions', this conference

63. J.B.Callis, M.Gouterman, J.D.S.Danielson, 'Flash Calorimeter for
    measuring triplet yields', Rev. Sci. Instr. 40, 1599 (1969)

64. J.B.Callis, 'The Calorimetric Detection of Excited States',
    J. Res. Natl. Bur. Stand. 80A, 413 (1976)

65. K.Gollnick, this conference
    -D.R.Ort, W.W.Parson, 'Flash-induced volume changes of bacterio-
    rhodopsin-containing membrane fragments and their relationship
    to proton movements and absorbance transients', J. Biol. Chem.
    253, 6158 (1978)
    -'The quantum yield of flash-induced proton release by bacterio-
    rhodopsin-containing membrane fragments', Biophys. J. 25, 341
    (1979)
    -'Enthalpy changes during the photochemical cycle of bacterio-
    rhodopsin', ibid. 355

66. A.C.Tam, C.K.N.Patel, R.J.Kerl, 'Measurement of small absorption in
    liquids', Opt. Lett .4, 81 (1979)
    -C.K.N.Patel, A.C.Tam, 'Optoacoustic spectroscopy of liquids', Appl.
    Phys. Lett. 34, 467 (1979)

-C.K.N.Patel, A.C.Tam, 'Quantitative spectroscopy of micron-thick liquid films', Appl. Phys. Lett. 36, 7 (1980)
-'Optical absorption coefficients of water', Nature 280, 302 (1979)
-'Absorption profile of the seventh harmonic of the C-H stretch in liquid benzene', Chem. Phys. Lett. 62, 511 (1979)
-A.C.Tam, C.K.N.Patel, 'Optical absorption of light and heavy water by laser optoacoustic spectroscopy', Appl. Opt. 18, 3348 (1979)
-'Ultimate corrosion resistant optoacoustic cell for spectroscopy of liquids', Opt. Lett. 5, 27 (1980)
-S.Rentsch, 'Eine photoakustische Meßmethode zur Bestimmung geringer linearer und nichtlinearer Absorptionen in Flüssigkeiten', Exp. Technik der Physik 27, 571 (1979)

67. C.K.N.Patel, E.T.Nelson, R.J.Kerl, 'Optoacoustic study of weak optical absorption of liquid methane', Nature 286, 368 (1980)

68. W.Lahmann, H.J.Ludewig, 'Opto-acoustic determination of absolute quantum yields in fluorescent solutions', Chem. Phys. Lett. 45, 177 (1977)

69. M.Hass, J.W.Davisson, P.H.Klein, L.L.Boyer, 'Infrared absorption in low-loss KCl single crystals near 10.6 µm', J. Appl. Phys. 45, 3959 (1974)
-J.A.Harrington, L.Bobbs, M.Braunstein, R.K.Kim, R.Stearns, R.Braunstein, 'Ultraviolet-visible absorption in highly transparent solids by laser calorimetry and wavelength modulation spectroscopy', Appl. Opt. 17, 1541 (1978)

70. M.B.Salamon, P.R.Garnier, B.Golding, E.Bruehler, 'Simultaneous measurement of the thermal diffusivity and specific heat near phase transitions', J. Phys. Chem. Solids 35, 851 (1974)

71. M.Bass, E.W.VanStryland, A.F.Stewart, 'Laser calorimetric measurement of two-photon absorption', Appl. Phys. Lett. 34, 142 (1979)
-T.J.Moravec, E.Bernal, G. 'Automation of a laser absorption calorimeter', Appl. Opt. 17, 1938 (1978)
-D.C.Johnson, 'Measurement of low absorption coefficient in crystals', Appl. Opt. 12, 2192 (1973)

72. W.J.Parker, R.J.Jenkinds, C.P.Butler, G.L.Abbott, 'Flash method of determining thermal diffusivity, heat capacity and thermal conductivity', J. Appl. Phys. 32, 1679 (1961)

73. G.H.Brilmeyer, A.Fujishima, K.S.V.Santhanam, A.J.Bard, 'Photothermal spectroscopy', Anal. Chem. 49, 2057 (1977)

74. M.Furuta, 'A two-layer slab method for measuring thermal diffusivity of polymer by irradiated light heat-wave I', Japan. J. Appl. Phys. 12, 1143 (1973)

75. H.Parker, K.W.Hipps, A.H.Francis, 'Thermo-optical spectroscopy of low temperature solids', Chem. Phys. 23, 117 (1977)

76. R.B.Bailey, P.L.Richards, 'Infrared measurements of molecules adsorbed on metal surfaces, a low-temperature thermal detection technique', Proc. Int. Conf. Infrared Phys. 2nd (1979), 292, E.Affolter, F.Kneubühl (Ed.)

-A.Bubenzer, D.Bimberg, 'Calorimetric absorption spectroscopy (CAS) of GaP:N,S', Proc. 15th Int. Conf. Phys. of Semiconductors, Kyoto, 1980, J. Phys. Soc. Japan 49 (1980) Suppl. A, p. 255
-D.Bimberg, A.Bubenzer, 'Calorimetric absorption spectroscopy of nonradiative recombination processes in GaP', Appl. Phys. Lett. 38, 803 (1981)

77. A.Bubenzer, S.Hunklinger, K.Dransfeld, 'New method for measuring extremely low optical absorptions', J. Non-Crystalline Solids 40, 605 (1980)

78. J.B.Smith, G.A.Laguna, 'Second sound spectroscopy: A new method for studying optical absorption in solids', Phys. Lett. 56A, 223 (1976)

79. M.B.Robin, 'Direct measurements of radiationless transitions in molecules and crystals', J. Luminescence 12/13, 131 (1976)

80. M.B.Robin, N.A.Kuebler, 'Radiationless relaxation in solids as measured by a heat pulse technique', J. Chem. Phys. 66, 169 (1977)

81. F.Vidal, 'Sur la thermométrie acoustique et ses applications à la spectroscopie electromagnétique dans les solides aux basses températures', C.R.Acad. Sc., Paris B285, 93 (1977)

82. J.Blitz, 'Elements of Acoustics', Butterworths, London (1964)

83. J.Pelzl, 'Frequency dependent photoacoustic spectroscopy of condensed matter', this conference

84. J.Pelzl, K.Klein, O.Nordhaus, 'The extended Helmholtz resonator in low temperature PAS', to be published.

PHOTOACOUSTIC CELL FOR LOW TEMPERATURE PAS

K.Klein, J.Pelzl and H.Fütterer
Institut f. Experimentalphysik VI
Ruhr-Universität, 463 Bochum, FRG

ABSTRACT

A photoacoustic (PA) cell is described, which can be used in
combination with a standard optical cryostat for variable
temperatures. The PA cell consists of a sample and a microphone chamber which are separated by a cylindrical tube. The
frequency dependence and the temperature variation of the PA
amplitude and of the PA phase angle from a graphite sample
have been studied in the range 35Hz < ν < 800Hz and 5K < T <
300K, respectively. The results are discussed in the context
of the extended Helmholtz resonator model.

INTRODUCTION

Photoacoustic spectroscopy of solids principally suffers
from a rather low sensitivity which is a result of the heat
generation involved in the PA process. However, with decreasing temperature the intensity of the PA signal becomes
considerably enhanced due to two effects: (a) The heat capacity
of solids decreases rapidly ($\sim T^3$) at temperatures below the
Debye temperature Θ; (b) according to the Rosencwaig-Gersho
theory /1/ the pressure response of the transducer gas is
proportional to μ_g/T which varies roughly as $T^{-1/4}$.
Descending from room temperature to liquid helium temperature
an enhancement of the order of magnitude of 10^3 can be expected. This sensitivity gain increases the attractivity of
a low temperature PA spectroscopy.

Two different designs of low temperature PA cells have
already been tested. One possibility is to use a microphone,
which can be cooled down with the whole PA cell /2,3/ .
An alternative construction proceeds from a spatial sepa -
ration of the sample and the microphone. The microphone is
contained in one cavity and is held at room temperature
whereas the temperature of the sample in the other cavity is
changed /4,5,6/.

EXPERIMENTS AND RESULTS

The photoacoustic cell presented here belongs to the
latter type. Two conditions had to be satisfied by the construction: 1. The cell arrangement should fit into a commercial optical cryostat with an available inner diameter of
21.5mm. 2. The temperature of the sample should be variable
in the range between 5K to 300K whereas the buffer gas is
maintained at a constant pressure.

The constructed PA cell is sketched in Fig.1. The sample
chamber at the lower part has an inner diameter of 7mm and
is made out of aluminium. The quartz window is sealed on the
cell body. The sample support opposite the window consists
of a copper plate. The temperature of the sample can be
measured by a thermocouple at the rear of the sample support.
The tube which interconnects the sample chamber and the microphone chamber is made out of stainless steel and has an inner
diameter of 1.5mm and a length of 400mm. The microphone is
mounted at the top of the microphone chamber. The gas inlet
at the left side can be closed by the needle valve. The lower
part of the cell fits into an optical cryostat (CryoVac) and
is sealed against vacuum by the connector. The sample can be
illuminated through the two windows of the cryostat and that
of the PA cell.

First measurements are performed in order to study the
frequency dependence of the cell transfer function at
different temperatures. Experiments were carried out with an
vector lock-in (Ithaco) and an electret microphone (Knowles).
The sample was a graphite foil and the buffer gas was helium.
Measurements of the PA signal were performed at constant
pressure (1atm) and with the sample at constant temperature
as a function of chopping frequency ν. The amplitude S and
the phase angle φ were recorded simultaneously.

Some selected spectra are shown in Fig.2 . The frequency
response of the cell exhibits a multiple resonance pattern
characteristic for an extended Helmholtz resonator (EHR) /7,8/.
At lower temperatures the amplitude S becomes considerably
enhanced and the frequencies of the resonance peaks are
shifted to lower values. A second resonance pattern can be observed below 50K.

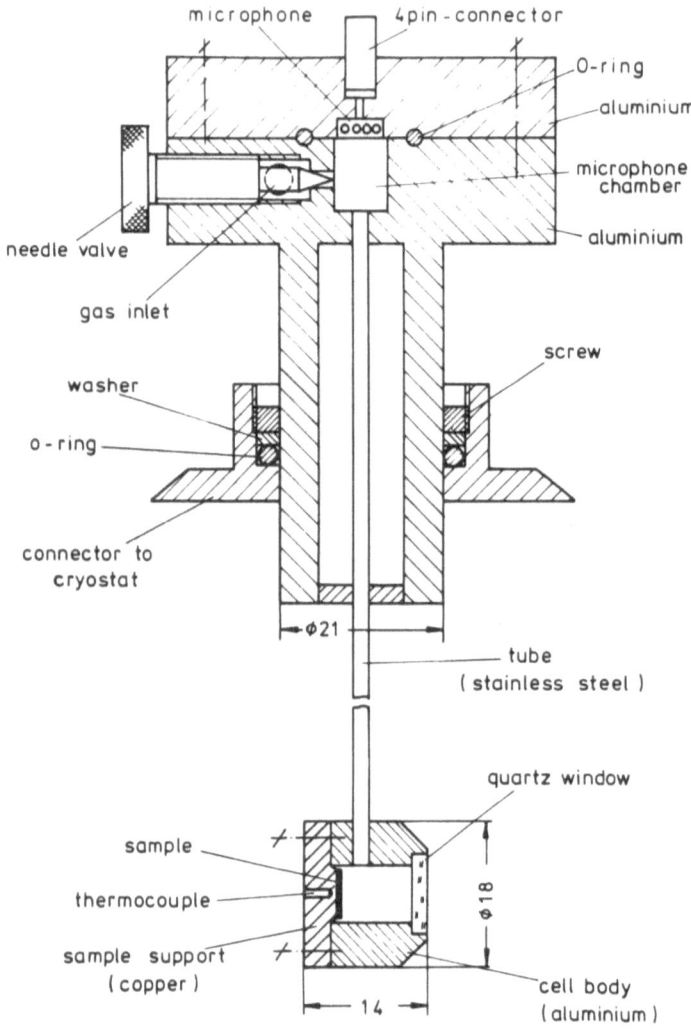

FIGURE 1: Design of the cell for low temperatures. The scales
are given in mm.

DISCUSSION

The frequency dependence of S and φ can be explained in
the context of the EHR model. This model takes into account
that the length of the interconnecting tube is not negligible
compared to the wave length of the sound in the cell. The tube
is treated as ·an acoustic transmission line whereas the two
chambers V_1 and V_2 (Fig.3) are kept as discrete elements.

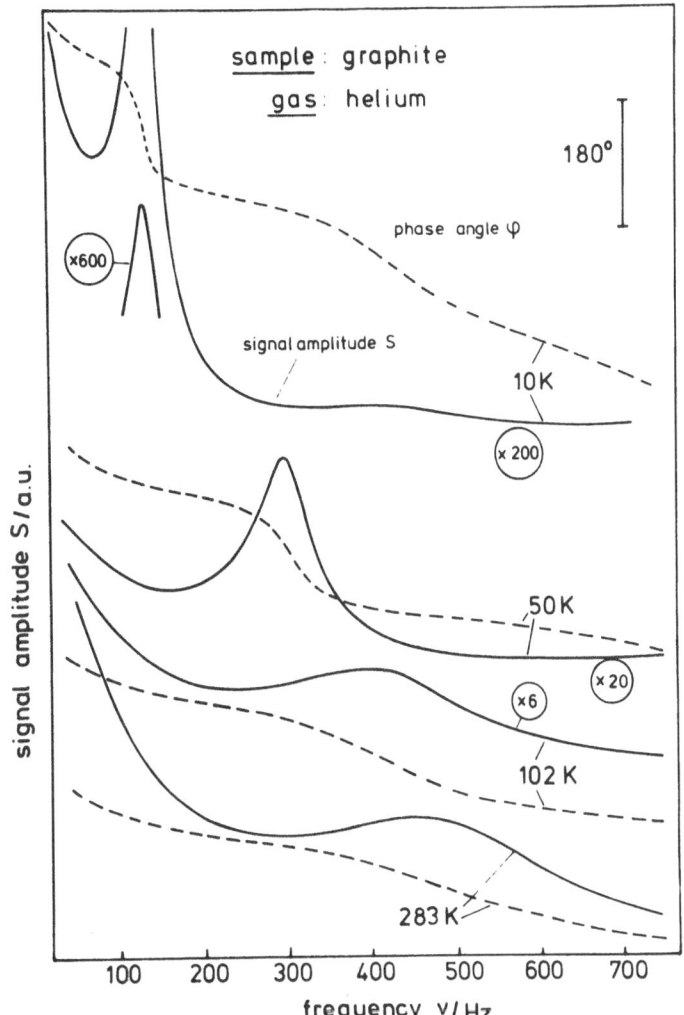

FIGURE 2: Amplitude A_2 and phase angle φ_2 at different temperatures as a function of the chopping frequency ν. The sample was a graphite foil and the buffer gas was helium.

The frequency response of such an acoustic resonance cell can be determined by an electrical analogy similar to that used for conventional Helmholtz resonators .Fig.3 shows schematically the cell arrangement and the analogous electrical equivalent circuit on the basis of the EHR model. The sample chamber and the microphone chamber have the two volumes V_1 and V_2, respectively. They are interconnected by the conduit with the cross-sectional area A. The electrical analogies to

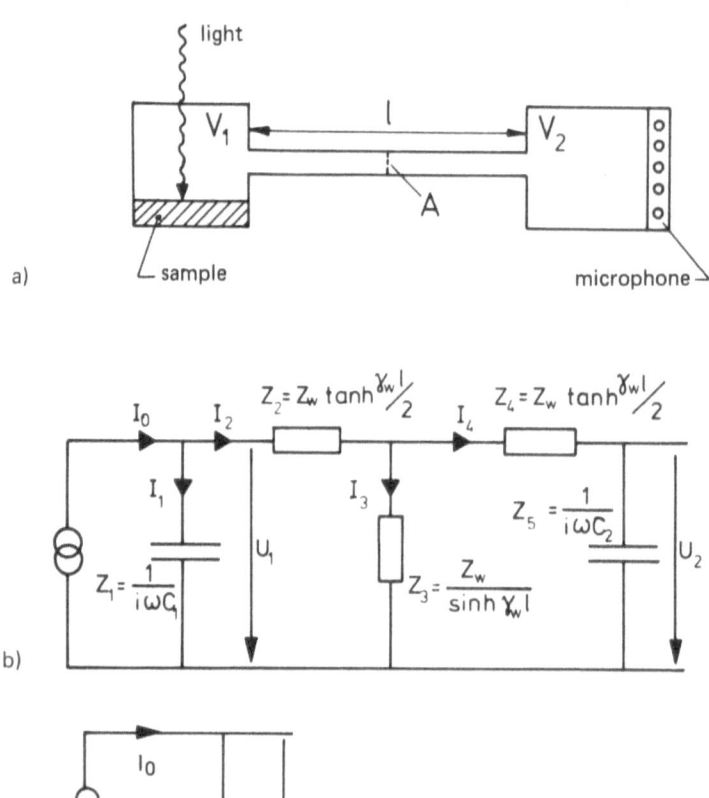

FIGURE 3: Acoustic cell and the electrical equivalent circuit
analogies. (a) cell arrangement, (b) equivalent circuit of
the extended Helmholtz resonator, (c) equivalent circuit of the
nonresonant cell.

the volumes V_1 and V_2 constitute the two capacitances C_1 and
C_2. The voltage U_1 and U_2 corresponds to the pressure p_1 and
p_2 in the sample and in the microphone cell, respectively.
The interconnecting tube is replaced by the T-network consisting of Z_2, Z_3 and Z_4 which are functions of the charcte-
ristic impedance Z_w and the propagation constant γ_w . These
quantities are defined as follows:
$Z_w = (Z/Y)^{1/2}$ and $\gamma_w = (Z \cdot Y)^{1/2}$ where $Z = R + i\omega L$ represents the series impedance and $Y = G + i\omega C$ the shunt ad-

mittance of the tube per unit length. R takes into account the
energy loss in the tube due to frictional forces whereas L
and C describe the kinetic energy and the storage of energy
of the gas in the tube, respectively. The shunt admittance G
considers the heat loss in the tube due to nonadiabatic expansion. The quantities R, L, G and C which are eaqually distributed on the tube are functions of system parameters only
/9/ and can be determined   theoretically or semiempiri-
cally /7/.

The photoacoustic signal is generated in the cavity V_1
while it is detected in the microphone chamber V_2. The
measured quantities are the amplitude A_2 and the phase angle
φ_2 of the acoustic pressure in V_2 which is represented in the
analogous electrical circuit by the voltage $U_2 = A_2\, e^{i(\omega t - \varphi_2)}$.
In order to draw a conclusion from the measureable amplitude
A_2 to the photoacoustic signal in the sample chamber it is
convenient to define an amplitude transfer factor of the EHR
by the ratio (A_2/A_0). A_0 is the acoustic pressure amplitude in
the "isolated" sample cell which is represented by the nonresonant electric circuit shown in Fig.3c. The factor A_2/A_0
depends only on acoustic properties of the resonance cell
and it can be evaluated as a function of the chopping frequency on the basis of the extended Helmholtz resonator model.
In order to be able to apply the equivalent circuit of Fig.3b
to those cases where a temperature gradient exists between
the sample and the microphone chamber, in a first approximation a homogeneous temperature distribution T_{eff} within the
whole cell was assumed. The effective temperature T_{eff} had to
meet the condition that at a given temperature the peak frequencies of the resonance pattern can be reproduced with the
EHR equivalent circuit of Fig.3b. This procedure works reasonably for temperature differences less than 250K between
both ends of the helium filled duct. Deviations at low temperature become evident from a plot of A_0^{exp} versus the frequency in Fig.4. $A_0^{exp} = A_2^{exp}/(A_2/A_0)^{theo}$ is the acoustic
pressure amplitude in the "isolated" sample chamber (Fig.3c)
which is deduced from the experimental amplitude A_2^{exp}
measured in the volume V_2 by means of the theoretical amplitude factor $(A_2/A_0)^{theo}$ at $T = T_{eff}$.

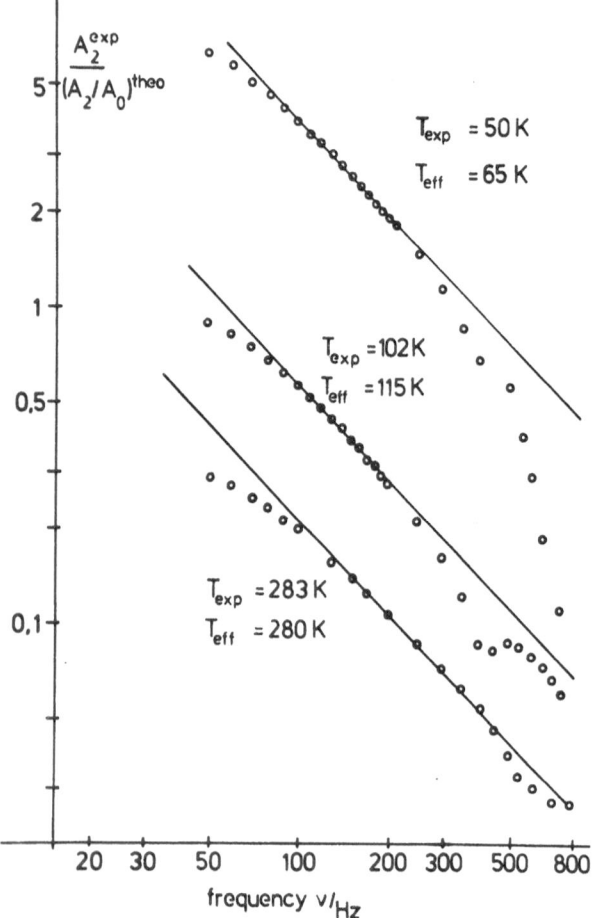

FIGURE 4: $A^{exp}/(A_2/A_0)^{theor}$ as a function of chopping frequency ν at three selected sample temperatures T_{exp} and the
corresponding effective temperature T_{eff}, respectively. The
circles are the experimental points. The solid line shows
a ν^{-1} dependence.

In the context of the Rosencwaig-Gersho theory A_0 should
obey a ν^{-1} dependence (full line in Fig.4) as we are concerned with the PA effect from an optically opaque and thermally thick sample. The discrepancies between experiment and
theory below 100Hz are consequence of the low frequency limit
of the electret microphone. The deviations at high frequencies
are temperature dependent indicating that the temperature gradient along the duct is no more negligible at very low temperatures. Despite these discrepancies the effective tempera-

ture approach yields an estimate of the temperature variation of the PA amplitude A_0 at intermediate frequencies. A_0 increases by a factor of 10^3 and follows roughly a T^{-3} behaviour between room temperature and 10K.

ACKNOWLEDGEMENTS

The authors would like to thank D.Krüger for technical assistance and K.Junge and H.Lerchner for valuable discussions.

REFERENCES

1   A.Rosencwaig and A.Gersho, 'Theory of the photoacoustic effect with solids', J.Appl.Phys. $\underline{47}$(1), 64 (Jan 76)

2   C.Pichon, M.LeLiboux, D.Fournier and A.C.Boccara, 'Variable temperature photoacoustic effect: Application to phase transitions', Appl.Phys.Letters $\underline{35}$(6), 435 (Sep 79)

3   D.Fournier and A.C.Boccara, 'Photoacoustic spectroscopies', this conference

4   J.C.Murphy and L.C.Aamodt, 'Photoacoustic spectroscopy of luminescent solids: Ruby ', J.Appl.Phys. $\underline{48}$(8), 3502 (Aug 77)

5   P.S.Bechthold, M.Campagna and J.Chatzipetros, 'Variable temperature photoacoustic spectroscopy, 1) Instrumentation', Opt.Comm. $\underline{36}$(5), 369 (1981)

6   H.Coufal, U.Möller and S.Schneider, 'Photoacoustic cell for measurements at various temperatures and pressures -Design and characterisation-', this conference

7   O.Nordhaus and J.Pelzl, 'Frequency dependence of resonant and nonresonant photoacoustic cells: The extended Helmholtz resonator', Appl.Physics, in print

8   J.Pelzl, 'Frequency dependent photoacoustic spectroscopy of condensed matter', this conference

9   A.H.Benade, 'On the propagation of sound waves in a cylindrical conduit', J.Ac.Soc.Am $\underline{44}$, 616 (1978)

PHOTOACOUSTIC CELLS FOR MEASUREMENTS AT

VARIOUS TEMPERATURES AND PRESSURES

- DESIGN AND CHARACTERISATION -

H. Coufal

Physik Department E13, Technische Universität München,
D 8046 Garching, FRG

U. Möller and S. Schneider

Institut für Physikalische und Theoretische Chemie
Technische Universität München, D 8046 Garching, FRG

A B S T R A C T

Five different types of photoacoustic cells are described
and their   characteristics are discussed with
respect to their possible application: resonant and
non-resonant cells for use at ambient temperature, resonant cells for high and low temperature samples as well
as one allowing for various gas pressures above the
sample (10 mbar to 10 bar).

INTRODUCTION

Photoacoustic cells which can be used for measurements
at various temperatures and pressures are described.
The periodic pressure fluctuations generated in these
gas filled cells are detected with conventional condenser
or electret microphones. To characterize the cells the
frequency-dependence of both the amplitude (S_{PAS}) and the
phase (ϕ_{PAS}) of the PA-signal from carbon black samples
were measured.

As light source an Ar^+ -laser pumped tunable dye
laser was used. The laserbeam was modulated with an acousto-
optic modulator. More details of the experimental set up
are given in reference 1. The data points were taken in

2 or 3 intervals, respectively (10-100 Hz, 100 - 1000 Hz, 1- 5 kHz). In each interval the amplitudes and phases were measured for 11 equally spaced frequencies. All tests were performed with a laser power $p \leq 8$ m W at a wavelength of $\lambda \approx 604$ nm.

CELL DESIGN

The light enters the cell through a quartz window and illuminates the sample (fig. 1). The distance between the sample and the window is approximately equal to the calculated thermal active layer in the gas ($\pi \cdot \mu_g$) at the lowest modulation frequency (f ≈ 5 Hz). Typical volumes of the sample chamber (V_{SC}), the microphone chamber (V_{MC}) and the connecting tube (V_T) are:

$$V_{SC} \approx 0,2 \ cm^3$$
$$V_{MC} \approx 0,03 \ cm^3$$
$$V_T \geq 0,14 \ cm^3$$

V_{SC} is kept very small to achieve a high sensitivity and a good signal/noise-ratio (SNR), V_{MC} is used to tune the helmholtz resonance frequency of the system well above the highest working frequency. The diameter of the connecting tube (ϕ_T) is chosen such that no thermoviscous damping effects have to be taken into account ($\phi_T \geq 1$ mm). The lengths of the tubes (l_T) are kept as short as possible. Depending on the sample type, different sample holders can be inserted in the cell bottom plates.

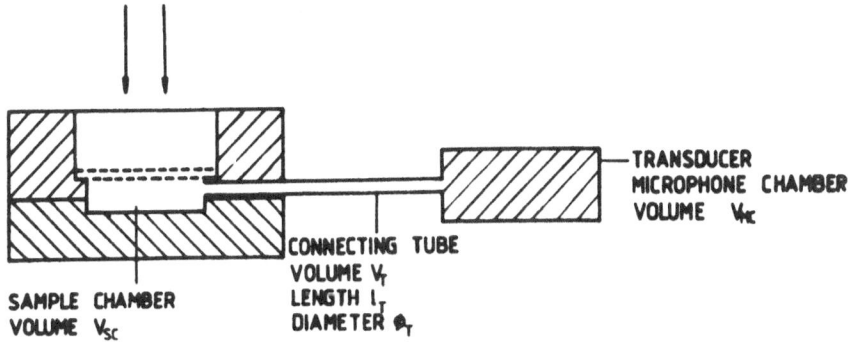

TRANSDUCER
MICROPHONE CHAMBER
VOLUME V_{MC}

CONNECTING TUBE
VOLUME V_T
LENGTH l_T
DIAMETER ϕ_T

SAMPLE CHAMBER
VOLUME V_{SC}

Figure 1   Principle of the cells

The nominal resonance frequencies mentioned below
are calculated by means of the following formula without any
corrections

$$f_R = \frac{c_o}{4\sqrt{\pi}} \cdot \phi_T \cdot \sqrt{\frac{V_{MC} + V_{SC}}{V_{SC} \cdot V_{MC} \cdot (l_T + 0.85 \cdot \phi_T)}}$$

c_o: velocity of sound

a) Helmholtz resonant cell with condenser microphone

The Brüel & Kjaer microphone cartridge type 4166 with
preamplifier type 2619 has a flat response ($\pm$ 2dB)
from 3 Hz to 9 kHz. From ambient noise the microphone

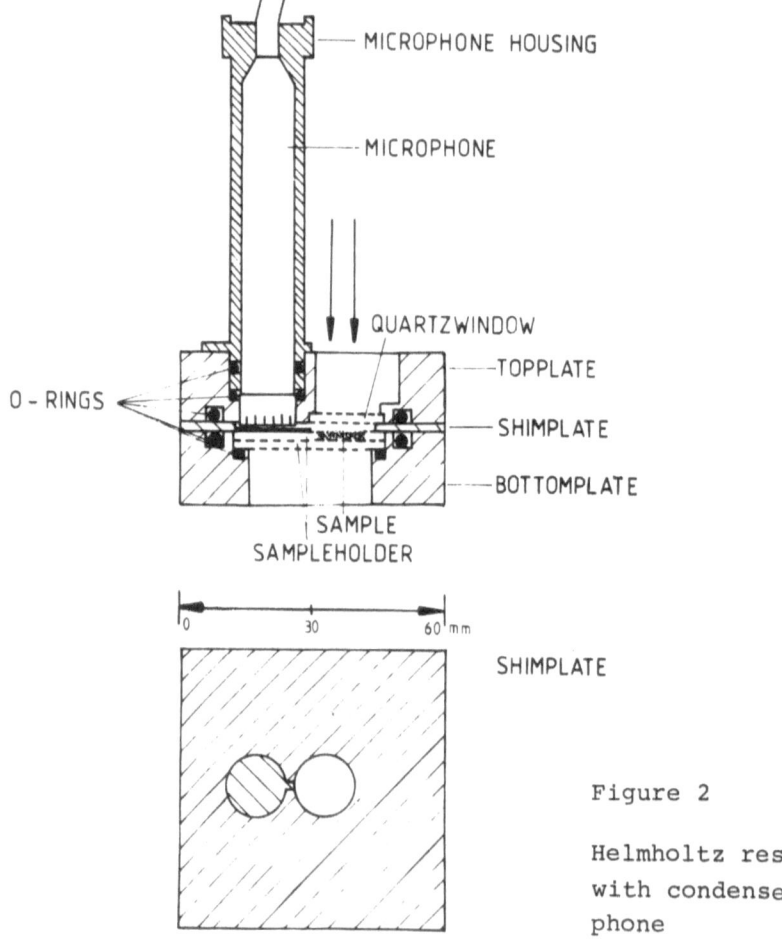

Figure 2

Helmholtz resonant cell
with condenser micro-
phone

is shielded with an additional microphone housing
(fig. 2). The cellvolume can be varied by
plates with different geometries according to the
experimental requirements. The total volume of the
cell ($V_t = V_{SC} + V_{MC} + V_T$) is $V_t \approx 0,5$ cm^3,
the resonance frequency is calculated to $f_R \approx 6.5$ kHz.

b) Nonresonant cell with electret microphone

In this type of cell an electret and integrated MOS-
FET preamplifier is used (fig. 3). This microphone can be used with a simple battery, instead of a
special power supply; grounding or noise problems
can therefore be conveniently avoided. Both electret and
preamplifier are potted with epoxy   directly into
the cell housing, thereby suppressing external sound
sources. The gas volume is $V_t \approx 0.25$ cm^3.

c) Helmholtz resonant cell with detachable transducer

In this PA-cell built for work at room temperature,
both the electret and the MOS-FET preamplifier (transducer) are potted with epoxy into a separate housing
and can be coupled with a capillary to the cell
(fig. 4). With this kind of transducer it is possible
to work in a pressure range of 10   mbar to 10 bar.
This was tested with air at room temperature. The vo-

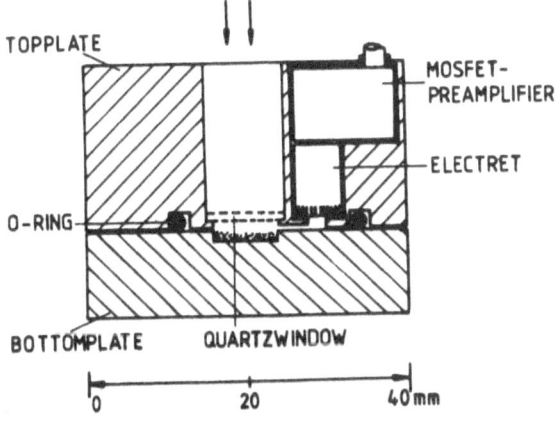

Figure 3   Non-resonant cell with electret microphone

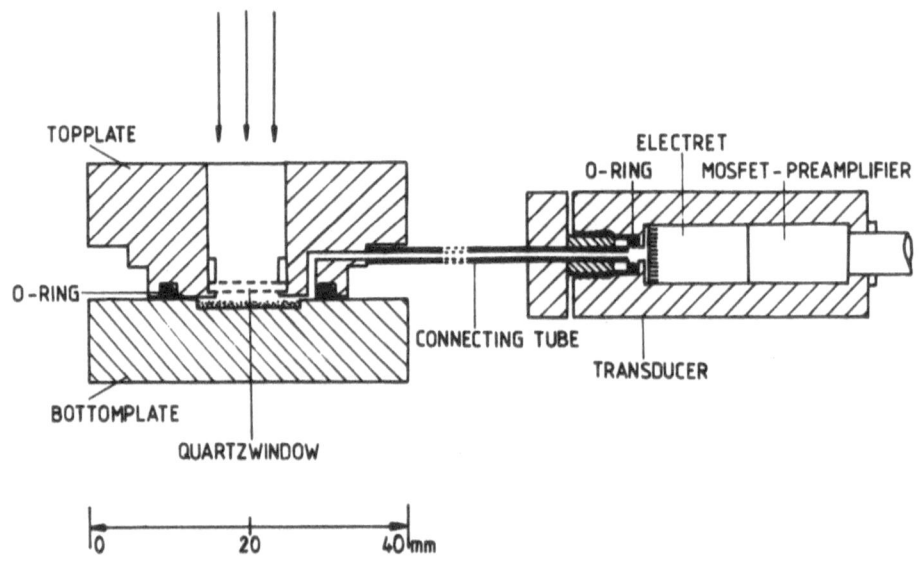

Figure 4   Helmholtz resonant cell with removable trans-
            ducer for room temperature measurements.

lume of the cell is $V_t \approx 0.33$ cm^3 and the calculated
resonance frequency $f_R \approx 2.1$ kHz.

d) Helmholtz resonant low temperature cell

The cell itself is inside of a vacuum vessel (fig. 5).
The top part of the latter is a plexiglass window to
enable illumination and fast exchange of samples.
On the front side there are feed throughs for the
electronics and the liquid coolant in- and outlet.
Platinum resistance thermometers in the cell are
used to measure the temperature and keep it constant
by combination of a continous flow of liquid nitrogen
and a gated resistance heater. The PA-cell itself is
constructed similarly to the helmholtz resonant room
temperature cell discussed above, except that the O-
ring seal is replaced by an indium wire. The connecting
tube is slightly longer; this leads to a total gas-
volume of $V_t \approx 0.42$ cm^3 with a calculated resonance
frequency of $f_R \approx 1.5$ kHz. The cell design allows
measurements in the temperature range of 85 K to 375 K.

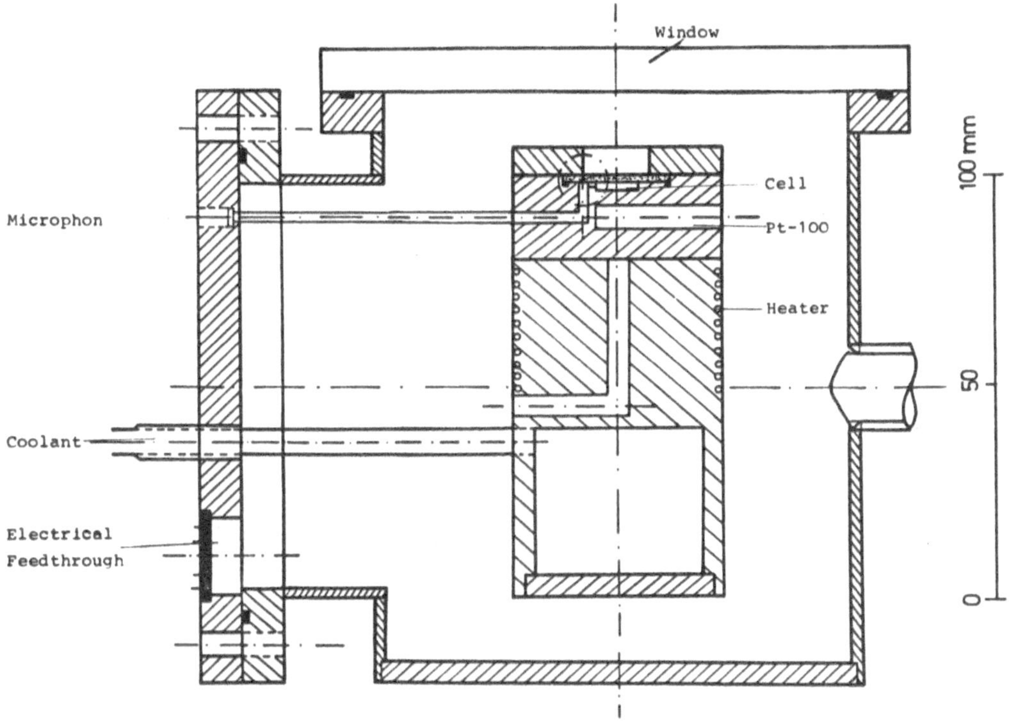

Figure 5   Helmholtz resonant low temperature cell

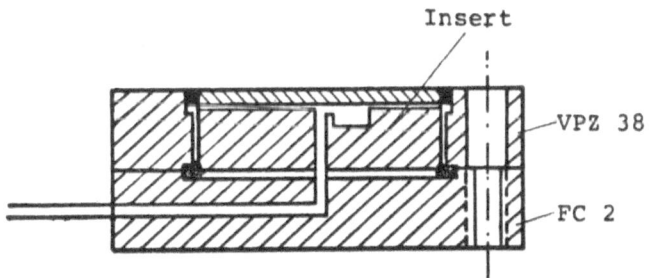

Figure 6   Helmholtz resonant high temperature cell

e) Helmholtz resonant high temperature cell

The cell, which has been tested in the temperature
range of 290 K to 800 K can be placed in the vacuum
vessel instead of the low temperature PA-cell. Special
attention has been paid to choose a construction which
makes use of three standard UHV-parts only (fig. 6):

one UHV-conflat flange
one UHV-conflat flange with viewing port
one OFHC-cooper gasket.

An insert reduces the volume to approximately
$V_t \approx 0.6$ cm^3 resulting in a resonance frequency
of $f_R \approx 1.4$ kHz.

RESULTS AND DISCUSSION

Figure 7 and 8 show the modulation frequency dependence
of the amplitude and phase response of the PA-signal

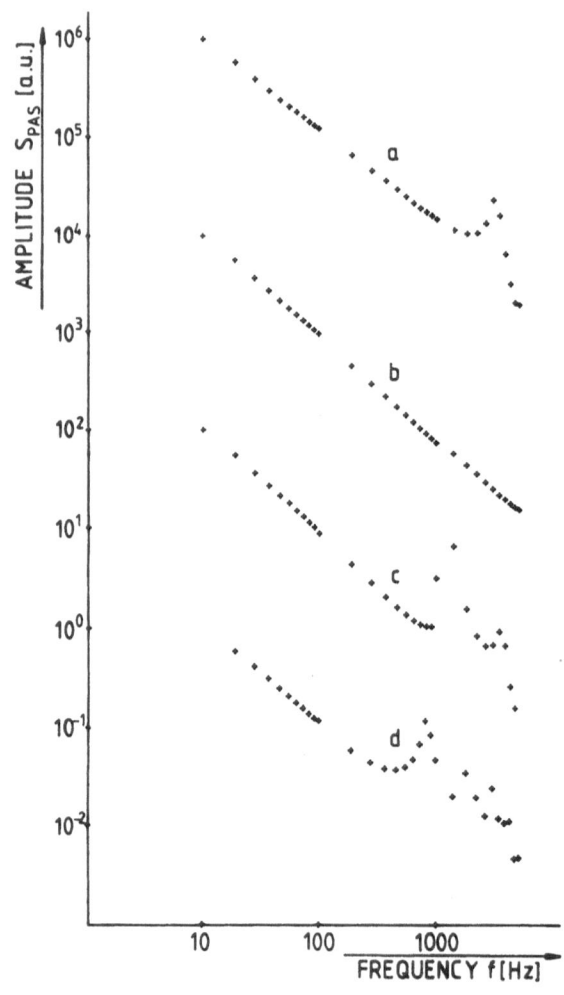

Figure 7

Modulation frequency
dependence of the ampli-
tude of the PA-signal for
cells a) to d) (carbon
black sample)

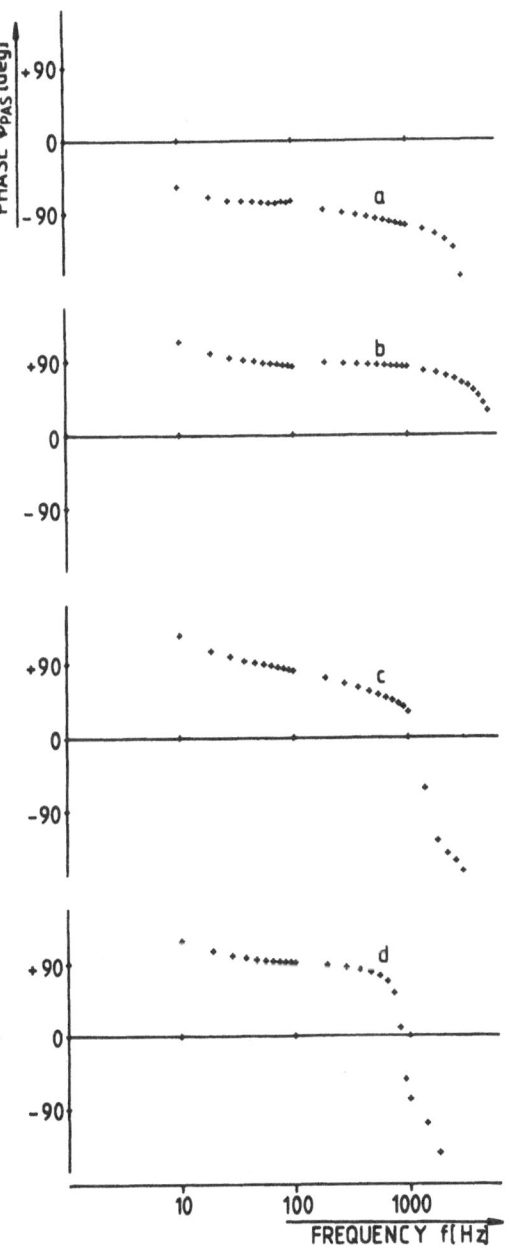

Figure 8   Modulation frequency dependence of the phase
of the PA-signal for cells a) to d) (carbon black
sample)

produced by carbon black samples in the cells a) to d).
Well below the first resonance frequency, for each cell
an excellent f^{-1}-dependence of the amplitude is measured.
The observed resonance frequency is lower by a factor of
2 to 3 than calculated by the simple formula given above.
For an exact calculation of the first resonance frequency
and the explanation of the appearance of several higher
resonances, more refined models have to be considered [3].
The f-dependence of the phase displayed in figure 8 is
typical for helmholtz resonant cells.

The data presented in the figures were taken at constant
gas pressure in the cells. Those made at constant volume
were nearly identical with the exception of a small shift
of the resonances to lower frequencies.

With increasing length of the connecting tube the resonances are shifted towards lower frequencies. A similar
low frequency shift of about 300 Hz is observed upon
lowering the temperature in cell d) from 295 K to 90 K.

Measurements of the dependence of the PA-signal from
the incident laser power and the length of the gas column
in the cell a) showed in agreement with the RC-theory a
perfect linear relationship [4,5]. The laser power was
varied in the range from 0.03 to 13 mW, the length of the cell
volume in the range from 1 to 8.5 mm at chopping frequencies of about 27 Hz and 157 Hz, respectively. Figure
9 displays the dependence of the PA-signal on temperature
(carbon black sample). As one might expect, there is no
simple law governing this dependence since a series of
effects is influencing the signal amplitude. According
to the RG-theory a dependence like μ_g / T is predicted
under the assumption of constant pressure. In reality,
we have a temperature gradient along the coupling capillary, besides of the fact that pressure equalization
was done only once during the course of the experiment.

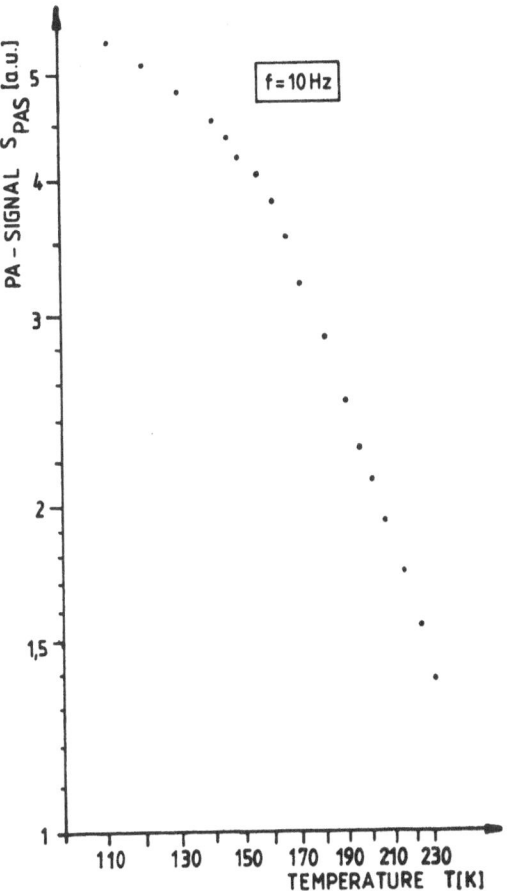

Figure 9   Amplitude variation of PA-signal during warm
            up of the cell. Pressure equalization was per-
            formed at T = 110 K.

CONCLUSION

It has been shown that the helmholtz resonant systems
show a f^{-1} dependence below the resonance frequencies.
The PA-cells with electret transducer and pressure equalization possibility are less sensitive to ambient noise
and mechanical vibrations resulting in a superior S/N-ratio.
With one type of cell and detector PA-spectra of samples
at low, room and high temperatures with different gases and
variable gas pressures are easily accessible.

ACKNOWLEDGEMENT

Final support by the "Deutsche Forschungsgemeinschaft"
and the "Fonds der Chemischen Industrie" is gratefully
acknowledged. Furthermore, we wish to thank Miss E. Kudler,
Mrs. M. Reiche and Mrs. H. Harlandt for their valuable
help in preparing the figures.

REFERENCES

/1/   S. Schneider, U. Möller, H. Coufal
      Photoisomerisation of DODCI studied by Photo-
      acoustic Spectroscopy
      This volume

/2/   L.L. Baranek
      Acoustics: Chapter 5, Part XIII, "Acoustic Elements"
      Mac Graw-Hill, New York (1954)

/3/   O. Nordhaus, J. Pelzl
      Frequency Dependence of resonant Photoacoustic-Cells:
      The extended Helmholtz-Resonator
      Appl. Phys. to be published

/4/   A. Rosencwaig
      Photoacoustic Spectroscopy: Chapter IV and V
      Theory of the PA-Effect
      Advances in Electronics and Electron Physics
      46, 207 (1979)

/5/   L.C. Aamodt, J.C. Murphy, J.G. Parker
      Size Considerations in the Design of Cells for
      Photoacoustic Spectroscopy
      J. Appl. Phys. **48**, 927 (1977)

SELF-SUPPORTING CARBON GLASS FILMS: A PAS REFERENCE SAMPLE

H. Coufal

Physikdepartment E13, Technische Universität München, D-8046-
Garching, West Germany

ABSTRACT

The preparation and use of self-supporting carbon glass films as
reference samples for PAS are described. Spectral, structural
and photoacoustic parameters of these films are presented and
compared with other standard samples.

INTRODUCTION

Photoacoustics being an effect due to optical and thermal
processes within a sample require reference samples with well-
defined spectroscopic, thermal and geometric properties. Reproducibility, stability over extended periods of time and, last but
not least, a high dynamic range combined with an excellent signal
to noise ratio are desirable.

Standard samples for the visible spectral range that are
widely used in the field (such as carbon black) boast extremely
low reflectivities and high extinction coefficients. Their
optical properties are optimal and hard to beat; but their poor
reproducibility and the rapid aging process /1/ due to the large
surface, as well as their need of a backing material cause problems that should be solved by a reference sample whose overall
photoacoustic properties are optimised. Self-supporting carbon
glass films seem to fulfill these requirements for the visible
spectral range for all modulation frequencies used in the gas-
coupled microphone technique.

DESIGN CONSIDERATIONS

A maximum signal at a given light intensity will be observed using an optically opaque, thermally thin sample /2/.

One prerequisite a standard sample should have is, therefore, a high extinction coefficient α . Certainly, carbon with α in the order of 3×10^7 m^{-1} /3/ is an excellent choice. But, of course, a high extinction coefficient implies a high reflectivity that can only be overcome by multiple reflections - as is done in carbon black - thus increasing the surface area that is prone to aging and contamination. A moderately rough carbon sample should be a good compromise.

Assuming typical thermal properties for this sample - a thermal conductivity k = $8{,}3\times10^{-1}$ cal/m°C /4/, a density ϱ of amorphous graphite of 1.5×10^6 g/m^3 /5/ and a specific heat of 0.178 cal/g°C /6/ - the thermal diffusion length μ_s at 10 kHz is approximately 2.5×10^{-5} m. To qualify as thermally thin even at this rather high modulation frequency - typical frequencies usually employed with gas-coupled microphones are of lower orders of magnitude - a sample thickness in the 1μm-regime would be sufficient. Of course, with thermally thin samples the backing material plays an important role. Its thermal conductivity should be small, whereas its thermal diffusion length should be as great as possible for maximum signal strength /2/. Plain air as a backing material is unsurpassed in this respect. A self-supporting carbon film with a thickness in the order of 1μm should make an excellent reference sample. Graphite foils that have been used to test, for example, photoacoustic theories /8/ or cells /9/ suffer from the strongly anisotropic optical and thermal properties of graphite /1, 10/. An amorphous carbon glass with isotropic properties is therefore preferable.

SAMPLE PREPARATION

From all the samples that have been examined, those prepared according to a prescription from P. Maier-Komor /11/ performed best. A clean microscope coverglass is coated with Betaine: carbon is deposited under UHV-conditions with an electron beam

evaporator, but care has to be taken not to anneal the carbon
glass during this deposition; 100 µg/cm^2, corresponding to
0.7 µm thickness, give optimum optical and thermal properties.
The parting agent is dissolved by flooding the substrate with
de-ionised water; the film is then floated on to its final mount.
Films deposited on other types of parting agents, or prepared by
pyrolysis, or sputtering, displayed insufficient reproducibility
or optical properties.

SAMPLE PROPERTIES

About 30 samples prepared according to the above prescription
in different laboratories and in different production runs varied
in their optical and photoacoustic properties by only about 0.3%.
Furthermore, an 18-month old sample showed practically identical
behaviour to that of a freshly prepared one.

The reflection coefficient R (Fig.1) of 100 µg/cm^2 samples
under 0°/diffuse conditions using a BaSO$_4$ reference can be described by

$$R/\% = 20.6 + 1.175 \times 10^{-2} \times \lambda \ /nm$$

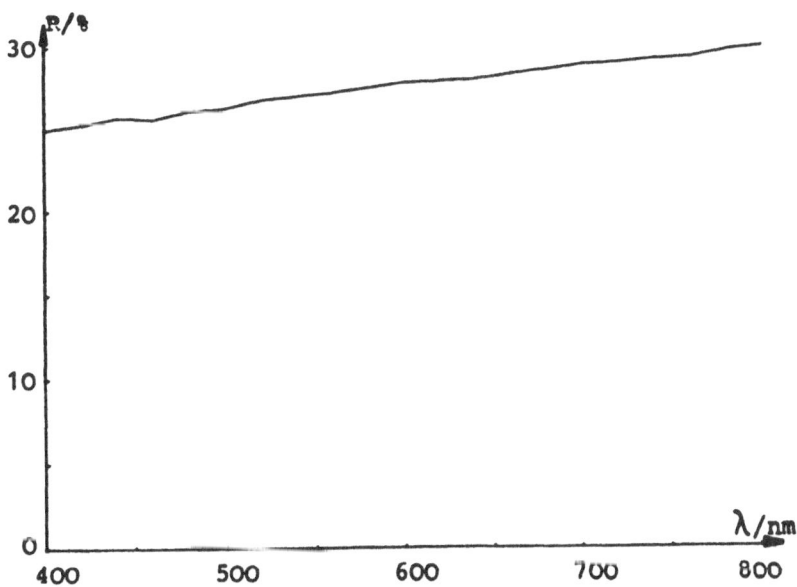

FIGURE 1: Reflection coefficient of a 100 µg/cm^2 carbon glass
film versus wavelength, under 0°/diff. conditions against BaSO$_4$.

in the wavelength range

$$400 \text{ nm} \leq \lambda \leq 800 \text{ nm}.$$

Regular reflection was found to be independent of the angle of incidence for all the angles between normal incidence and 45°. Black absorbers in this spectral range, like fresh soot from a parafin candle with R ≈ 2%, or Nextel velvet coating 101-C10 black with R ≈ 4%, do not show the wavelength behaviour of this particular carbon glass. An SEM picture of the carbon glass (Fig.2) shows that the wavelength dependence is due to the sur- face structure of the sample, which is itself just a genuine replica of the parting agent. This underlines the importance of the parting agent.

Transmission spectra (Fig.3) of 5 μg/cm^2 samples corrected for reflection losses give an absorbance

$$\alpha \, x1 = 1.516 - 1.115 \times 10^{-3} \times \lambda \, /\text{nm}$$

in the above-mentioned wavelength range; with an optical absorp- tion length in the order of 30 nm, therefore, a 600 nm thick

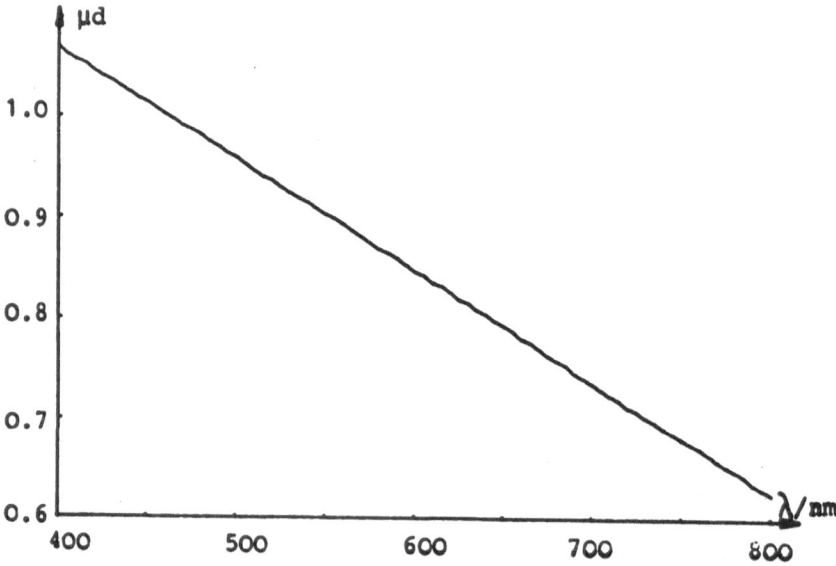

FIGURE 2:   SEM photograph of 100 μg/cm^2 carbon glass film, floated from Betaine as parting agent.

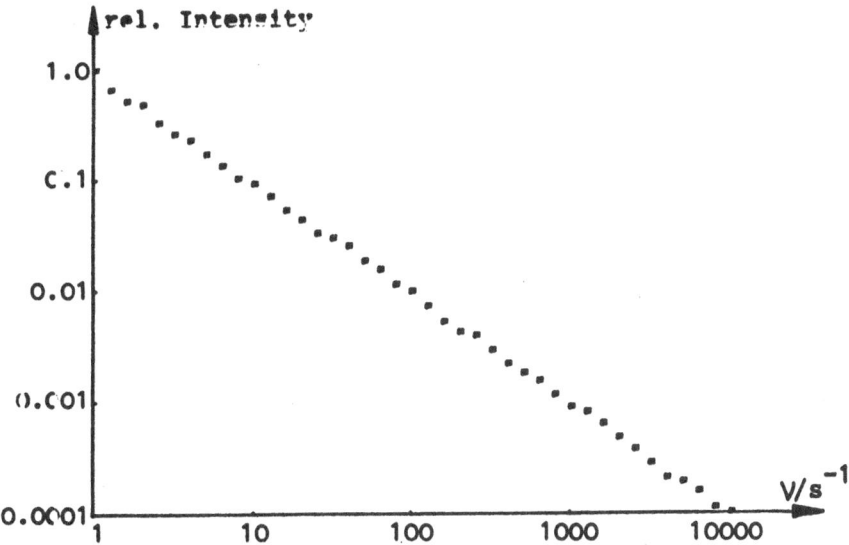

FIGURE 3: Transmission spectrum of a 5 μg/cm^2 carbon glass film
versus wavelength, corrected for reflection.

sample with 100 μg/cm^2 is an excellent approximation for an
optically opaque sample, regardless of wavelength.

To demonstrate the spectral features of the sample, the
photoacoustic signal of an EIMAC VIX-300 UV lamp vs. wavelength
was recorded. The throughput of the monochromator was calibrated
by using a PTB-calibrated tungsten ribbon lamp Wi 17/G as light
source and an NBS-calibrated Coblentz type thermopile by Eppley
as a detector. Taking the reflection correction and the throughput of the monochromator into account, the spectral power distribution of the light source was calculated from the PA-spectra. A
comparison of this spectrum with data from the manufacturer /12/
shows (Fig.4) excellent agreement; spectra were matched at 600 nm.

To characterise the photoacoustic properties, the frequency
dependence of a sample mounted directly on a calibrated microphone
was measured. As one expects for an optically opaque, thermally
thin sample, the result   exhibits a clear f^{-1} modulation
frequency dependence, thus providing a convenient means for frequency calibration of other photoacoustic cells.

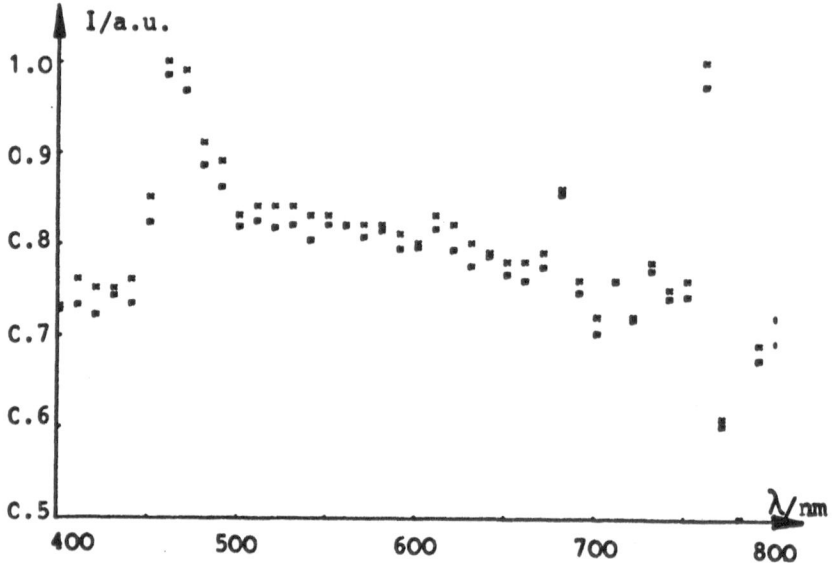

FIGURE 4:   Power density of EIMAC-lamp versus wavelength: x data
provided by manufacturer, ■ data from PAS.

The signal was proportional to the incident light-power within the power-range 1 μW to 1 W. About 1 W/cm², a slow degradation,
and above 10 W/cm² luminance, the onset of recrystallisation ir-
reversibly changing the sample properties, was observed.

ACKNOWLEDGEMENT

The author would like to thank P. Maier-Komor, Target Laboratory of the Technical University of Munich, and F. Sequeda,
Materials Technology, IBM Research Laboratory, San Jose, for
providing invaluable advice and samples.

REFERENCES

/1/   C.L. Mantell, Carbon and Graphite Handbook, Interscience,
      New York (1968)

/2/   A. Rosencwaig and H. Gersho, J.Appl.Phys. 47, 64 (1970)

/3/   M. Blue and G. Danielson, J.Appl.Phys. 28, 583 (1957)

/4/   S. Yamada, H. Sato, T. Ishii, Carbon $\underline{2}$, 253 (1964)

/5/   T. Tsuzuku, H. Kobayashi, Proc.5 Conf.Carbon $\underline{2}$, 539 (1961)

/6/   C.F. Lucks and H.W. Deem, WADC TR 55-496, 1 (1956)

/7/   A. Rosencwaig, Adv.Electron.Electron Phys. $\underline{46}$, 218 (1978)

/8/   J. Baumann, private communication

/9/   K. Klein, J. Pelzl, H. Fütterer, this volume

/10/   Gmelins Handbuch der anorg. Chemie $\underline{14}$, B2 (1968)
                               ibid $\underline{14}$, B3 (1968)

/11/   P. Maier-Komor, Proc.INTDS, to be published, Plenum, New York

/12/   G. Liljegren, EIMAC, private communication

A PHOTOACOUSTIC DEMONSTRATION EXPERIMENT: CHEAP AND EASY

H. Coufal, Chr. Stein

Physikdepartment E13, Technische Universität München, D-8046-
Garching, West Germany

ABSTRACT

An experiment at undergraduate level, demonstrating the detection
of heat waves with a minimum of instrumentation, is described.
The characteristics of the photoacoustic signal-like thermal
damping within the sample, the influence of the gas within the
cell, the length of the cell and the modulation frequency dependence - can easily be shown.

INTRODUCTION

One hundred years ago, Alexander Graham Bell discovered
the photoacoustic effect using just a simple hearing tube. Modern
photoacoustic laboratory set-ups use highly sophisticated instrumentation that is expensive and tends to disguise the simple
physics involved in the photoacoustic effect. To overcome this
handicap and give everybody the chance to gain hands-on experience with photoacoustics, an experiment was designed that allows
the demonstration of the essential processes and variables involved in the photoacoustic effect with a minimum of instruments
and specialised hardware. A signal generator, a Hifi amplifier
and approximately DM 10,-- of hardware from a local hobby shop,
and half an hour, is all one needs for this experiment.

EXPERIMENT

The sample is enclosed in a simple cell (Fig.1) made from
a 5 cm long, 14 mm diameter glass tube, a fitting glass rod and
a low-cost electret microphone with integrated preamplifier. In a
first series of experiments, resistance heating of a thin aluminised mylar sheet - available for example as rescue blanket - is

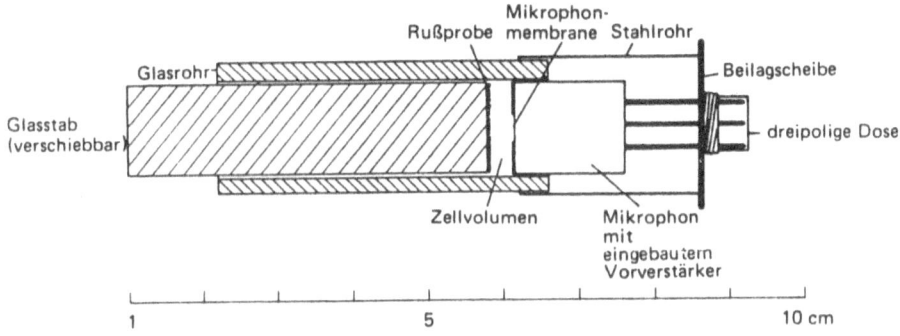

Figure 1:   Diagram of low-cost photoacoustic demonstration cell.

used to produce a thermoacoustic effect. Signal generator and
Hifi-amplifier are used to heat the aluminum coating periodically.
The thermally induced acoustic signal can be conveniently detected
with the microphone and a second Hifi-amplifier driving a loud-
speaker. The influence of cell length or filling gas, as well as
the modulation frequency dependence of the thermoacoustic signal,
can be shown. Having the aluminised side of the sample first
facing into the cell and afterwards directly attached to the glass
rod demonstrates the damping of the thermal waves.

    Replacing the resistance-heated sample within the cell by
a piece of carbon typewriter ribbon and connecting the modulation
amplifier output to a tungsten halogen lamp whose light is focussed on to the black sample now allows the demonstration of the
same effects as those caused by the photoacoustic effects. For
quantitative results, the output voltage of the detection amplifier can be measured using an AC-Voltmeter or an oscilloscope.

    Typical results obtained with that set-up are shown in
Figs.2-5. The results are in excellent agreement with photoacoustic theories.

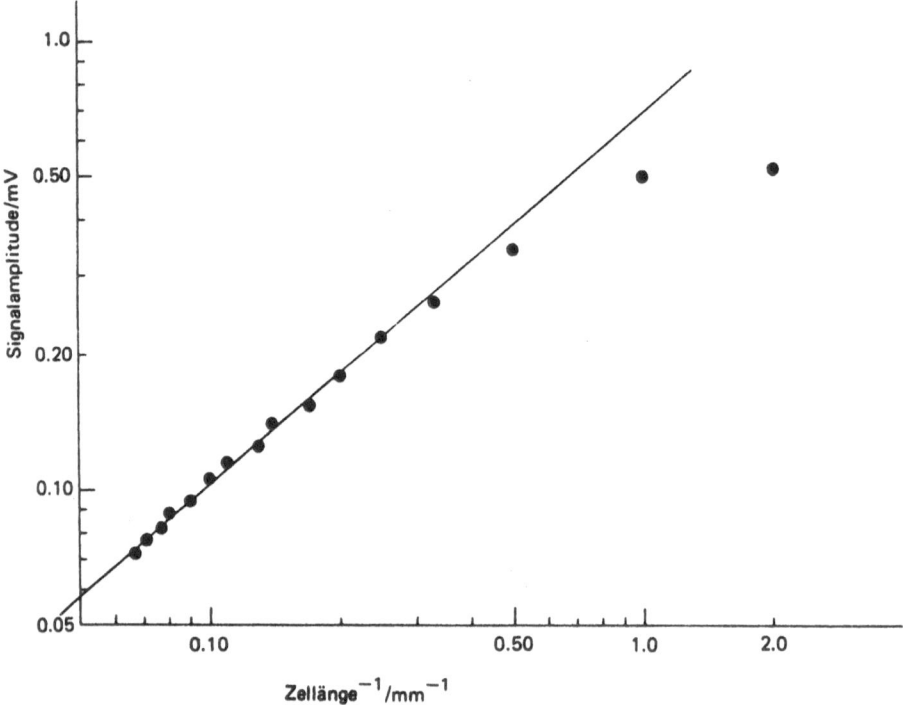

Figure 2:   Signal amplitude versus Cell-length^{-1} using periodic
            illumination with a frequency of 189 Hz.

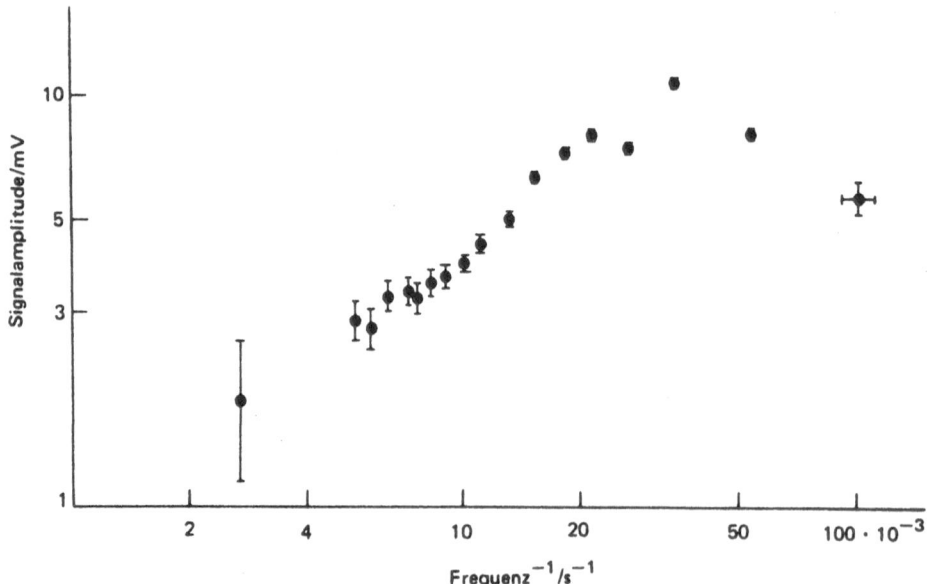

Figure 3:   Signal amplitude versus modulation-frequency^{-1} at a
            cell length of 0.5 mm.

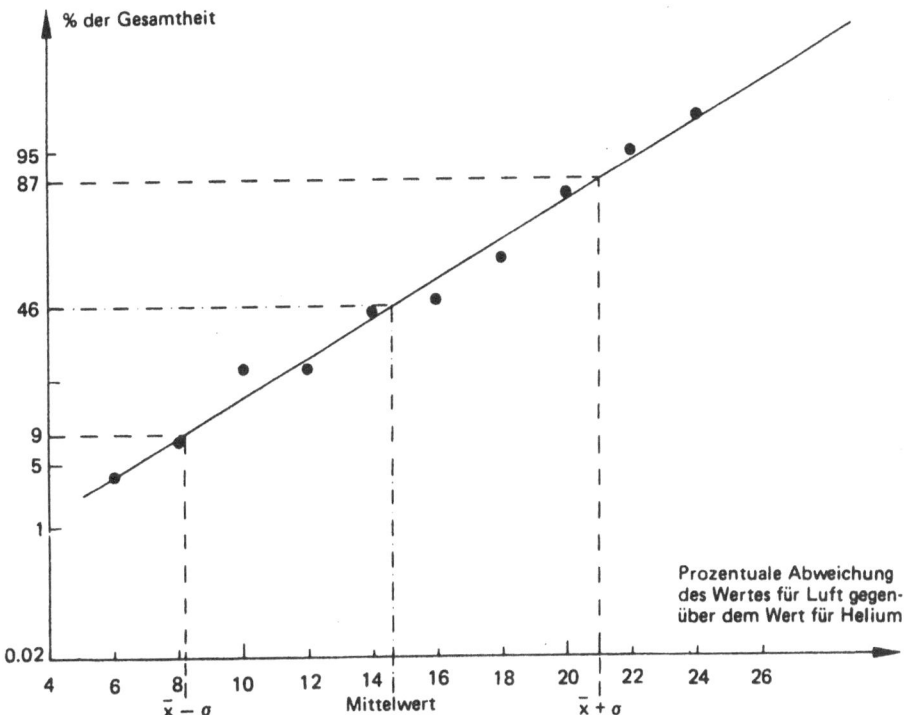

Figure 4:   Statistical evaluation of the signal decrease upon
            replacement of the Helium filling gas by plain air.

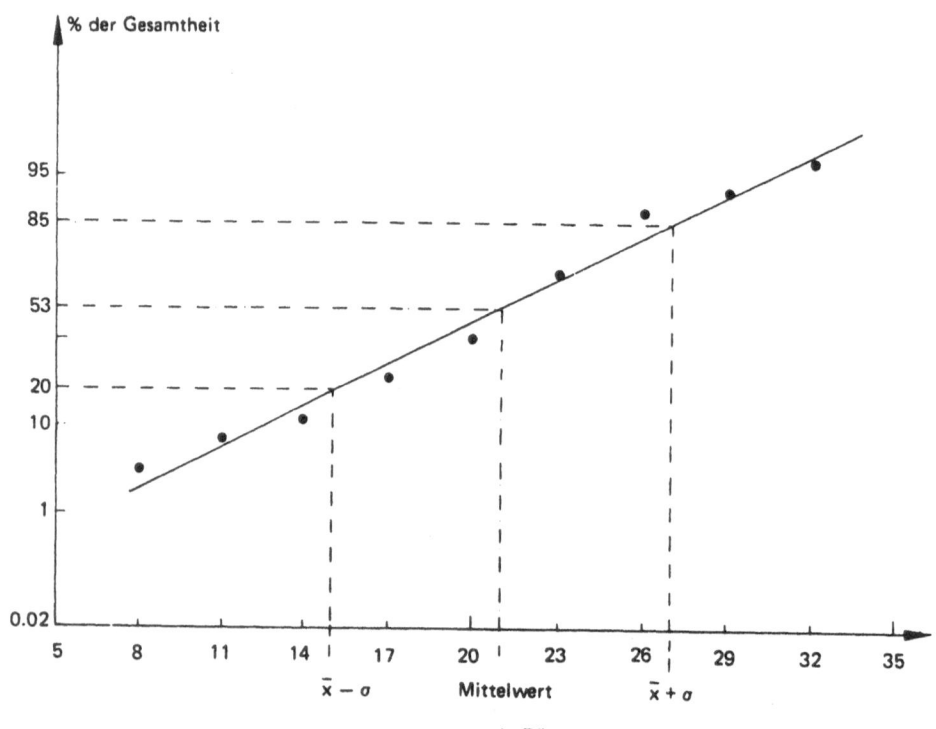

Figure 5: Statistical evaluation of the signal decrease due to thermal damping when the carbon surface is attached to the cell wall, compared to the effect of the carbon facing the cell gas.